Lithium-Ion Batteries

Lithium-Ion Batteries

Latest Advances and Prospects

Editor

Mohammad (Mim) Rahimi

MDPI • Basel • Beijing • Wuhan • Barcelona • Belgrade • Manchester • Tokyo • Cluj • Tianjin

Editor
Mohammad (Mim) Rahimi
Department of Chemical Engineering,
Massachusetts Institute of Technology
USA

Editorial Office
MDPI
St. Alban-Anlage 66
4052 Basel, Switzerland

This is a reprint of articles from the Special Issue published online in the open access journal *Batteries* (ISSN 2313-0105) (available at: https://www.mdpi.com/journal/batteries/special_issues/Lithium_Ion_Batteries_Latest_Advances_Prospects).

For citation purposes, cite each article independently as indicated on the article page online and as indicated below:

LastName, A.A.; LastName, B.B.; LastName, C.C. Article Title. *Journal Name* **Year**, *Volume Number*, Page Range.

ISBN 978-3-0365-0584-8 (Hbk)
ISBN 978-3-0365-0585-5 (PDF)

© 2021 by the authors. Articles in this book are Open Access and distributed under the Creative Commons Attribution (CC BY) license, which allows users to download, copy and build upon published articles, as long as the author and publisher are properly credited, which ensures maximum dissemination and a wider impact of our publications.

The book as a whole is distributed by MDPI under the terms and conditions of the Creative Commons license CC BY-NC-ND.

Contents

About the Editor vii

Mohammad Rahimi
Lithium-Ion Batteries: Latest Advances and Prospects
Reprinted from: *Batteries* 2021, 7, 8, doi:10.3390/batteries7010008 1

Marco Fritsch, Matthias Coeler, Karina Kunz, Beate Krause, Peter Marcinkowski, Petra Pötschke, Mareike Wolter and Alexander Michaelis
Lightweight Polymer-Carbon Composite Current Collector for Lithium-Ion Batteries
Reprinted from: *Batteries* 2020, 6, 60, doi:10.3390/batteries6040060 7

Pavel L'vov and Renat Sibatov
Effect of the Particle Size Distribution on the Cahn-Hilliard Dynamics in a Cathode of Lithium-Ion Batteries
Reprinted from: *Batteries* 2020, 6, 29, doi:10.3390/batteries6020029 25

Bharat Balagopal and Mo-Yuen Chow
The Physical Manifestation of Side Reactions in the Electrolyte of Lithium-Ion Batteries and Its Impact on the Terminal Voltage Response
Reprinted from: *Batteries* 2020, 6, 53, doi:10.3390/batteries6040053 37

S M Rakiul Islam, Sung-Yeul Park and Balakumar Balasingam
Unification of Internal Resistance Estimation Methods for Li-Ion Batteries Using Hysteresis-Free Equivalent Circuit Models
Reprinted from: *Batteries* 2020, 6, 32, doi:10.3390/batteries6020032 51

Seyed Saeed Madani, Erik Schaltz and Søren Knudsen Kær
Effect of Current Rate and Prior Cycling on the Coulombic Efficiency of a Lithium-Ion Battery
Reprinted from: *Batteries* 2019, 5, 57, doi:10.3390/batteries5030057 71

Hendrik Zappen, Georg Fuchs, Alexander Gitis and Dirk Uwe Sauer
In-Operando Impedance Spectroscopy and Ultrasonic Measurements during High-Temperature Abuse Experiments on Lithium-Ion Batteries
Reprinted from: *Batteries* 2020, 6, 25, doi:10.3390/batteries6020025 81

Christiane Essl, Andrey W. Golubkov, Eva Gasser, Manfred Nachtnebel, Armin Zankel, Eduard Ewert and Anton Fuchs
Comprehensive Hazard Analysis of Failing Automotive Lithium-Ion Batteries in Overtemperature Experiments
Reprinted from: *Batteries* 2020, 6, 30, doi:10.3390/batteries6020030 101

William Yourey
Theoretical Impact of Manufacturing Tolerance on Lithium-Ion Electrode and Cell Physical Properties
Reprinted from: *Batteries* 2020, 6, 23, doi:10.3390/batteries6020023 129

Tobias Jansen, Maja W. Kandula, Sven Hartwig, Louisa Hoffmann, Wolfgang Haselrieder and Klaus Dilger
Influence of Laser-Generated Cutting Edges on the Electrical Performance of Large Lithium-Ion Pouch Cells
Reprinted from: *Batteries* 2019, 5, 73, doi:10.3390/batteries5040073 139

Sören Hollatz, Sebastian Kremer, Cem Ünlübayir, Dirk Uwe Sauer, Alexander Olowinsky and Arnold Gillner
Electrical Modelling and Investigation of Laser Beam Welded Joints for Lithium-Ion Batteries
Reprinted from: *Batteries* **2020**, *6*, 24, doi:10.3390/batteries6020024 . . . . **159**

Nenad G. Nenadic, Thomas A. Trabold and Michael G. Thurston
Cell Replacement Strategies for Lithium Ion Battery Packs
Reprinted from: *Batteries* **2020**, *6*, 39, doi:10.3390/batteries6030039 . . . . **171**

Takumi Yamanaka, Daiki Kihara, Yoichi Takagishi and Tatsuya Yamaue
Multi-Physics Equivalent Circuit Models for a Cooling System of a Lithium Ion Battery Pack
Reprinted from: *Batteries* **2020**, *6*, 44, doi:10.3390/batteries6030044 . . . . **191**

Agnieszka Sobianowska-Turek and Weronika Urbańska
Future Portable Li-Ion Cells' Recycling Challenges in Poland
Reprinted from: *Batteries* **2019**, *5*, 75, doi:10.3390/batteries5040075 . . . . **211**

About the Editor

Mohammad (Mim) Rahimi is a postdoctoral associate at the Department of Chemical Engineering at Massachusetts Institute of Technology (MIT). He conducts research on developing various electrochemical processes for carbon capture and utilization. Dr. Rahimi obtained his Ph.D. in chemical engineering from the Pennsylvania State University in 2017. Dr. Rahimi has authored and coauthored more than 18 peer-reviewed articles and has served as a reviewer for more than 30 scientific journals. He has also served as a guest editor for several journals on topics related to electrochemical processes as well as carbon capture and utilization.

Editorial

Lithium-Ion Batteries: Latest Advances and Prospects

Mohammad Rahimi

Department of Chemical Engineering, Massachusetts Institute of Technology, Cambridge, MA 02139, USA; rahimi@mit.edu

Received: 15 January 2021; Accepted: 19 January 2021; Published: 20 January 2021

The anthropogenic release of greenhouse gases, especially carbon dioxide (CO_2), has resulted in a notable climate change and an increase in global average temperature since the mid-20th century [1,2]. To arrive at the margin of a 2 °C global temperature rise, it is essential to design and execute a multiscale comprehensive action plan to effectively mitigate climate change before its impacts overwhelm our ability to manage the situation [3–5]. Electrochemistry is a powerful tool for designing diverse CO_2 mitigation approaches that can effectively help prevent dangerous anthropogenic interference with the climate system. Several implementations of electrochemical systems are being considered within the electrochemistry and climate change framework. Besides emerging tasks such as CO_2 capture [6–8] and conversion [9–11], electrochemical systems are mainly being developed to help integrate renewable energy into electricity systems, through developing electrochemical energy storage systems such as batteries. Batteries are currently being developed to power an increasingly wide range of applications, including electrification of transportation [12,13] and grid-scale energy storage [14,15]. Large-scale developments and implementations of batteries offer sustainable energy supply based on renewables, which is a major step toward reducing CO_2 emissions associated with the energy sector and ultimately assisting in climate change mitigation.

Among the developed batteries, lithium-ion batteries (LIBs) have received the most attention, and have become increasingly important in recent years. Compared with other batteries, LIBs offer high energy density, high discharge power, high coulombic efficiencies, and long service life [16–18]. These characteristics have facilitated a remarkable advance of LIBs in many frontiers, including electric vehicles, portable and flexible electronics, and stationary applications. Since the field of LIBs is advancing rapidly and attracting an increasing number of researchers, it is necessary to often provide the community with the latest updates. Therefore, this Special Issue was designed to focus on updating the electrochemical community with the latest advances and prospects on various aspects of LIBs. Researchers were invited to submit their original research as well as review/perspective articles for publication in the Special Issue "Lithium-Ion Batteries: Latest Advances and Prospects".

In response to this call, twelve research papers [19–30] and one case report [31] were thoroughly peer-reviewed and published. The published research papers covered advances in several fronts of the technology, including detailed fundamental studies of the electrochemical cell and investigations to better improve parameters related to battery packs. In the domain of fundamental studies, various components of the electrochemical cell, including electrodes and electrolyte, were investigated. In this context, a lightweight dense polymer–carbon composite-based current collector foil for applications in LIB was developed and evaluated in comparison to the state-of-the-art aluminum foil collector [19]. It was found that the resistance of the developed electrode based on this current collector to be by a factor of five lower compared to the aluminum-based collector, which was attributed to the low contact resistance between the proposed current collector and the other elements of the electrode. In addition, due to a 50% lower material density, the developed lightweight current collector offers the possibility to significantly decrease the mass loading of the electrode, which can be of special interest for bipolar battery architectures. In another study [20], the lithium intercalation dynamics in a cathode electrode with particles of distributed size was studied using the phase-field model, aiming to better understand the effect of this particle size distribution on the LIBs' dynamic performance.

To advance the electrolyte-related research, a first principle-based modeling framework was developed to identify the physical manifestation that electrolyte degradation has on the battery and the response observed in the terminal voltage [21]. The developed framework relates the different kinds of side reactions in the electrolyte to the material properties affected due to these side reactions—these material property changes directly impact the electrochemical reactions, and ultimately the voltage across the terminals of the battery.

Internal resistance is one of the important parameters in LIBs, which requires developing precise experimental procedures and/or theoretical frameworks to accurately evaluate this parameter. In this context, two different methods were investigated in the research paper published in this Special Issue: electrochemical impedance spectroscopy (EIS) and parameter estimation based on equivalent circuit model (ECM) [22]. It was found that unlike the conventional parameter estimation method that yields a different value than EIS, internal resistances estimated based on ECM match the values obtained from EIS. The proposed methods will be supplementary in tracking the internal resistance properly which can improve the accuracy of battery performance prediction.

An experimental investigation was performed to evaluate the effect of the current rate and prior cycling on the coulombic efficiency of LIBs [23]. The determination of coulombic efficiency of LIBs can contribute to comprehend better their degradation behavior. Therefore, a detailed understanding of the effect of these parameters would be beneficial to further optimize the cell charge/discharge procedures.

Experiments were performed at high temperatures to provide better insights regarding battery performance at elevated temperatures. In this context, advanced in-operando measurement techniques such as fast impedance spectroscopy and ultrasonic waves as well as strain-gauges were employed to evaluate the cell performance at these temperatures [24]. These methods have the potential to be integrated into the battery management system in the future, making it possible to achieve higher battery safety even under the most demanding operating conditions. In addition, comprehensive hazard analysis of failing LIBs used in electric vehicles was evaluated experimentally at elevated temperatures [25]. In this investigation, several hazard-relevant parameters were quantified, including the temperature response of the cell, the maximum reached cell surface temperature, the amount of produced vent gas, the gas venting rate, the composition of the produced gases, and the size and composition of the produced particles. The results are valuable for all who deal with batteries, including firefighters, battery pack designers, and cell recyclers.

The effect of cell manufacturing parameters was also investigated on the performance of the produced LIBs. For example, a theoretical framework was developed to highlight the considerable impact of electrode porosity, electrode internal void volume, cell capacity, and capacity ratio that result from electrode coating and calendering tolerance (as the manufacturing parameters) on the cell-to-cell and lot-to-lot performance variation [26]. In another study published in this Special Issue, the impact of manufacturing parameters in laser cutting, which is a promising technology for the singulation of conventional and advanced electrodes for LIBs, was investigated [27]. In specifics, it was shown how cutting edge characteristics affect electrochemical performance. These types of information would be beneficial to manufacture better LIBs.

In addition to improving individual LIB cells, several researches were focused on strategies to obtain better battery packs. Every single cell in the battery pack needs a contact for its cell terminals, which raises the necessity of an automated contacting process with low joint resistances to reduce the energy loss in the cell transitions. A capable joining process suitable for highly electrically conductive materials like copper or aluminum is laser beam welding. In the research paper published in this Special Issue, a theoretical examination of the joint resistance and a simulation of the current flow dependent on the contacting weld's position in an overlap configuration was performed [28]. This investigation highlighted the influence of the shape and position of the weld seams as well as the laser welding parameters, and how these parameters can be leveraged to further reduce the cell-to-cell joint resistances.

For LIB packs, it is necessary to understand how to best replace poorly performing cells to extend the lifetime of the entire battery pack. In a comprehensive investigation [29], cell replacement strategies were studied considering two scenarios: early life failure, where one cell in a pack fails prematurely, and building a pack from used cells for less demanding applications. Early life failure replacement found that a new cell can perform adequately within a pack of moderately aged cells. The second scenario for the reuse of lithium-ion battery packs examines the problem of assembling a pack for less-demanding applications from a set of aged cells, which exhibit more variation in capacity and impedance than their new counterparts. The cells used in the aging comparison part of the study were deeply discharged, recovered, assembled in a new pack, and cycled. The criteria for selecting the aged cells for building a secondary pack were discussed in the paper and the performance and coulombic efficiency of the secondary pack were compared to the pack built from new cells and the repaired pack. The results showed that the pack that employed aged cells performed well, but its efficiency was reduced.

The cooling system of LIB packs used in electric vehicles was also investigated in the Special Issue. Thermal management systems of LIBs play an important role as the performance and lifespan of the batteries are affected by the temperature. A detailed study published in this Special Issue proposed a framework to establish equivalent circuit models that can reproduce the multi-physics phenomenon of Li-ion battery packs, which includes liquid cooling systems with a unified method. The developed equivalent circuit models were found to be very accurate and computationally cost-effective [30].

Besides the detailed research papers, one case report on the future portable LIB recycling challenges in Poland was published in this Special Issue [31]. This case report presents the market of portable LIBs in the European Union (EU) with particular emphasis on the stream of used cells in Poland by 2030. The report also draws attention to the fact that, despite a decade of efforts in Poland, it has not been possible to create an effective management system for waste batteries and accumulators that would include waste management, waste disposal, and component recovery technology for reuse. This report highlights the critical role of recycling strategies and challenges that need to be investigated to effectively deal with used LIBs.

Funding: This research received no external funding.

Acknowledgments: At the end of this editorial, I would like to express my sincere gratitude to the authors for their valuable contributions, and the reviewers for their efforts in analyzing the relevance and quality of the papers that were submitted to this special issue. I would also like to thank the editorial staff of *Batteries* for their help and support during the review process.

Conflicts of Interest: The author declares no conflict of interest.

References

1. Rogelj, J.; Huppmann, D.; Krey, V.; Riahi, K.; Clarke, L.; Gidden, M.; Nicholls, Z.; Meinshausen, M. A new scenario logic for the Paris Agreement long-term temperature goal. *Nature* **2019**, *573*, 357–363. [CrossRef] [PubMed]
2. Luderer, G.; Vrontisi, Z.; Bertram, C.; Edelenbosch, O.Y.; Pietzcker, R.C.; Rogelj, J.; De Boer, H.S.; Drouet, L.; Emmerling, J.; Fricko, O. Residual fossil CO_2 emissions in 1.5–2 C pathways. *Nat. Clim. Chang.* **2018**, *8*, 626–633. [CrossRef]
3. O'Neill, B.C.; Carter, T.R.; Ebi, K.; Harrison, P.A.; Kemp-Benedict, E.; Kok, K.; Kriegler, E.; Preston, B.L.; Riahi, K.; Sillmann, J. Achievements and needs for the climate change scenario framework. *Nat. Clim. Chang.* **2020**, *10*, 1–11. [CrossRef] [PubMed]
4. Fawzy, S.; Osman, A.I.; Doran, J.; Rooney, D.W. Strategies for mitigation of climate change: A review. *Environ. Chem. Lett.* **2020**, *18*, 1–26. [CrossRef]
5. Rahimi, M. Public Awareness: What Climate Change Scientists Should Consider. *Sustainability* **2020**, *12*, 8369. [CrossRef]
6. Rahimi, M.; Catalini, G.; Hariharan, S.; Wang, M.; Puccini, M.; Hatton, T.A. Carbon Dioxide Capture Using an Electrochemically Driven Proton Concentration Process. *Cell Rep. Phys. Sci.* **2020**, *1*, 100033. [CrossRef]

7. Rahimi, M.; Diederichsen, K.M.; Ozbek, N.; Wang, M.; Choi, W.; Hatton, T.A. An Electrochemically Mediated Amine Regeneration Process with a Mixed Absorbent for Postcombustion CO_2 Capture. *Environ. Sci. Technol.* **2020**, *54*, 8999–9007. [CrossRef]
8. Rahimi, M.; Zucchelli, F.; Puccini, M.; Hatton, T.A. Improved CO_2 Capture Performance of Electrochemically Mediated Amine Regeneration Processes with Ionic Surfactant Additives. *ACS Appl. Energy Mater.* **2020**, *3*, 10823–10830. [CrossRef]
9. Zheng, Y.; Vasileff, A.; Zhou, X.; Jiao, Y.; Jaroniec, M.; Qiao, S.-Z. Understanding the roadmap for electrochemical reduction of CO2 to multi-carbon oxygenates and hydrocarbons on copper-based catalysts. *J. Am. Chem. Soc.* **2019**, *141*, 7646–7659. [CrossRef]
10. Lee, G.; Li, Y.C.; Kim, J.-Y.; Peng, T.; Nam, D.-H.; Rasouli, A.S.; Li, F.; Luo, M.; Ip, A.H.; Joo, Y.-C. Electrochemical upgrade of CO_2 from amine capture solution. *Nat. Energy* **2020**, *6*, 46–53.
11. Sun, Z.; Ma, T.; Tao, H.; Fan, Q.; Han, B. Fundamentals and challenges of electrochemical CO2 reduction using two-dimensional materials. *Chem* **2017**, *3*, 560–587. [CrossRef]
12. Cano, Z.P.; Banham, D.; Ye, S.; Hintennach, A.; Lu, J.; Fowler, M.; Chen, Z. Batteries and fuel cells for emerging electric vehicle markets. *Nat. Energy* **2018**, *3*, 279–289. [CrossRef]
13. Harper, G.; Sommerville, R.; Kendrick, E.; Driscoll, L.; Slater, P.; Stolkin, R.; Walton, A.; Christensen, P.; Heidrich, O.; Lambert, S. Recycling lithium-ion batteries from electric vehicles. *Nature* **2019**, *575*, 75–86. [CrossRef] [PubMed]
14. Posada, J.O.G.; Rennie, A.J.; Villar, S.P.; Martins, V.L.; Marinaccio, J.; Barnes, A.; Glover, C.F.; Worsley, D.A.; Hall, P.J. Aqueous batteries as grid scale energy storage solutions. *Renew. Sustain. Energy Rev.* **2017**, *68*, 1174–1182. [CrossRef]
15. Jiang, L.; Lu, Y.; Zhao, C.; Liu, L.; Zhang, J.; Zhang, Q.; Shen, X.; Zhao, J.; Yu, X.; Li, H. Building aqueous K-ion batteries for energy storage. *Nat. Energy* **2019**, *4*, 495–503. [CrossRef]
16. Manthiram, A.; Yu, X.; Wang, S. Lithium battery chemistries enabled by solid-state electrolytes. *Nat. Rev. Mater.* **2017**, *2*, 1–16. [CrossRef]
17. Manthiram, A. A reflection on lithium-ion battery cathode chemistry. *Nat. Commun.* **2020**, *11*, 1–9. [CrossRef]
18. Xiao, J.; Li, Q.; Bi, Y.; Cai, M.; Dunn, B.; Glossmann, T.; Liu, J.; Osaka, T.; Sugiura, R.; Wu, B. Understanding and applying coulombic efficiency in lithium metal batteries. *Nat. Energy* **2020**, *5*, 561–568. [CrossRef]
19. Fritsch, M.; Coeler, M.; Kunz, K.; Krause, B.; Marcinkowski, P.; Pötschke, P.; Wolter, M.; Michaelis, A. Lightweight Polymer-Carbon Composite Current Collector for Lithium-Ion Batteries. *Batteries* **2020**, *6*, 60. [CrossRef]
20. L'vov, P.; Sibatov, R. Effect of the Particle Size Distribution on the Cahn-Hilliard Dynamics in a Cathode of Lithium-Ion Batteries. *Batteries* **2020**, *6*, 29. [CrossRef]
21. Balagopal, B.; Chow, M.-Y. The Physical Manifestation of Side Reactions in the Electrolyte of Lithium-Ion Batteries and Its Impact on the Terminal Voltage Response. *Batteries* **2020**, *6*, 53. [CrossRef]
22. Islam, S.; Park, S.-Y.; Balasingam, B. Unification of Internal Resistance Estimation Methods for Li-Ion Batteries Using Hysteresis-Free Equivalent Circuit Models. *Batteries* **2020**, *6*, 32. [CrossRef]
23. Madani, S.S.; Schaltz, E.; Knudsen Kær, S. Effect of Current Rate and Prior Cycling on the Coulombic Efficiency of a Lithium-Ion Battery. *Batteries* **2019**, *5*, 57. [CrossRef]
24. Zappen, H.; Fuchs, G.; Gitis, A.; Sauer, D.U. In-Operando Impedance Spectroscopy and Ultrasonic Measurements during High-Temperature Abuse Experiments on Lithium-Ion Batteries. *Batteries* **2020**, *6*, 25. [CrossRef]
25. Essl, C.; Golubkov, A.W.; Gasser, E.; Nachtnebel, M.; Zankel, A.; Ewert, E.; Fuchs, A. Comprehensive Hazard Analysis of Failing Automotive Lithium-Ion Batteries in Overtemperature Experiments. *Batteries* **2020**, *6*, 30. [CrossRef]
26. Yourey, W. Theoretical Impact of Manufacturing Tolerance on Lithium-Ion Electrode and Cell Physical Properties. *Batteries* **2020**, *6*, 23. [CrossRef]
27. Jansen, T.; Kandula, M.W.; Hartwig, S.; Hoffmann, L.; Haselrieder, W.; Dilger, K. Influence of Laser-Generated Cutting Edges on the Electrical Performance of Large Lithium-Ion Pouch Cells. *Batteries* **2019**, *5*, 73. [CrossRef]
28. Hollatz, S.; Kremer, S.; Ünlübayir, C.; Sauer, D.U.; Olowinsky, A.; Gillner, A. Electrical Modelling and Investigation of Laser Beam Welded Joints for Lithium-Ion Batteries. *Batteries* **2020**, *6*, 24. [CrossRef]

29. Nenadic, N.G.; Trabold, T.A.; Thurston, M.G. Cell Replacement Strategies for Lithium Ion Battery Packs. *Batteries* **2020**, *6*, 39. [CrossRef]
30. Yamanaka, T.; Kihara, D.; Takagishi, Y.; Yamaue, T. Multi-physics equivalent circuit models for a cooling system of a lithium ion battery pack. *Batteries* **2020**, *6*, 44. [CrossRef]
31. Sobianowska-Turek, A.; Urbańska, W. Future Portable Li-Ion Cells' Recycling Challenges in Poland. *Batteries* **2019**, *5*, 75. [CrossRef]

Publisher's Note: MDPI stays neutral with regard to jurisdictional claims in published maps and institutional affiliations.

 © 2021 by the author. Licensee MDPI, Basel, Switzerland. This article is an open access article distributed under the terms and conditions of the Creative Commons Attribution (CC BY) license (http://creativecommons.org/licenses/by/4.0/).

Article

Lightweight Polymer-Carbon Composite Current Collector for Lithium-Ion Batteries

Marco Fritsch [1],*, Matthias Coeler [1], Karina Kunz [2], Beate Krause [2], Peter Marcinkowski [1], Petra Pötschke [2], Mareike Wolter [1] and Alexander Michaelis [1]

1. Fraunhofer Institute for Ceramic Technologies and Systems (IKTS), 01277 Dresden, Germany; matthias.coeler@ikts.fraunhofer.de (M.C.); peter.marcinkowski@ikts.fraunhofer.de (P.M.); mareike.wolter@ikts.fraunhofer.de (M.W.); alexander.michaelis@ikts.fraunhofer.de (A.M.)
2. Leibniz Institute of Polymer Research Dresden e.V. (IPF), 01069 Dresden, Germany; karinakunz2015@gmail.com (K.K.); krause-beate@ipfdd.de (B.K.); poe@ipfdd.de (P.P.)
* Correspondence: marco.fritsch@ikts.fraunhofer.de; Tel.: +49-351-2553-7869

Received: 16 October 2020; Accepted: 30 November 2020; Published: 8 December 2020

Abstract: A hermetic dense polymer-carbon composite-based current collector foil (PCCF) for lithium-ion battery applications was developed and evaluated in comparison to state-of-the-art aluminum (Al) foil collector. Water-processed $LiNi_{0.5}Mn_{1.5}O_4$ (LMNO) cathode and $Li_4Ti_5O_{12}$ (LTO) anode coatings with the integration of a thin carbon primer at the interface to the collector were prepared. Despite the fact that the laboratory manufactured PCCF shows a much higher film thickness of 55 µm compared to Al foil of 19 µm, the electrode resistance was measured to be by a factor of 5 lower compared to the Al collector, which was attributed to the low contact resistance between PCCF, carbon primer and electrode microstructure. The PCCF-C-primer collector shows a sufficient voltage stability up to 5 V vs. Li/Li^+ and a negligible Li-intercalation loss into the carbon primer. Electrochemical cell tests demonstrate the applicability of the developed PCCF for LMNO and LTO electrodes, with no disadvantage compared to state-of-the-art Al collector. Due to a 50% lower material density, the lightweight and hermetic dense PCCF polymer collector offers the possibility to significantly decrease the mass loading of the collector in battery cells, which can be of special interest for bipolar battery architectures.

Keywords: lithium-ion battery; bipolar battery; polymer-carbon composite; current collector; water-based electrode slurries; carbon primer; CNTs; LMNO; LTO

1. Introduction

Lithium-ion batteries play an important role in the development of electric vehicles and portable electronic devices. Bipolar battery concepts [1,2] utilize the connection of multiple cells in series to form a battery stack. This approach avoids the use of numerous passive components and parts usually required for packaging as well as external electrical wiring, which lowers the overall electrical resistance, volume, weight, complexity and cost of the battery.

In a bipolar battery architecture, anode and cathode electrodes are coated on both sides of the same current collector (bipolar plate). To avoid internal short-circuits between the unit cells, this collector has to be pore free. Since anode and cathode operate in different cell potential ranges, the collector material has to be stable against corrosion in a wide voltage range (e.g., 0 to >5 V vs. Li/Li^+). State-of-the-art lithium-ion batteries use thin aluminum (Al) and copper (Cu) foils as current collectors for cathode and anode, respectively [3,4]. Al shows a destructive alloying reaction below 1 V vs. Li/Li^+, which falls within the potential window of state-of-the-art carbon anodes [5]. That is why Al is only used as cathode collector or in combination with high voltage anodes like $Li_4Ti_5O_{12}$ (~1.5 V vs. Li/Li^+). On the other hand, Cu is dissolved above 3.5 V vs. Li/Li^+, the potential window of common oxide cathodes, which limits the applicability of Cu collector to the anode side. To overcome this issue, bimetal collectors with a combination of Al-Ni or Al-Cu were developed [6–10]. However, this leads to increased costs and there are considerable difficulties in industrial implementation. These comprise residual pores, which can lead to short circuits [7]. During welding processes above 120 °C, Al and Cu form intermetallic compounds, which are brittle and lead to a poor strength and high electrical resistance [11–13].

In state-of-the-art lithium-ion batteries, metallic collectors can represent a significant percentage of an electrode weight [14] and they exhibit corrosion problems during processing of water-based electrode slurries [15] as well as with electrolyte components [5,16]. Especially for novel mechanical flexible battery concepts, alternative current collector materials were developed based on carbon, coated paper, textiles and conductive polymers [17,18]. Carbon collectors based on graphite, carbon fibers, carbon nanotubes and graphene offer a low density and high stability over a wide range of electrode potentials [19]. However, their packing density and mechanical stability is limited, which makes it difficult to achieve benefits on the macro-scale of battery performance. Residual porosity in such carbon collectors is the main hindrance to use them for bipolar battery concepts, since short circuits can occur.

The use of electrically conductive polymer composites, that remain electrochemically stable in the whole potential window of the battery, would greatly simplify the process of manufacturing of bipolar collectors and the processing of lithium-ion batteries [20]. Further, polymer composites can lead to more lightweight collectors, since Al and Cu have correspondingly moderate and high densities (2.7 g/cm^3 and 8.9 g/cm^3 respectively) in comparison to polymers like poly(vinylidene fluoride) (PVDF) (1.8 g/cm^3) or polyethylene (PE) and polypropylene (PP) (0.9 g/cm^3). Due to applicability, in terms of a required mechanical strength and secure handling during electrode production, the thickness of metal collectors is limited to approx. 10 to 30 µm. Today, several technical applications use much thinner polymer foils and there is a high probability that even battery-compatible polymer collectors can be developed.

We recently published the attractive electronic properties of polymer-carbon composite foils (PCCF) based on PVDF polymer [21–23]. In this study, we present the processing, electrochemical stability and performance of a PVDF polymer carbon nanocomposite current collector, which can be extruded to a thin hermetic dense collector foil and processed in a roll-to-roll process. The applicability for lithium-ion battery applications was studied based on water processed $LiNi_{0.5}Mn_{1.5}O_4$ (LMNO) cathode and $Li_4Ti_5O_{12}$ (LTO) anodes coatings with the integration of a thin carbon primer at the interface to the collector. For comparison reasons comparable electrodes were also fabricated on a thin Al collector. We used LMNO and LTO active materials, since they offer a more environmentally friendly approach (no cobalt component) and can be charged to high voltages. The applicability of different current collector materials depends, beside aspects of processing and costs, on a low electrical resistance (influence on the overall cell resistance and capacity losses with increasing C-rates) and the chemical compatibility to other cell components (e.g., active materials and electrolyte). These aspects will be discussed in this paper for PCCF in comparison to Al-foil collector.

2. Results and Discussion

2.1. Characterization of the Polymer-Carbon Composite and Electrodes

Figure 1a shows the developed polymer-carbon composite foil (PCCF) after film extrusion. The process allows the handling in roll-to-roll process, which is state of the art in today´s battery electrode manufacturing. Due to the extrusion process, the PCCF shows at the edges a small shiny strip (~1 cm) with a different thickness and roughness, which can be cut off prior to the electrode coating process. Figure 1b shows the PCCF microstructure, where the carbon nanotube and carbon black filler particles are generally homogeneously distributed in the PVDF polymer matrix. In the SEM-CCI image, the 3 wt.-% CB dominates the appearance; however, an orientation of the CNT particles in the extrusion direction occurs. SEM-CCI images of foils with lower CB content show the discussed CNT filler orientation more clearly (Figure 5 in [23]). Due to the high aspect ratio of the CNTs, the particles rotate in the viscous melt during flowing through the extrusion die and are oriented in plane, whereas the small spherical carbon black particles are hardly oriented and thus connect the CNTs in the through-plane direction (see schematic Figure 1c). This particular microstructure will lead to differences in mechanical and electrical properties in dependence on PCCF orientation as discussed below and in Section 2.2.1.

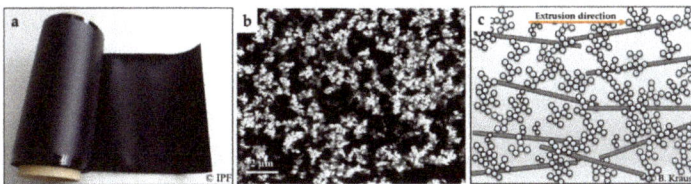

Figure 1. (**a**) Polymer-carbon composite foil (PCCF, width 22 cm); (**b**) top view of PCCF microstructure in SEM-CCI, composition PVDF/1.0 wt.-% b-MWCNT + 3 wt.-% CB; and (**c**) schematic of arrangement of the two kinds of carbon fillers in the PVDF matrix (cross-sectional view).

The microstructure in Figure 1b may create the impression, that the PCCF is porous. However, this is not the fact, since the PCCF proved to be hermetically dense based on results of gas leakage test with helium (1.8×10^{-7} mbar·L/s) and air (1×10^{-12} mbar·L/s, equals detection limit of equipment). With this hermetic density, the PCCF can separate cathode and anode half cells in bipolar cell architectures, where no liquid electrolyte will penetrate trough.

The mechanical properties of the PCCF collector in comparison to those of the battery grade Al collector are summarized in Table 1. The results show for PCCF higher values of the elastic modulus and tensile strength parallel to the extrusion direction compared to perpendicular to it, which is attributed to the orientation effect of the CNT particles during film extrusion. The elongation at break is comparable and independently of orientation. The Al collector, which is a strain-hardened aluminum foil (H18), shows higher tensile strength (>135 MPa according to [24]) and elastic modulus, but its elongation at break is lower compared to PCCF. Nevertheless, we demonstrated successfully the applicability of this PCCF foil in an industrial manufacturing machine (roll-to-roll coating) for battery electrodes [25]. In comparison to standard Al-foil, one adjustment needed was the implementation of special electrically driven rolls at the front end of the coater that push the PCCF without applying high pulling forces during the coating process, which will otherwise cause an unwanted elongation of the PCCF.

Table 1. Mechanical properties of PCCF collector (55 μm) in comparison to Al-foil (19 μm).

Collector Type	E_t (MPa)	σ_B (MPa)	ε_B (%)
PCCF collector ‖	2105 ± 158	39 ± 1	3.9 ± 0.3
PCCF collector ⊥	1557 ± 518	31 ± 3	4.1 ± 0.9
Al-foil *[1]	45205 ± 2760	146 ± 3	1.2 ± 0.1

‖: Parallel to extrusion direction; ⊥: perpendicular to extrusion direction; E_t: E modulus; σ_B: tensile strength; ε_B: elongation at break. *[1] Own measurement. Results are comparable to [24].

The surface topography and roughness of the extruded PCCF collector are shown in Figure 2. Over the scanned 600 μm × 600 μm area the maximum differences in height is 6 μm, which equals to 10% of thickness. The overall homogeneity of the PCCF surface is quite good; the roughness can be correlated to the existence of small agglomerates of CNT and carbon black filler particles. This residual roughness can be beneficial for the later electrode coating process, since highly smooth polymer surfaces are usually difficult to coat with functional films. Especially, the adhesion of coated films on smooth and dense fluorine-containing polymer surfaces is challenging, due to their low surface energy [26]. The positive effect of an increased collector roughness is also known from Al collector foils, where special etching techniques are applied to increase the Al surface roughness, which leads to a better electrode adhesion [27].

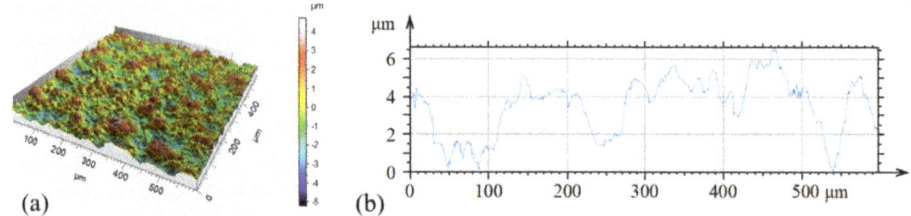

Figure 2. (a) 3D-image of roughness of PCCF (600 × 600 µm) and (b) height profile.

Table 2 summarizes the PCCF and Al collector materials used in this study. Prior to coating with electrodes, the PCCF was coated with a thin C-primer on both sides to reduce the electrical contact resistance between electrode and PCCF and to homogenize the electrical conductivity in plane at the interface to the active electrode layer. By comparing uncoated and C-primer coated PCCF, the C-primer film thickness and area weight were determined to be 7.5 ± 1.0 µm and 1.1 ± 0.1 mg/cm^2 (both sides in sum).

Table 2. Specification of PCCF and Al collector.

Sample	Material	Size (cm^2)	Thickness (µm)	Area Weight (mg/cm^2)
PCCF	PVDF-carbon composite	roll	55 ± 5	8.7 ± 0.8
PCCF + C-primer	batch of 9 sheets	12 × 16	70 ± 2	9.8 ± 0.1
Al collector	Al-alloy EN AW 1085-L H18	roll	19 ± 1	4.8 ± 0.3

Figure 3 shows the cross section of the C-primer coated on both sides of the PCCF and the interface microstructure between PCCF, C-primer and LTO electrode. In Figure 3a a variation of the PCCF thickness and C-primer layer thickness is visible. Such variations are common by using R&D laboratory equipment, but they can usually be avoided if industry relevant scale production and machinery are applied. Figure 3b,c show an excellent surface coverage and composite formation between C-primer and PCCF or active electrode layer, respectively.

Figure 3. (a) Cross section of double side C-primer coated PCCF, (b) interface of C-primer to PCCF and (c) interface between LTO electrode to C-primer.

Table 3 summarizes geometrical parameters and properties of the prepared LTO and LMNO electrode coatings on both types of current collectors. The initial porosity of the dried electrodes was 63 to 71 vol.-%, which was significantly reduced by the lamination densification down to ~40 vol.-%. Based on the information of the uncoated collector substrates (Table 2), the active electrode layer thickness was calculated to be 69 to 79 µm (LTO) and 63 to 64 µm (LMNO).

Table 3. Parameters of prepared electrode films (12 × 12 cm² size).

Sample	Thickness (μm)				Area Weight (mg/cm²)		Electrode Film				
densification (1) before (2) after	Sample		Electrode film		Sample	Electrode film	Density (g/cm³)			Porosity [*3] (vol.-%)	
	(1)	(2)	(1)	(2)	(1,2)	(1,2)	theor.	(1)	(2)	(1)	(2)
LTO on Al [*1]	127	88	108	69	15.94	11.13	2.77	1.03	1.61	63	42
LTO on PCCF [*2]	210	149	140	79	23.11	13.31	2.77	0.95	1.68	66	39
LMNO on Al [*1]	140	83	121	64	16.98	12.17	3.29	1.01	1.90	69	42
LMNO on PCCF [*2]	194	133	124	63	21.88	12.03	3.29	0.97	1.91	71	42

[*1] Al collector with 19 μm thickness and 4.81 mg/cm² area weight; [*2] PCCF/C-primer with 70 μm thickness and 9.80 mg/cm² area weight; [*3] calculated based on difference between geom. electrode film density and theoretical density of raw materials.

Figure 4 shows the interface between LTO and LMNO active electrodes on both current collectors. The interface microstructure between the C-primer of the PCCF and the active electrodes seems more interlocked in each other (Figure 4c,d) compared to the electrodes on the smooth Al collector (Figure 4a,b). Usually such carbon primer is not used in standard lithium-ion batteries with metal foil collectors, since it leads to additional processing steps and costs.

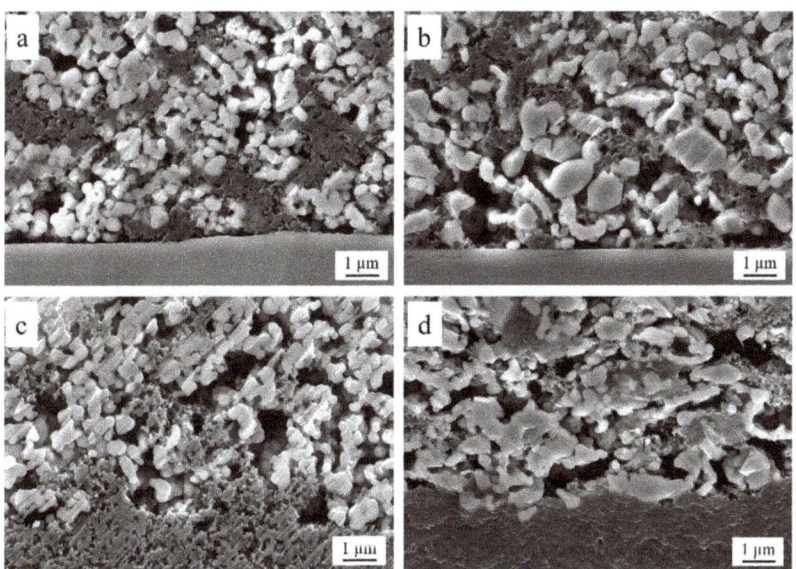

Figure 4. Cross section of (**a**) LTO on Al, (**b**) LMNO on Al, (**c**) LTO on PCCF-C-primer and (**d**) LMNO on PCCF-C-primer.

2.2. Electrical and Electrochemical Characterization

2.2.1. Electrical Properties and Advantage of Carbon Primer

Table 4 summarizes the electrical properties of the current collectors and prepared electrodes before the battery cell test.

The polymer composite shows low values of electrical resistivity of 0.7 Ω·cm in-plane parallel to the film extrusion direction. The resistivity in-plane perpendicular to the film extrusion direction was 2.7 Ω·cm. The resistivity measured through the film thickness, which is important for bipolar battery application, was 26 Ω·cm. As described in [22], carbon nanotubes are oriented mainly in in-plane direction due to their high aspect ratio. Therefore, the resistivity in-plane is much lower compared to through-plane. Especially for the MWCNTs used in this study it could be shown that the in-plane orientation in an extruded film is more pronounced than in a compressed molded plate. This is due to the melt flow and take-off forces during the film extrusion. In polymer composites filled solely with carbon black, only marginal anisotropy is determined due to the spherical shape of carbon black. No significant difference in CB orientation between pressed plate and extruded film was found. The quotient of in-plane and through-plane conductivity σ was calculated in [22] for the purpose of quantification of the different orientation degrees (see Table 5).

Table 4. Electrical properties of current collectors and electrodes.

Component	Direction	Resistivity ($\Omega \cdot$cm)
PCCF [*1]	in-plane parallel and	0.7
	perpendicular to extrusion direction	2.7
	through-thickness	26
C-primer [*2]	in-plane	0.3
Al-foil [*3]	in-plane and through-thickness	5.7×10^{-6}
LTO on Al		460
LTO on PCCF [*4]	through-thickness	90
LMNO on Al		500
LMNO on PCCF [*4]		100

[*1] measured with Ag-paste to reduce contact resistance of measurement; [*2] evaluated on 40 µm reference specimen on ceramic substrates; [*3] evaluated on 1 cm wide and 20 cm long stripe, equals two times Al-bulk value; [*4] with C-primer between PCCF and electrode coating.

Table 5. Comparison of the quotients of electrical conductivities σ measured in three directions for extruded films.

Filler Content	σ_x/σ_z (-) *	σ_y/σ_z (-) *	σ_x/σ_y (-) *
1 wt% b-MWCNT [22]	166	42	4
4 wt% CB [22]	5	4	1
PCCF (1 wt% b-MWCNT + 3 wt% CB) [23]	26	8	3

* x—parallel to extrusion direction (in-plane), y—perpendicular to extrusion direction (in-plane), z—through-plane.

In the PCCF composite used in the present study with a mixed filler system, the advantages of both fillers can now be combined. Thus, for a composite with highly conductive CNTs, the formation of a conductive network can be expected even at a low CNT content. However, since this network develops primarily in-plane, the carbon black is supposed to form bridges between neighboring CNTs due to the low particle orientation and thus generates conductive paths through the plate (see Figure 1c). The higher in-plane conductivity compared to the through-plane values show that the formed conductive network is slightly more oriented in-plane even when using a mixed filler system. The importance of the CB for the conductivity through the plane (z) was described in Krause et al. [23]. The z-values of conductivity increase significantly with increasing addition of CB (1–4 wt.-%) to PVDF/1 wt.-% b-MWCNT. The quotients of electrical conductivities (Table 5, Table 2 in [23]) indicate that the addition of CB to CNTs leads to a significant lower orientation of the whole conductive network in the film. However, even after addition of CB, the orientation in the extrusion direction (x) is higher than that perpendicular to the extrusion direction (y), whereby the CB addition results in a slight decrease in σ_x/σ_y from 4 to 3.

The resistivity of the Al collector is several orders in magnitude lower compared to the PCCF, which is reasonable for a metal material (Table 4). Surprisingly, the through-thickness resistivity of LTO and LMNO electrodes on PCCF were by a factor of five lower compared to electrodes on Al collector (see comparison in Figure 5).

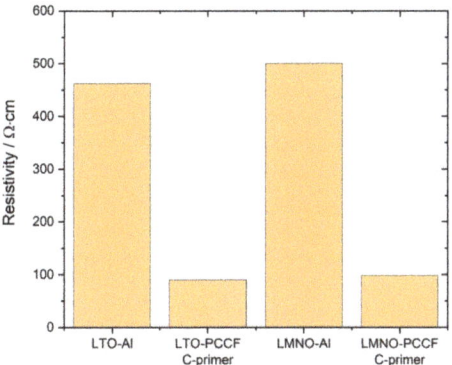

Figure 5. Resistivity of LTO and LMNO electrodes on Al and PCCF-C-primer collector (based on through-thickness measurement).

This was somewhat unexpected, since the measured resistivity of the bare PCCF was far higher compared to the Al collector. The measured electrode "through-thickness" resistance comprises a sum of collector bulk resistance, contact resistance between the coated films and electrode bulk resistance. The resistivity of the C-primer alone, prepared and measured as a bulk film, was determined to be 0.3 Ω·cm. This leads to the conclusion that the lateral in plane resistance on the surface of the C-primer coated PCCF-foil is rather low. The observed difference in electrical resistivity between electrodes based on Al or PCCF collector can be explained by a far lower contact resistance when using the developed C-primer in the case of PCCF collector. According to [28] the interfacial resistance accounts for a large portion of the whole impedance of an electrode without any treatments for the interfacial resistance reduction. One common procedure to reduce this interfacial resistance is the reduction of the electrode thickness by using a pressing technique, usually lamination or calander compression. However, this procedure leads to the reduction in pore size and volume, which causes the lithium-ion diffusivity resistance to increase [29,30]. Therefore, an optimized electrode porosity for most electrode material systems is around 35 vol.-% after densification, which is near to the 40 vol.% of the samples used in this study (see Table 3). Further, [28] demonstrated that a thin carbon under-coating layer, between the collector foil and the electrode film, can effectively decrease the impedance of the whole electrode. The microstructure of the electrodes on the Al collector in a cross section view (Figure 4a,b) shows, that at the interface between Al and electrode layer larger areas of "gaps", with only limited and more isolated contact, are visible. On the other side, the interface region between C-primer film and electrodes on PCCF collector (Figure 4c,d) is much more cohesively and interlocked. The rather soft C-primer film should lead to a better compression behavior with the electrode microstructure during the lamination densification, which leads to an overall lower contact and electrode resistance, which is highly important for bipolar battery concepts. In state-of-the-art battery manufacturing with Al-foil collector calandering, instead of lamination technique, is normally used for electrode densification. Since some issues for PCCF during calandering were observed (crack formation due to foil thickness variation), a lamination technique was used in this work. The authors suspect, that lamination technique can be a proper densification method for bipolar battery electrodes and forming a battery stack of bipolar plates, since two different active materials will be coated on one collector foil. However, we admit that calandering could lead to better results for metal Al-foil collector, since it could benefit from deformation ability of Al-metal. Nevertheless, the observed difference in through-thickness resistance of this study is quite remarkable.

2.2.2. Electrochemical Stability of PCCF Collector between 0 V to 5 V

One important requirement for the use of current collector materials in lithium-ion batteries is their voltage stability in a wide range of potential window during battery cycling. In case of a cathode-anode combination like Lithium Nickel Manganese Cobalt Oxide (NMC) and graphite this is 3.0 to 4.3 V vs. Li/Li$^+$ and in case of LMNO vs. graphite this is 3.0 to 5.0 V vs. Li/Li$^+$. LSV and CV tests regarding reduction and oxidation stability were conducted to verify the electrochemical stability of PCCF collector. Figure 6a shows a LSV curve of the uncoated PCCF collector cycled versus metallic lithium.

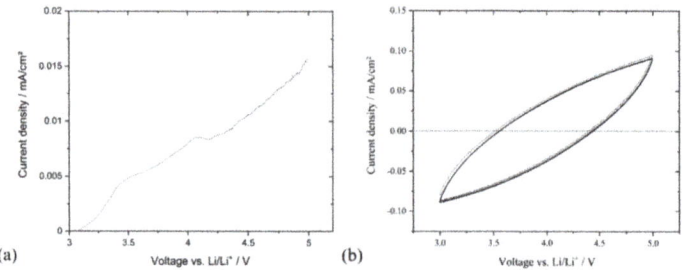

Figure 6. Electrochemical stability of PCCF collector: (**a**) LSV with 0.5 mV/s and (**b**) CV scan in liquid electrolyte (LP40) up to 5.0 V vs. Li/Li$^+$ 5 mV/s (plotted from 2nd cycle; additional drawn zero line to guide the readers eye).

The LSV curve in Figure 6a shows a near linear increase in current density up to 5.0 V vs. Li/Li$^+$. The visible two slight humpbacks at 3.5 and 4.0 V seem not to have a great effect on the further trend. We conclude that such anomalies arise from a small capacitance charge at the PCCF and not from an electrochemical degradation reaction. Moreover the measured highest current density with 0.02 mA/cm^2 is very low, underlining the fact that the PCCF is electrochemically stable up to 5.0 V vs. Li/Li$^+$. The CV-curve in Figure 6b shows the current density of the PCCF collector cycled 10 times between 3.0 and 5.0 V vs. Li/Li$^+$. The symmetrical shape of the curve indicates a capacity and no faradic reaction. The results demonstrate that the developed PCCF collector is compatible to the voltage range of NMC cathodes (approx. 4.3 V charging end potential) and also compatible to higher voltage materials like LMNO cathodes (5.0 V charging end potential), if long-term stable high-voltage liquid electrolytes are available.

Tests concerning the Li-ion intercalation into bare PCCF compared to C-primer coated PCCF collector were conducted to evaluate possible irreversible capacity losses due to Li-ion intercalation into carbon ingredients of PCCF (CNT-CB in PCCF as well as carbon black in C-primer layer).

In Figure 7a, the PCCF collector coated with C-primer shows two cathodic (reductive) peaks, which can be attributed to a solid electrolyte interface (SEI) formation (0.7 V) and a beginning of lithium intercalation (0 V). The first peak disappears after the first cycle, which supports the thesis for SEI formation. The following cycles show that the second peak slightly declines. The CV curve of PCCF collector without C-Primer (Figure 7b) shows a small intercalation peak at 0 V, but no affiliated de-intercalation peak. The comparison shows that the overall current density of the PCCF collector without C-primer coating is one order of magnitude lower (µA/cm^2 range) than the C-primer coated PCCF (low mAh/cm^2 range). It is suggested that a beginning of lithium intercalation into the C-Primer is visible, whereas the bare PCCF collector shows no intercalation behavior at all. To classify these observations, the CV-curves were integrated and the total amount of transported charge was calculated. Figure 8 shows these integrated negative and positive charge values from the oxidation and reduction parts of the CV curve.

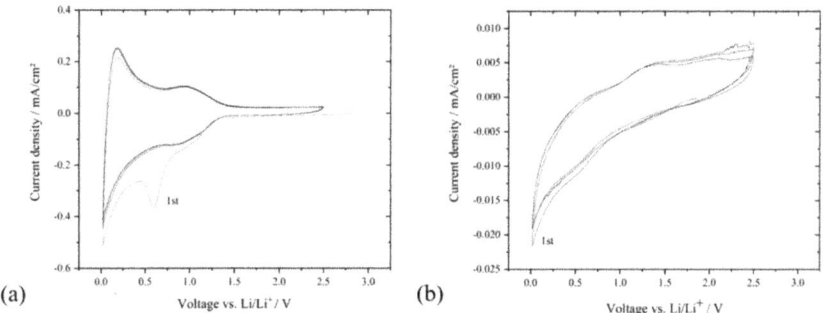

Figure 7. Electrochemical stability of (**a**) PCCF-C-primer collector and (**b**) bare PCCF collector. CV-curves between OCV and 0 V vs. Li/Li$^+$ 1 mV/s.

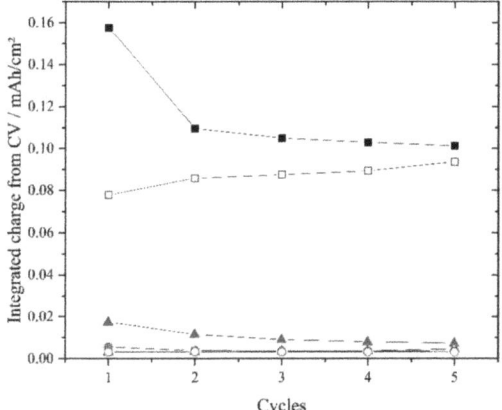

Figure 8. Sum of oxidation and reduction charges from CV test (Figure 7); black square: reduction of PCCF foil with C-primer; white square: oxidation of PCCF foil with C-primer; blue triangle: reduction of PCCF foil without C-primer, green triangle (is hiding under white dot curve): oxidation of PCCF foil without C-primer; green dot: reduction of copper foil; white dot: oxidation of copper foil.

For the PCCF collector with C-primer the charge values from reduction and oxidation tend to come close to each other after four cycles. A reversible intercalation and de-intercalation of 0.10 mAh/cm^2 in each cycle was observed. The bare PCCF collector without C-primer shows charge values almost identical to a tested copper foil collector under identical conditions, with no intercalation effects. In summary we conclude that the C-primer coating is more dominant for a small Li-ion intercalation compared to the bare PCCF collector. The overall observed capacity losses due to Li-ion intercalation into C-primer are below 0.2 mAh/cm^2. This is one magnitude lower compared to the area capacity of high energy (3–4 mAh/cm^2) or high-power electrodes (0.5–1.5 mAh/cm^2). A reduction of this effect can be expected by reducing the C-primer film thickness. Further, if an optimization of the surface microstructure of the PCCF could make the use of C-primer obsolete, it will avoid the observed capacity loss.

2.2.3. C-Rate Performance Test of LMNO and LTO on Al and PCCF Collector

In order to validate the functionality of the developed PCCF collector as an alternative current collector for Al foil, cycling tests with LMNO and LTO electrodes in monopolar half-cell configuration were conducted with performance tests from 0.1 to 5 C.

Figure 9 shows the cycling performance and of LTO on Al collector in comparison to LTO on PCCF-C-primer collector. The electrodes on Al collector show a reproducible capacity ranging from 167 mAh/g (0.1 C), 162 mAh/g (1 C) and 139 mAh/g (5 C) for higher C-rates. The cells with LTO on PCCF collector show 159 to 166 mAh/g (0.1 C), 153 to 164 mAh/g (1 C) and 108–127 mAh/g (5 C, with a slight decreasing trend). The coulombic efficiency in both tests, LTO on aluminum as well as LTO on PCCF, show high values above 0.99, which illustrates that no major side reaction is occurring. Single efficiency drops in both experiments, after changing to a 5 C cycle rate, are visible, which are attributed to mathematical artefacts from the efficiency calculation. During a constant 1 C cycling the coulombic efficiency stays constantly above 0.99. We observed that capacity values from the LTO electrodes on PCCF collector scatter more compared to Al collector. The reason is a larger deviation of the calculated electrode weight in the individual test cells, since the thickness of the prepared PCCF collector with nominal 70 μm thickness shows a higher thickness tolerance compared to the industrial 19 μm thick Al collector. This includes a possible thickness variation of 2 to 5 μm of the C-primer coating on PCCF. In sum, these deviations due to laboratory preparation methods will add up and lead to a slight variation of the calculated LTO mass, which was used to derive the collector cell capacities. It can be assumed, that by using scaled industrial manufacturing machines such deviations in collector thickness precision will be limited. The results demonstrate, that LTO electrodes on PCCF collector show comparable cell performance compared to electrodes on Al collector.

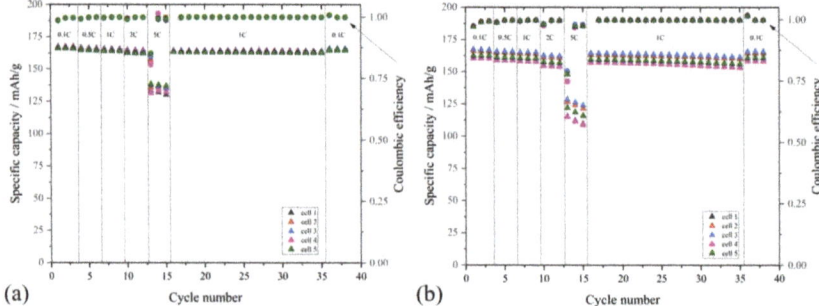

Figure 9. Cycling test of LTO electrodes on (**a**) Al (5 cells) and (**b**) PCCF-C-primer collector (5 cells).

Figure 10 shows the cycling performance test of LMNO on Al collector in comparison to PCCF-C-primer collector. The electrodes on Al collector show capacity values of 114 mAh/g (0.1 C), 103 mAh/g (1 C) and 52 to 78 mAh/g (5 C). The cells of LMNO on PCCF collector show 103 to 116 mAh/g (0.1 C), 95 to 106 mAh/g (1 C) and 67 to 90 mAh/g (5 C). The prepared LMNO test cells show a scattering of capacities by a given C-rate and a capacity decline within 20 cycles at 1 C, independently of collector type. Since a non-commercial, self-developed LMNO active material was used in this study, we attribute the capacity fade to the degradation of the active material [31]. The coulombic efficiency in the beginning of each cycle after changing the C-rate shows noticeably scattering values, which are attributed to artefacts from the mathematical efficiency calculation. Additionally, we did not observe any differences in the voltage curves from the cycling experiments, between electrodes on Al collector or PCCF. This underlines the fact that the PCCF is equivalent to an Al collector in terms of the here tested cycling behavior of the electrodes. The detailed voltage profiles of one coin cell out of five from the measurements in Figures 9 and 10 are shown in Figure S1 (in supplementary). For further development we propose long-term cycling studies to exclude effects, which might occur after many cycles and were out of scope for this work.

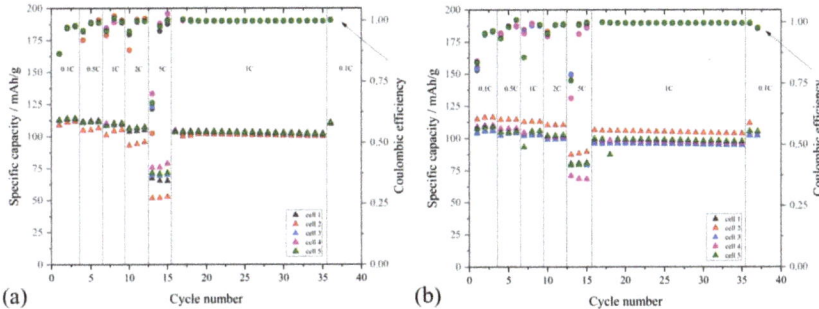

Figure 10. Cycling test of LMNO electrodes on (**a**) Al (5 cells) and (**b**) PCCF-C-primer collector (5 cells).

Table 6 summarizes the cell capacities at 0.1 C before and after a 20-cycle test at 1 C. The capacities as well as the cell resistance of the prepared samples are comparable between Al collector and PCCF collector. The observed lower electrode resistances for PCCF collector sheet samples (Table 4) has no visible influence on the overall cell resistance compared to Al collector, since the proportion to the total cell resistance is low (single digits of Ohms). The post-mortem disassembled cells show no visible degradation of the PCCF collector (Figures S2 and S3 in Supplementary).

Table 6. Average capacities at 0.1 C and cell resistances (based on 3 test cells) of LTO and LMNO electrodes before and after cycling test (20 cycles at 1 C).

Sample	Average Capacity at 0.1 C Begin of Cycling (mAh/g)	Average Capacity at 0.1 C End of Cycling (mAh/g)	Average Cell Resistance Begin of Cycling (Ω)	Average Cell Resistance End of Cycling (Ω)
LTO on Al	167 ± 0.6	166 ± 0.5	39 ± 7.1	18 ± 1.0
LTO on PCCF	165 ± 3.8	163 ± 3.9	40 ± 10.6	18 ± 1.9
LMNO on Al	114 ± 0.4	102 ± 1.0	24 ± 6.1	15 ± 2.2
LMNO on PCCF	108 ± 2.4	104 ± 2.0	28 ± 2.4	20 ± 2.8

Further, the observed capacity losses due to intercalation effects into the C-primer of PCCF collector (Figure 8) seem to be so small, that they show no visible influence on the overall cell capacities, which are several orders in magnitudes higher. Even possible electrochemical side reactions at high voltage of 5 V (see LSV test in Figure 6) seem to have no pronounced effect on the cycling performance of LMNO on PCCF collector.

The results show that the developed PCCF collector fulfills the electrochemical requirements to be used as an alternative current collector for lithium-ion batteries.

2.3. Discussion of PCCF as an Alternative Current Collector for Li-Ion Batteries

Table 7 compares the mass loading between standard Al and Cu collector and the developed PCCF polymer collector. The bulk density of the PCCF-C-primer collector is 1.4 g/cm^3 and 44 to 48% lower in comparison to state-of-the-art Al foil with 2.5 to 2.7 g/cm^3 and 84% lower compared to Cu foil with 8.9 g/cm^3. However, due to a much lower Cu- and Al foil thickness, the mass loading is higher for the PCCF collector manufactured in our laboratory. It can be assumed that progress in the manufacturing technology of such polymer collectors can lead to PCCF thicknesses in the range of 25 to 40 µm, which will decrease the mass loading of the collector to approx. 24% below Al foil. Further, if one day PCCF collectors with optimized surfaces without additional carbon primer coating are available, a potential saving in mass loading of 36% compared to state-of-the-art Al collector is possible. Compared to Cu foil, the developed PCCF of this study is comparable in mass loading and a PCCF optimization will lead to an even greater potential for mass saving compared to Al foil.

Table 7. Comparison of mass loading of Al- and Cu-foil compared to PCCF polymer collector.

Collector Material	Density (g/cm^3)	Typical Thickness (µm)	Mass Loading (mg/cm^2)	Mass Loading Relative to Al [*3] (%)
Al-foil	2.5–2.7	20–30 [*1]	5.0–8.1	100
Cu-foil	8.9	9–18 [*1]	8.0–16.0	160
PCCF-C-primer (this study)	1.4	70 [*2]	9.8	196
PCCF-C-primer (potential)	1.4 to 1.5 [*4]	25 to 40	3.8 to 6.0	76
PCCF composite (potential) [*5]	1.58	20 to 30	3.2 to 4.7	64

[*1] State-of-the-art thickness of Cu- and Al collector in industrial pouch cell [32]; [*2] PVDF-collector with 55 µm coated on both sides by carbon primer with 7.5 µm thickness; [*3] Based on thinnest version of each collector type; [*4] depending on thickness and porosity of C-primer; [*5] surface optimized PVDF-collector, which makes C-primer unnecessary.

Concerning material costs for both current collector types, a rough estimation can be done based on prices for laboratory scale developments: Purchasing prices of Al-foil of 26.85 EUR/kg, Cu-foil of 66.74 EUR/kg and raw materials costs of PCCF collector with PVDF polymer powder and carbon additives of 42.20 EUR/kg and approx. 80 EUR/kg for processing of PCCF due to composite compounding and film extrusion steps (5 kg batch). Table 8 shows, that the developed PCCF polymer collector exceeds the price of a commercial Al-foil by a factor of 9 and compared to Cu-foil by a factor of 2. However, since only small-scale laboratory consumables and equipment were used in this study, it can be expected, that by industrial scaling the raw material prices, compounding and extrusion costs as well as the PCCF thickness can be significantly reduced (38 EUR/kg and 25 µm thickness) to achieve a comparable and competitive price competitive with Al-and Cu-foil.

Table 8. Comparison of estimated collector costs based on laboratory scale consumables.

Component	Price (EUR/kg)	Thickness (µm)	Density (g/cm^3)	Volume (cm^3 @ 1 m^2)	Mass (g @ 1 m^2)	Price (EURct for 1 m^2)	Price Factor Rel. to Al
Al collector	26.85	20	2.5	20	50	1.34	1.0
Cu-collector	66.74	10	8.9	10	89	5.94	4.4
PCCF-C-primer (this study)	122.20	70	1.4	70	98	11.98	8.9
PCCF-C-primer (potential)	38.0	25	1.4	25	35	1.33	1.0

In comparison to metal collectors, the developed PCCF collector is compatible to a large variety of anode and cathode materials due to his wide potential range stability (0 to ~5 V vs. Li/Li$^+$). Further, since the PCCF is hermetically dense, it is suitable to be used in bipolar battery architectures, where alternative bimetal or carbon based collectors are usually facing issues of residual porosity and the possibility of internal battery short circuits [7,19]. One concern of using our developed PCCF collector is the ecologically impact for scaling to mass production. Regarding recycling of battery cell components, polyvinylidene fluoride (PVDF) normally decomposes during thermal treatment in volatile hydrogen fluoride, which can cause equipment corrosion and creates a potential environmental hazard. But recent studies indicate that the use of CaO as a reaction medium can avoid the release of hydrogen fluoride and reduce the processing costs during recycling [33]. Estimates about energy consumption of PCCF vs. Al collector for battery cell manufacturing are hard to assess. In state-of-the-art 10 Ah NMC cells, the Al cathode collector weights 6 to 16 times more than the PVDF binder in the cathode, depending if the cell is energy or power optimized. Another study regarding the energy consumption of Li-ion battery materials and production process [34], gives the information that the proportionally ratio in energy consumption for NMC-cathodes between Al (collector) and PVDF (binder for cathode) is 10 to

1 [35]. Based on these assumptions, we suppose that energy consumption for manufacturing of battery cells, where traditional Al collector is replaced by the developed PCCF collector (consists mainly of PVDF polymer), will be comparable.

3. Materials and Methods

3.1. Polymer-Carbon Collector Foil (PCCF)

A commercially available poly (vinylidene fluoride) (PVDF) was applied, namely Kynar720 (Arkema, Colombes Cedex, France) with a melt flow index of 5–29 g/10 min at 5.0 kg loading (230 °C). As the electrically conductive fillers, mixtures of branched multi-walled carbon nanotubes (b-MWCNTs) and carbon black (CB) were chosen. The b-MWCNT "CNS flakes" (Applied NanoStructured Solutions LLC, Baltimore, MD, USA) are coated with 3 wt.-% poly (ethylene)glycol and have a diameter of 14 ± 4 nm and length of ~70 µm (aspect ratio ~5000) [22]. The CB is a highly structured type of Ketjenblack EC600JD (Akzonobel, Cologne, Germany) with a BET surface area value of 1200 m^2/g and a primary particle size d_{50} of 34 nm (according to the supplier). For the polymer-carbon composite a combination of 1.0 wt.-% b-MWCNT with 3 wt.-% CB was used to achieve an optimized electrical conductivity.

Compounding was done via melt mixing by using a laboratory twin-screw extruder ZE 25 (KraussMaffei Berstorff GmbH, Hannover, Germany) with a screw with L/D ratio of 48. The pre-mixed PVDF powder with both carbon-fillers was compounded at a temperature of 210–230 °C, a rotation speed of 200 rpm and a throughput of 5 kg/h. For homogenization, the composite was extruded again under the same conditions. The extruded strands were granulated into approx. 2 mm diameter pellets. Cast film extrusion was performed with these composite granules using a 30 mm single-screw extruder (DAVO GmbH & Co. Polyrema KG, Troisdorf, Germany) in combination with a cast film line (Dr. Collin GmbH, Maitenbeth, Germany). The width of the flat die was 30 cm, the gap width was set to 100 µm, and the mass temperature was 290 °C. The take-off velocity was set to 3.7 m/min. Rolls of polymer-carbon collector foil (PCCF) with a width of 22 cm, a thickness of 55 µm and lengths of 50 m were achieved.

3.2. LMNO, LTO and C-Primer Electrode Coatings

$LiNi_{0.5}Mn_{1.5}O_4$ (LMNO) powder (PU110, synthesized by Fraunhofer IKTS described elsewhere [31]) and $Li_4Ti_5O_{12}$ (LTO) powder (HOMBITEC® LTO5, Huntsman Pigments and Additives, Duisburg, Germany) were used along with carbon black (SUPER P® Li, Imerys Graphite & Carbon, Bironico, Switzerland) and water-based polyacrylate (PAA) cathode binder BA-310C or anode binder BA-210S (15 wt.-% polymer solid content, both from Fujian Blue Ocean & Black Stone Techn. Co LTD, Zhangzhou, China). For the carbon primer, the BA-210S binder was used. Coatings were applied on Al foil (Li-ion battery grade, EN AW 1085-L H18, Hydro Aluminum Rolled Products GmbH [24]) or PCCF (see 3.3.1) as current collectors Prior to coating both current collector samples were cut to 12 × 12 cm^2 size.

For preparation of the carbon primer (C-primer) BA-210S binder was mixed with carbon black powder in deionized water with a dissolver (Dispermat LC, VMA Getzmann, Reichshof, Germany) up to a powder solid content of 6.2 wt.-%. The ratio of carbon black/BA-210S was set to 68.4/31.6 wt.-%. The slurry showed a viscosity of 4.0 Pas at a shear rate of 20 s^{-1} at 20 °C. PCCF collector was coated on both sides by a manual film applicator (Model 360, ERICHSEN GmbH & Co. KG, Hemer, Germany) with 120 µm blade gap and 90 mm coating width. In between both coating steps, a drying step at 60 °C for 24 h was set.

For the LMNO slurry, carbon black was first dispersed by a dissolver (Dispermat LC, VMA Getzmann, Germany) in BA-310C and additional deionized water (powder solid content of 2.8 wt.-%) followed by mixing with LMNO powder up to a powder solid content of 30.5 wt.-%. The quantity ratio of LMNO/carbon black/BA-310C binder was set to 85/6/9 wt.-%. The LMNO slurry pH was 6.2 and viscosity of 0.9 Pas at a shear rate of 20 s^{-1} at 20 °C. For the LTO slurry, carbon black was first dispersed by dissolver in BA-210S and additional deionized water (powder solid content of 3.0 wt.-%) followed by mixing with LTO powder up to a powder solid content of 31.6 wt.-%. The quantity ratio of LTO/carbon black/BA-210S binder was set to 85/6/9 wt.-%. The LTO slurry pH was 8.3 and viscosity of 5.4 Pas at a shear rate of 20 s^{-1} at 20 °C. Both LMNO and LTO slurries were casted on Al-foil and C-primer pre-coated PCCF collector by a manual doctor blade film applicator (LBT304 from Jokob Weiß & Söhne, Sinsheim, Germay) with 400 μm blade gap. The films were dried at 60 °C in a laboratory drying cabinet for 24 h. After drying, half of the electrodes were densified by pressing in an isostatic laminator at 250 bar, 70 °C for 10 min (IL-4012PC from Pacific Trinetics Corp, Fremont, USA).

3.3. Material Characterization

3.3.1. Polymer-Carbon Collector Foil (PCCF)

Tensile tests for determination of mechanical properties were performed with a tensile universal testing machine Z010 (ZwickRoell, Ulm, Germany) based on cut stripes of PCCF (length 115 mm, width 10 mm) with foil extrusion direction perpendicular or in extrusion direction and a displacement rate of 5 mm/min (according to DIN 53504/1A/5). The roughness and topography of the PCCF was investigated using a confocal 3D microscope μsurf (Nanofocus, Oberhausen, Germany), which derives a 3D image and the height profile. The characterization of filler dispersion in the PVDF composite by scanning electron microscopy (SEM) was performed on the foil surface using a Zeiss Ultra Plus microscope (Carl Zeiss AG, Oberkochen, Germany) in charge contrast imaging mode (CCI). To characterize the gas-tightness of the PCCF, a gas leak detector (air and helium) with PCCF sample size of 5 × 5 cm^2 was used (PhoenixL300 Leybold GmbH, Collogne, Germany).

3.3.2. Electrode Coatings

The collector foils and the prepared electrodes were characterized regarding thickness and weight using a mechanical thickness gauge for films and paper and a precision balance. Based on the measured thickness (D in μm), weight (in mg) and sample area size (in cm^2), values for area weight (AW = mass/area in mg/cm^2) and density (ρ = AW·10/D in g/cm^3) were calculated for current collectors and electrode coated samples. By subtraction of thickness and area weight of the current collector from the electrode coated samples, specific parameters (D, AW) were derived explicitly for the electrode film. By comparison of the electrode film density (ρ_{film} in g/cm^3) with the theoretical density of the electrode raw materials (sum of active material, carbon black and binder: $\rho_{film,th}$ in g/cm^3) the electrode film porosity was estimated ((1-$\rho_{film,th}$/ρ_{film})·100 in vol.%). The electrode cross-section, prepared by ion polishing was characterized by SEM (Crossbeam NVISION 40, Carl Zeiss SMT, Oberkochen, Germany).

3.3.3. Electrical Measurements

The electrical resistance of the PCCF collector in-plane was measured by a 4-point measurement using an device developed by the authors (see detailed description of this method in [22]). Therefore, PCCF samples of 30 by 25 mm size where metallized on the sample surfaces by a thin film of silver paste to reduce the contact resistance during measurement.

To evaluate electrical performances of electrodes coated on the PCCF foil two-point through plane resistance measurements were carried out. For this, electrode samples of 4 × 4 cm2 size were clamped between two copper plungers with a graphite fleece in between. The plungers were pressed with a force of 402 N (equals to 2.5 bar) against each other, measured by a force measuring sensor. A constant voltage of 10.0 V was used to supply the force sensor and the output voltage was measured with a Keithley 2700 multimeter. The resistance was measured with a milliohmmeter (HP 4338A). The resistance is internally calculated by applying a 1 kHz alternating current, and the sample impedance at 1 kHz was measured, which is in that case equal to the ohmic resistance. Values of uncoated as well as with electrodes coated PCCF current collectors were compared.

To characterize the film resistance of the thin C-primer, reference specimens of 40 µm thick and laminated C-primer films were prepared on ceramic substrates by using the carbon primer slurry and a manual film applicator (procedure is described elsewhere [36]).

3.4. Electrochemical Characterization

The PCCF current collector as well as the prepared LMNO and LTO electrodes on Al-foil and PCCF were electrochemically tested in cycling charge-discharge and cyclic voltammetry experiments by using a coin cell setup. The assembly was conducted in an argon filled glove box with an atmosphere of $O_2 < 2$ ppm and $H_2O < 2$ ppm. All materials used for this assembly, as well as the electrodes were pre-dried in a vacuum oven at 105 °C at 40 mbar for 24 hrs to ensure a complete removal of water residues. Lithium chips of thickness 300 µm (Xiamen Tob New Energy Technology, Xiamen, China) were used as a counter electrode and 150 µL of the electrolyte LP40 (BASF, Ludwigshafen, Germany) was soaked into two separators (FS3002-23, Freudenberg Performance Materials Holding SE & Co. KG, Weinheim, Germany). The assembly of the coin cell was carried out with a crimp machine (MT-160D, MTI Corp., Richmont, CA, USA). All electrode samples used for cyclic voltammetry (CV) were assembled into El-Cell test cells (ECC-Standard, El-Cell GmbH, Hamburg, Germany) under the same conditions as the coin cells in the glove box. In CV tests, the PCCF-foil was measured against metallic lithium with a LP 40 soaked separator (Freudenberg). CV experiments were carried out by a potentiostat (VMP3, BioLogic, Seyssinet-Pariset, France) in a climate chamber at 30 °C. Charge and discharge cycling experiments were conducted with a Basytec CTS potentiostat (Basytec GmbH, Asselfingen, Germany) in the same climate chambers (CTS T-40/50, CTS GmbH, Hechingen, Germany) at 30 °C. LMNO electrodes were cycled between 5.0 and 3.5 V vs. Li/Li$^+$ and LTO electrodes were cycled between 2.5 V and 1.0 V vs. Li/Li$^+$ in coin cells. For the charge and discharge experiments, 5 identical coin cells were manufactured and measured simultaneously (referenced as cell 1–5 in each experiment), adding up to 20 coin cells. To evaluate a possible degradation of PCCF collector after cell cycling, a post-mortem analysis was done by disassembling of test cells and visual inspection.

4. Conclusions

A polymer-carbon composite current collector foil (PCCF) for bipolar lithium-ion battery applications is developed and evaluated in comparison to state-of-the-art Al-foil collector. The PCCF shows sufficient mechanical properties, which allow the processing of the PCCF collector in a roll-to-roll industrial electrode coater. The PCCF proved to be hermetical dense, which is important to avoid liquid electrolyte penetration through the collector. The applicability for lithium-ion batteries was studied based on water-processed $LiNi_{0.5}Mn_{1.5}O_4$ (LMNO) cathode and $Li_4Ti_5O_{12}$ (LTO) anode coatings with the integration of a thin carbon primer at the interface to the collector. Despite the fact that the laboratory-manufactured PCCF shows a much higher film thickness of 70 µm compared to Al-foil of 19 µm, the electrode resistance was measured to be by a factor of five lower compared to Al collector, which was attributed to the low contact resistance between PCCF, carbon primer and electrode microstructure. The PCCF-C-primer collector shows a sufficient voltage stability up to 5 V vs. Li/Li+ and low Li-intercalation losses into the carbon primer of the PCCF (~0.1 mAh/cm^2), which makes him compatible to a wide range of anode and cathode active materials. Electrochemical cell tests demonstrate the applicability of the developed PCCF for LMNO and LTO electrodes, with no obvious disadvantage compared to Al collector. The advantage of a nearly 50% lower raw material density of the PCCF polymer collector compared to metal Al-foil along with expected improvements in collector thickness reduction and cost savings, due to a scaled industry manufacturing approach, will offer the possibility to significantly reduce the mass loading of the collector in the battery cell. Overall, the developed PCCF collector appears to be advantageous, especially for bipolar battery architectures, where a combination of the abovementioned properties is needed which cannot be fulfilled by today´s metal-, bimetal- or carbon-based collectors.

Supplementary Materials: The following are available online at http://www.mdpi.com/2313-0105/6/4/60/s1, Figure S1: Voltage profiles of measured coin cells from Figures 9 and 10 (first three cycles at 0.1 C); Top left: LTO on PCCF, Top right LTO on Al-collector, Bottom left: LMNO on PCCF, Bottom right: LMNO on Al, Figure S2: Post-mortem picture of PCCF-foil of a LTO cell after cycling test according to Figure 9; Left: PCCF C-primer side in contact to LTO electrode after cycling, Right: Backside of the PCCF after cycling of LTO in coin cell, Figure S3: Post-mortem picture of PCCF-foil of a LMNO cell after cycling test according to Figure 10; Left: PCCF C-primer side in contact with LMNO electrode coating (some separator residue white) after cycling on coin cell; Right: Backside of the PCCF after cycling of LMNO in coin cell.

Author Contributions: Conceptualization, M.F. and M.C.; methodology, M.F., M.C., B.K., P.P.; validation, M.F. and M.C.; investigation, M.F., M.C., P.M., B.K.; writing—original draft preparation, M.F., M.C., K.K., B.K. and P.P.; writing—review and editing, M.F.; visualization, M.F., M.C., B.K. and P.P.; supervision, M.W. and A.M.; project administration, M.W. All authors have read and agreed to the published version of the manuscript.

Funding: This research was funded by German Federal Ministry of Education and Research (BMBF) project EMBATT2.0 No. 03XP0068G and No. 03XP0068E.

Acknowledgments: The authors thank the processing laboratory of IPF, headed by I. Kühnert, and especially F. Pursche for carrying out the melt compounding and film extrusion experiments. We thank O. Kobsch (IPF) for performing roughness measurements and S. Höhn (IKTS, Ceramography and Phase Analysis) for sample preparation and SEM analysis.

Conflicts of Interest: The authors declare no conflict of interest.

References

1. Marsha, R.A.; Russell, P.G.; Reddy, T.B. Bipolar lithium-ion battery development. *J. Power Sources* **1997**, *65*, 133–141. [CrossRef]
2. Jung, K.; Shin, H.; Park, M. Solid-State Lithium Batteries: Bipolar Design, Fabrication and Electrochemistry. *ChemElectroChem* **2019**, *6*, 3842–3859. [CrossRef]
3. Iwakura, C.; Fukumoto, Y.; Inoue, H.; Ohashi, S.; Kobayashi, S.; Tada, H.; Abe, M. Electrochemical characterization of various metal collectors as a current collector of positive electrode for rechargeable lithium batteries. *J. Power Sources* **1997**, *68*, 301–303. [CrossRef]
4. Whitehead, A.H.; Schreiber, M. Current Collectors for Positive Electrodes of Lithium-Based Batteries. *J. Electrochem. Soc.* **2005**, *152*, A2105–A2113. [CrossRef]
5. Myung, S.-T.; Hitoshi, Y.; Sun, Y.-K. Electrochemical behavior and passivation of current collectors in lithium-ion batteries. *J. Mater. Chem.* **2011**, *21*, 9891–9911. [CrossRef]
6. Stich, M.; Fritz, M.; Roscher, M.; Peipmann, R. Korrosionsverhalten von bipolaren Stromableitern für Lithium-Ionen-Batterien (Teil 1). *Galvanotechnik* **2018**, *12*, 2330–2336.
7. Stich, M.; Fritz, M.; Roscher, M.; Peipmann, R. Korrosionsverhalten von bipolaren Stromableitern für Lithium-Ionen-Batterien (Teil 2). *Galvanotechnik* **2019**, *1*, 67–72.
8. Beikai, Z. The Height that a Kind of Bipolar Template Transitionality Unit Lithium Battery of Cu-Al Bimetal and Its Series Connection Are Formed Forces down Changeable Internal Damp Battery Heap and Method for Packing. China Patent 104916864B, 16 January 2018.
9. Culver, D.; Dyer, C.K.; Epstein, M.L. Modular Battery with Battery Cell Having Bimetallic End. U.S. Patent 20110200867, 18 August 2011.
10. Hossain, S. Bipolar Lithium-Ion Rechargeable Battery. U.S. Patent 5595839, 21 January 1997.
11. Braunovic, M.; Aleksandrov, N. Effect of electrical current on the morphology and kinetics of formation of intermetallic phases in bimetallic aluminum-copper joints. In Proceedings of the IEEE Holm Conference on Electrical Contacts, Pittsburgh, PA, USA, 27–29 September 1993; pp. 261–268. [CrossRef]
12. Silveria, V.L.A.; Mury, A.G. Analysis of the behavior of bimetallic joints (Al /Cu). *J. Microstruct. Sci.* **1987**, *14*, 277–287.
13. Abbasi, M.; Taheri, A.K.; Salehi, M.T. Growth rate of intermetallic compounds in Al/Cu bimetal produced by cold roll welding process. *J. Alloy. Compd.* **2001**, *319*, 233–241. [CrossRef]
14. Fritsch, M.; Standke, G.; Heubner, C.; Langklotz, U.; Michaelis, A. 3D-cathode design with foam-like aluminum current collector for high energy density lithium-ion batteries. *J. Energy Storage* **2018**, *16*, 125–132. [CrossRef]

15. Li, S. Corrosion of Aluminum Current Collector in Cost Effective Rechargeable Lithium-Ion Batteries. Dissertation, No.1384, University of Wisconsin-Milwaukee, USA. 2016. Available online: https://dc.uwm.edu/etd/1384 (accessed on 4 December 2020).
16. Ma, T.; Xu, G.-L.; Li, Y.; Wang, L.; He, X.; Zheng, J.; Liu, J.; Engelhard, M.H.; Zapol, P.; Curtiss, L.A.; et al. Revisiting the Corrosion of the Aluminum Current Collector in Lithium-Ion Batteries. *J. Phys. Chem. Lett.* **2017**, *8*, 1072–1077. [CrossRef] [PubMed]
17. Kim, W.S.; Cho, K.Y. Current Collectors for Flexible Lithium Ion Batteries: A Review of Materials. *J. Electrochem. Sci. Technol.* **2015**, *6*, 1–6. [CrossRef]
18. Foreman, E.; Zakri, W.; Sanatimoghaddam, M.H.; Modjtahedi, A.; Pathak, S.; Kashkooli, A.G.; Garafolo, N.G.; Farhad, S. A Review of Inactive Materials and Components of Flexible Lithium-Ion Batteries. *Adv. Sustain. Syst.* **2017**, *1*, 1700061. [CrossRef]
19. Wang, K.; Wu, Y.; Wu, H.; Luo, Y.; Wang, D.; Jiang, K.; Li, Q.; Li, Y.; Fan, S.; Wang, J. Super-aligned carbon nanotube films with a thin metal coating as highly conductive and ultralight current collectors for lithium-ion batteries. *J. Power Sources* **2017**, *351*, 160–168. [CrossRef]
20. Evanko, B.; Yoo, S.J.; Lipton, J.; Chun, S.-E.; Moskovits, M.; Ji, X.; Boettcher, S.W.; Stucky, G.D. Stackable bipolar pouch cells with corrosion resistant current collectors enable high-power aqueous electrochemical energy storage. *Energy Environ. Sci.* **2018**, *11*, 2865–2875. [CrossRef]
21. Krause, B.; Barbier, C.; Kunz, K.; Pötschke, P. Comparative study of singlewalled, multiwalled, and branched carbon nanotubes melt mixed in different thermoplastic matrices. *Polymer* **2018**, *159*, 75–85. [CrossRef]
22. Kunz, K.; Krause, B.; Kretzschmar, B.; Juhasz, L.; Kobsch, O.; Jenschke, W.; Ulrich, M.; Pötschke, P. Direction dependent electrical conductivity of polymer/carbon filler composites. *Polymers* **2019**, *11*, 591. [CrossRef]
23. Krause, B.; Kunz, K.; Kretzschmar, B.; Kühnert, I.; Pötschke, P. Effect of filler synergy and cast film extrusion parameters on extrudability and direction-dependent conductivity of PVDF/carbon nanotube/carbon black composites. *Polymers* **2020**. Submitted for publication.
24. Hampel, U.; Denkmann, V.; Siemen, A.; Eckhard, K.; Schenkel, W.; Eberhard, S.; Bögershausen, D. Chemically Treated Current Collector Foil Made of Aluminium or an Aluminium Alloy. U.S. Patent 9160006B2, 13 October 2015.
25. Kunz, K.; de Limé, A.d.B.; Seeba, J.; Reuber, S.; Wolter, M.; Michaelis, A. Possibilities for Processing of All Solid State Batteries/Components as Pilot Plant Scale. Poster, Dresden Battery Days, 23–25 September 2019. Available online: https://www.energy-saxony.net/veranstaltungen/dresden-battery-days-23-25092019.html (accessed on 4 December 2020).
26. Sacher, E. (Ed.) *Metallization of Polymers 2*; Springer: Boston, MA, USA, 2002; ISBN 978-1-4615-0563-1. [CrossRef]
27. Nakanishi, S.; Suzuki, T.; Cui, Q.; Akikusa, J.; Nakamura, K. Effect of surface treatment for aluminum foils on discharge properties of lithium-ion battery. *Trans. Nonferr. Met. Soc. China* **2014**, *24*, 2314–2319. [CrossRef]
28. Nara, H.; Mukoyama, D.; Shimizu, R.; Momma, T.; Osaka, T. Systematic analysis of interfacial resistance between the cathode layer and the current collector in lithium-ion batteries by electrochemical impedance spectroscopy. *J. Power Sources* **2019**, *409*, 139–147. [CrossRef]
29. Heubner, C.; Nickol, A.; Seeba, J.; Reuber, S.; Junker, N.; Wolter, M.; Schneider, M.; Michaelis, A. Understanding thickness and porosity effects on the electrochemical performance of $LiNi_{0.6}Co_{0.2}Mn_{0.2}O_2$-based cathodes for high energy Li-ion batteries. *J. Power Sources* **2019**, *419*, 119–126. [CrossRef]
30. Heubner, C.; Schneider, M.; Michaelis, A. Diffusion-Limited C-Rate: A Fundamental Principle Quantifying the Intrinsic Limits of Li-Ion Batteries. *Adv. Energy Mater.* **2020**, *10*, 1902523. [CrossRef]
31. Seidel, M.; Kugaraj, M.; Nikolowski, K.; Wolter, M.; Kinski, I.; Jähnert, T.; Michaelis, A. Comparison of Electrochemical Degradation for Spray Dried and Pulse Gas Dried $LiNi_{0.5}Mn_{1.5}O_4$. *J. Electrochem. Soc.* **2019**, *166*, A2860–A2869. [CrossRef]
32. Kovachev, G.; Schröttner, H.; Gstrein, G.; Aiello, L.; Hanzu, I.; Martin, H.; Wilkening, R.; Foitzik, A.; Wellm, M.; Sinz, W.; et al. Analytical Dissection of an Automotive Li-Ion Pouch Cell. *Batteries* **2019**, *5*, 67. [CrossRef]
33. Wang, M.; Tan, Q.; Liu, L.; Li, J.; Facile, A. Environmentally Friendly, and Low-Temperature Approach for Decomposition of Polyvinylidene Fluoride from the Cathode Electrode of Spent Lithium-ion Batteries. *ACS Sustain. Chem. Eng.* **2019**, *7*, 12799–12806. [CrossRef]
34. Berga, H.; Zackrisson, M. Perspectives on environmental and cost assessment of lithium metal negative electrodes in electric vehicle traction batteries. *J. Power Sources* **2019**, *415*, 83–90. [CrossRef]

35. Emilsson, E.; Dahllöf, L. *Lithium-Ion Vehicle Battery Production—Status 2019 on Energy Use, CO2 Emissions, Use of Metals, Products Environmental Footprint, and Recycling*; Report number C 444; IVL Swedish Environmental Research Institute Ltd.: Stockholm, Sweden, 2019; ISBN 978-91-7883-112-8.
36. Nanomanufacturing—Key Control Characteristics—Part 4-3: Nano-Enabled Electrical Energy Storage—Contact and Coating Resistivity Measurements for Nanomaterials, DIN IEC/TS 62607-4-3:2018-07. 2018. Available online: https://dx.doi.org/10.31030/2534286 (accessed on 4 December 2020).

Publisher's Note: MDPI stays neutral with regard to jurisdictional claims in published maps and institutional affiliations.

© 2020 by the authors. Licensee MDPI, Basel, Switzerland. This article is an open access article distributed under the terms and conditions of the Creative Commons Attribution (CC BY) license (http://creativecommons.org/licenses/by/4.0/).

Article

Effect of the Particle Size Distribution on the Cahn-Hilliard Dynamics in a Cathode of Lithium-Ion Batteries

Pavel L'vov [1,2,*] and Renat Sibatov [1,2]

[1] Laboratory of Diffusion Processes, Ulyanovsk State University, 432017 Ulyanovsk, Russia; ren_sib@bk.ru
[2] Institute of Nanotechnology of Microelectronics of the Russian Academy of Science, 119991 Moscow, Russia
* Correspondence: LvovPE@sv.uven.ru

Received: 4 April 2020; Accepted: 11 May 2020; Published: 15 May 2020

Abstract: The phase-field model based on the Cahn-Hilliard equation is employed to simulate lithium intercalation dynamics in a cathode with particles of distributed size. We start with a simplified phase-field model for a single submicron particle under galvanostatic condition. We observe two stages associated with single-phase and double-phase patterns typical for both charging and discharging processes. The single-phase stage takes approximately 10–15% of the process and plays an important role in the intercalation dynamics. We establish the laws for speed of front propagation and evolution of single-phase concentration valid for different sizes of electrode particles and a wide range of temperatures and C-rates. The universality of these laws allows us to formulate the boundary condition with time-dependent flux density for the Cahn-Hilliard equation and analyze the phase-field intercalation in a heterogeneous cathode characterized by the particle size distribution.

Keywords: lithium-ionbattery; Cahn-Hilliard equation; intercalation; particle size distribution

1. Introduction

Currently, lithium-ion batteries (LIB) are one of the most common devices for energy storage [1]. Many works are devoted to understanding the lithium intercalation/deintercalation mechanisms into electrodes and to optimization of LIB components for better electrochemical characteristics (see reviews [2,3]). The most widely accepted electrochemical model of LIB is the so-called pseudo-two-dimensional (P2D) model [4,5], where the intercalation and deintercalation in electrode particles, ion transport in electrolyte and separator are described in terms of the normal diffusion equations. The generalization to the anomalous diffusion case is proposed in [6]. However, in several electrode materials, the mutual solubility of the lithiated and delithiated phases is very low, and two phases coexist in a wide range of the state of charge (SOC). Such behavior is typical for charging and discharging in cathode based on lithium iron phosphate and in anode materials based on lithium titanate [7]. For this reason, lithiation and delithiation of the cathode particles can be associated with reversible phase transition in quasi-binary system, where the intercalant concentration can vary in the range from 0 to 1 during charging or discharging. The corresponding theoretical approach is based on the phase-field theory [8–16] that is commonly employed in studies on phase transitions [17–19].

The fact that the electrodes consist of particles of various sizes is often neglected. For example, the classic single particle model (SPM) and P2D model assume the electrodes consisting of spherical particles of identical size. Nonetheless, electrodes are typically made of porous materials with particles of different sizes and shapes. As a result, the medium in LIB is a highly heterogeneous system. Neglecting the particle size distribution (PSD) largely underestimates the capacity in the case of elevated C-rates [20], and the SPM approaches poorly describe the battery performance at higher current densities. PSD is an important factor in the degradation of batteries [21,22], and it may be altered during battery operation due to cracking or agglomeration of particles [21]. In [23], the multiple-particle model indicates that the PSD broadening may lead to higher values of volumetric capacity and energy density. On the other hand, this broadening can cause amplification of electrode polarization [23].

There are several papers (see e.g., [20,23–26]) where the particle size dispersion is considered within the diffusion approach. In most cases, the actual PSD is described by finite number of particle groups [20,25,26]. The multiple-particle approach meets computational difficulties related to the need to solve diffusion equations for each particle bin [26].

In the present paper, we extend the phase-field model based on the Cahn-Hilliard equation to simulate lithium intercalation dynamics in a cathode with particles of distributed size. Due to PSD, during charging/discharging, number of active electrode particles and their interfacial flux density depend on time. The boundary condition with time-dependent flux density for the Cahn-Hilliard equation is formulated from the universal laws established for speed of front propagation and evolution of single-phase concentration in individual particles. Calculation of the time-dependent flux density is performed by the developed general approach that can be employed for analysis of intercalation process in particle ensemble with arbitrary PSD.

2. Phase-Field Intercalation in a Single Spherical Particle

We start with a simplified phase-field model for a single submicron particle under galvanostatic condition. Expecting to establish some general rules of phase-field intercalation in a spherical particle, we analyze the dynamics for wide ranges of particle size and C-rate. The process of intercalation is simulated in terms of the Cahn-Hilliard (CH) equation [14,16,18,19]:

$$\frac{\partial c}{\partial t} = M\nabla^2 \left[\frac{\partial f}{\partial c} - \kappa \nabla^2 c \right], \quad (1)$$

where $c \equiv c(r,t)$ is concentration field of intercalating atoms that defines the local state of charge of a cathode particle, M is the mobility, $f(c)$ is the free energy of mixing in a quasi-binary system, κ is the gradient energy coefficient. In this study, the free energy of mixing $f(c)$ of quasi-binary solid solution is considered in the regular solution approximation [17–19]:

$$f(c) = \Omega c(1-c) + k_B T [c \ln c + (1-c) \ln(1-c)],$$

Ω is the interaction parameter, T is the temperature, and k_B is the Boltzmann constant. The influence of elastic strain is neglected. Assuming spherical symmetry of cathode particles, we use the Laplace operator of the form: $\nabla^2 = \frac{\partial^2}{\partial r^2} + \frac{2}{r}\frac{\partial}{\partial r}$.

The insertion and extraction of intercalating atoms are simulated at the temperature of 300 K. The mobility of intercalated atom M in a cathode particle is related to the diffusion coefficient through the Einstein relation, $M = D/(k_B T)$. The diffusion coefficient of intercalating atom (e.g., lithium) for different cathode materials is $D = 10^{-13}$–10^{-15} m^2/s at the considered temperature [4,8,13]. The interaction parameter Ω for different cathode materials lies in the range of 0.059–0.193 eV/atom [8–16]. The gradient energy coefficient can be related to the width of equilibrium profile [17,18] that is usually of several nanometers [8,10]. In this study, we employ the diffusion coefficient $D = 10^{-14}$ m^2/s, interaction parameter $\Omega = 0.115$ eV/atom. The gradient energy coefficient is usually written as $\kappa = \eta \Omega r_0^2$ [8,14,17,18], where r_0 is the intermolecular distance and η is the dimensionless coefficient depending on form of interaction potential [18]. In this study, the value of gradient energy coefficient is $\kappa = 0.228$ eVnm^2.

The interaction parameter allows us to calculate phase diagram of the system (Figure 1). The equilibrium values of intercalating atom concentration are $c_{b1} = 0.013$ and $c_{b2} = 0.987$ at the temperature of 300 K. The unstable region lies in the interval between $c_{s1} = 0.129$ and $c_{s2} = 0.871$.

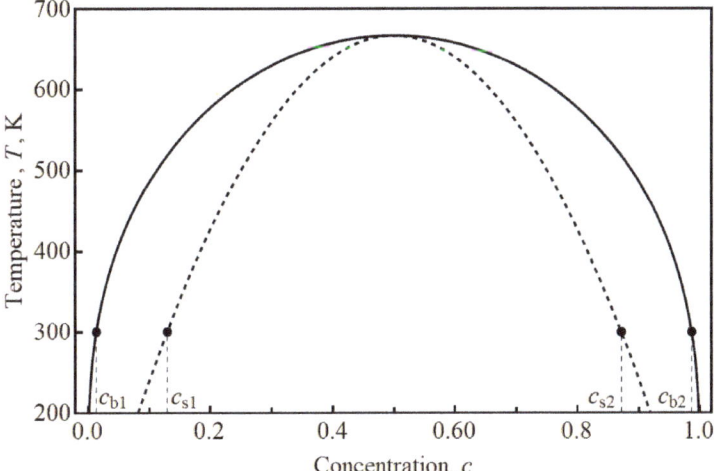

Figure 1. Phase diagram for regular binary solution with interaction parameter of $\Omega = 0.115$ eV/atom.

The CH Equation (1) is solved under natural boundary conditions corresponding to the galvanostatic mode:

$$\left.\frac{\partial c}{\partial r}\right|_{0,R_0} = 0, \quad \left.\frac{\partial \mu}{\partial r}\right|_0 = 0, \quad \left.\frac{\partial \mu}{\partial r}\right|_{R_0} = -\frac{j_r}{M}. \tag{2}$$

Here, $\mu = \partial f/\partial c - \kappa \nabla^2 c$ is the chemical potential defining the flux density of intercalating atoms. Initial distribution of the atoms is uniform and is characterized by equilibrium values of concentration, i.e., $c_0 = c_{b1}$ for insertion and $c_0 = c_{b2}$ for extraction process. The CH Equation (1) with boundary conditions (2) is solved numerically by using the explicit Euler time integration scheme [27].

Figure 2 demonstrates the evolution of concentration profile of intercalating atoms computed with the introduced values of the model parameter. The insertion and extraction processes are simulated for the cathode particle with radius of $R_0 = 0.1$ μm under C-rate of 1C and 10C. Also, we simulated the insertion and extraction dynamics for cathode particle with radius $R_0 = 0.2$ μm under C-rate equal to C/2.

Analyzing the results of insertion and extraction of intercalating atoms in cathode particles, we can identify some important features of these processes. At the very beginning of the processes, the concentration of intercalant decreases uniformly over the particle volume (see lines 1 and 2 in Figure 2a–f). At this stage, the particle corresponds to the single-phase pattern. The uniformity is explained by fast equilibration of atom distribution due to high diffusion coefficient and small size of the particle. The stage continues until the nearest value of metastability limit (c_{s1} or c_{s2}) is achieved. At this stage, the average intercalant concentration $\bar{c}$ changes linearly with time (Figure 3). The approximate dependence can be easily obtained from the CH Equation (1):

$$\bar{c} = c_0 - \frac{3j_r t}{R_0}, \tag{3}$$

that agrees well with the results of direct solution of CH Equation (Figure 3). The relation (3) is valid at the time interval $0 \leq t \leq t_s$. From Equation (3) we can calculate the duration of the single-phase stage as $t_s = (c_{b2} - c_{s2})R_0/(3|j_r|)$ for extraction and $t_s = (c_{s1} - c_{b1})R_0/(3|j_r|)$ for insertion process.

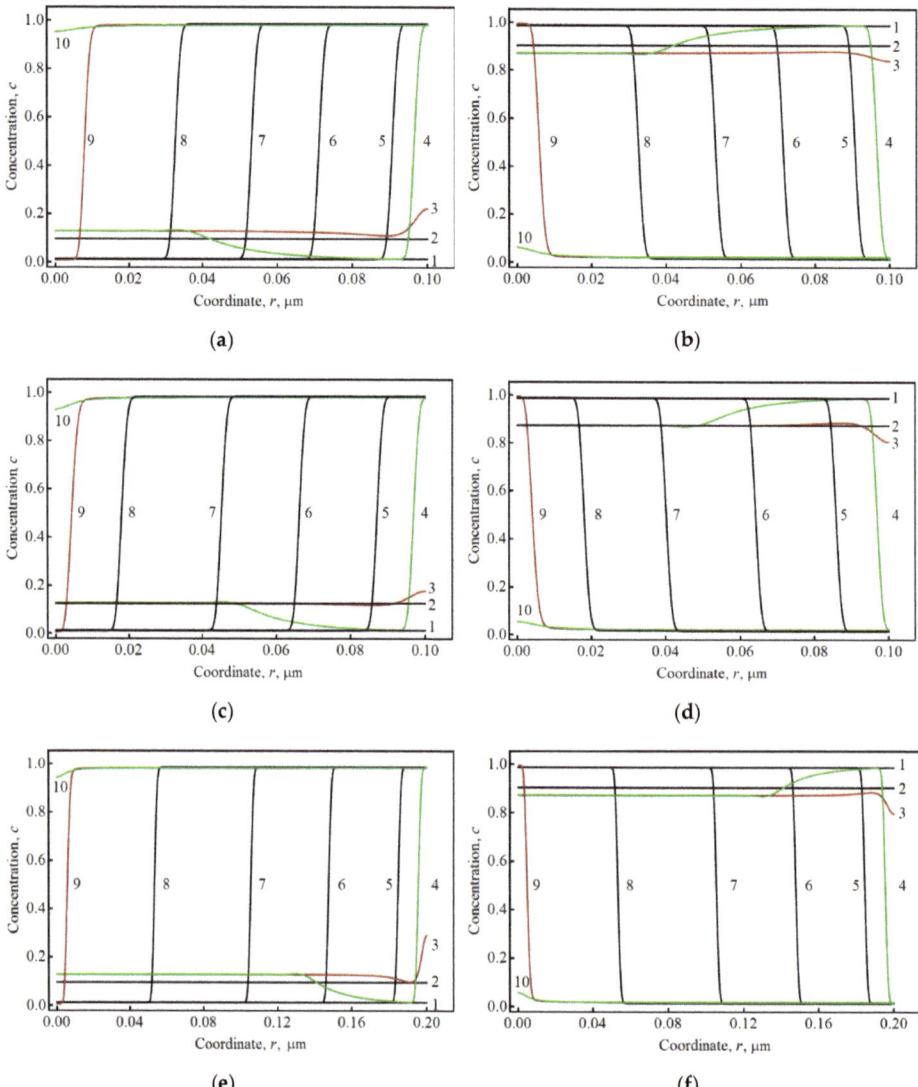

Figure 2. Concentration profiles of intercalating atoms for different time points at $T = 300$ K. Corresponding radius of particle and C-rate are: (**a**) 0.1 µm, 1C insertion, (**b**) 0.1 µm, 1C extraction, (**c**) 0.1 µm, 10C insertion, (**d**) 0.1 µm, 10C extraction, (**e**) 0.2 µm, 0.5C insertion, (**f**) 0.2 µm, 0.5C extraction. The lines 3, 4, 9 and 10 correspond to passing from single-phase stage to double-phase stage and vice versa.

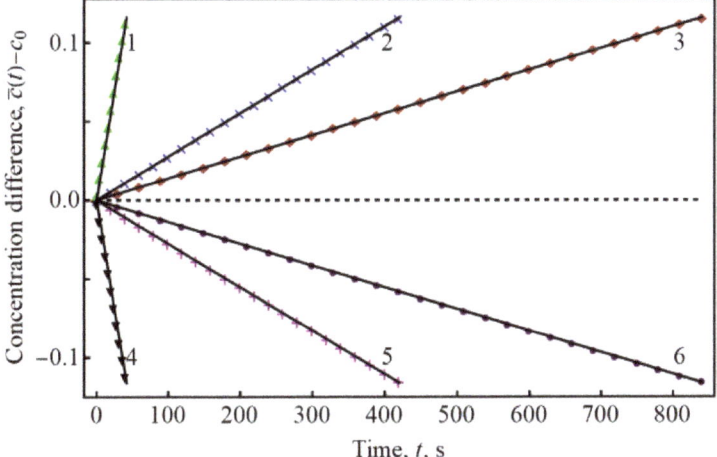

Figure 3. Kinetics of average concentration at the single-phase stage of insertion (1–3) and extraction (4–6) processes. Points are the solution of the CH equation, lines correspond to the linear dependence (3).

After that the concentration profile dramatically changes. Interfacial region depletes in very short time interval and the particle passes to double-phase pattern (see lines 3 and 4 in Figure 2a–f). At this stage, the intercalant redistributes over the particle volume and passes to the nearest equilibrium composition (see lines 4 and 5 in Figure 2a–f). Transition from single-phase to double-phase stage takes place for the short time interval estimated as ~0.2 s and ~0.5 s for particles with size of 0.1 µm (1C) and 0.2 µm, respectively. The longer time interval for larger particle is explained by the presence of additional concentration wave moving toward the particle center (see line 4 in Figure 2a–f) during equilibration of concentration profile. Overpotential of electrode depends on interfacial concentration of intercalating atoms [4,5,8]. Therefore, it is expected that passing from single-phase to double-phase regime can cause the abrupt change of overpotential. The effect could be used for determination of metastability limit.

The next stage is characterized by motion of the concentration wave from the particle interface to the center (lines 5–9 in Figure 2a–f). The position of concentration wave can be associated with coordinate $R_h(t)$ corresponding to concentration $c_h = (c_{b1} + c_{b2})/2$. The position of the wave front $R_h(t)$ satisfies the linear dependence

$$\frac{R_0^3 - R_h^3(t)}{R_0^2} = \frac{3|j_r|t}{(c_{b2} - c_{b1})}, \quad (4)$$

obtained from the solution of the CH Equation (1). The Equation (4) is valid for $t > t_s$ (after the single-phase stage). The results of simulation by the CH Equation (Figure 2a–f) confirm Equation (4) (see Figure 4). The end of the double-phase stage corresponds to the time point $t_d = R_0(c_{b2} - c_{b1})/(3|j_r|)$ for both insertion and extraction processes. The relation can be employed for calculation of the flux density j_r corresponding to 1C-rate, if duration of charging or discharging process equals to $t_d = 3600$ s.

At the end of insertion or extraction process, the system passes over unstable region to another single-phase state that corresponds to fully charged or discharged state of a particle (lines 9 and 10 in Figure 2a–f). The process of insertion or extraction is stopped, when the corresponding equilibrium composition is achieved.

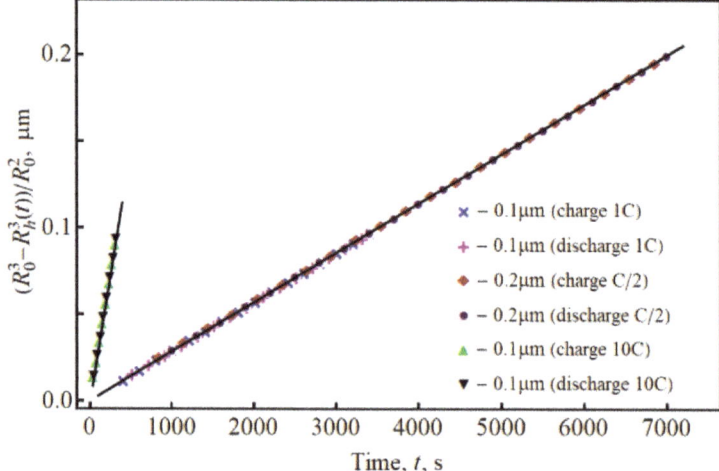

Figure 4. Kinetics of concentration wave propagation obtained by the CH Equation (points). Lines represent the linearized dependence (4). Size of particles and applied C-rate are indicated in Figure 2.

Contribution of the single-phase and double-phase stages can be estimated by ratio $\tau_s = t_s/t_d$. This ratio can be easily found from Equations (3) and (4):

$$\tau_{s1} = \frac{c_{s1} - c_{b1}}{c_{b2} - c_{b1}}, \quad \tau_{s2} = \frac{c_{b2} - c_{s2}}{c_{b2} - c_{b1}}$$

for charging and discharging, respectively. The value of τ_s is defined by the phase diagram and depends on temperature only. In case of symmetric phase diagram, this parameters are equal to each other $\tau_{s1} = \tau_{s2}$. Calculation of the ratio in the range 250–400 K gives contribution of single-phase stage about 10–15%. Simulation of insertion and extraction for lower diffusion coefficient reveals deviation from concentration profiles at the single-phase stage only if the diffusion coefficient is less than ~10^{-17} m^2/s.

Remarkably, variation of insertion and extraction rates in the range 1–10 C at the temperature of 300 K does not modify the two-stage mechanism and relations (3) and (4) remain valid (see Figures 2–4). Thus, Equations (3) and (4) can be used for determination of the composition profile at any time point of insertion or extraction process.

3. Intercalation in Particles of Distributed Size

Real cathodes consist of particles of various sizes and can be characterized by certain PSD. Under the assumption that the current is equally distributed over the active surface, smaller particles are charged or discharged for shorter time interval and withdraw earlier from the intercalation process. Larger particles have to balance the total current change in the galvanostatic mode. Therefore, flux density at the particle surface turns to depend on time $j_r = j_r(t)$.

Let us modify the obtained relations with respect to size distribution of cathode particles. If the flux density at the particle interface depends on time after integration of CH equation we obtain for single-phase and double-phase regimes:

$$\bar{c}(t) = c_0 - \frac{3}{R_0} \int_0^t j_r(t) dt, \quad 0 \leq t \leq t_s,$$

$$\frac{R_0^3 - R_h^3(t)}{R_0^2} = \frac{3}{(c_{b2} - c_{b1})} \int_0^t |j_r| dt, \quad t_s < t < t_d. \tag{5}$$

Whereas the total amount of intercalating atoms is conserved, the relation for total time of insertion or extraction process can be written in the form

$$\frac{1}{3}R_0(c_{b2} - c_{b1}) = \int_0^{t_d} |j_r(t)|dt. \tag{6}$$

Equation (6) can be used for definition of threshold value R_{min} defining the minimal size of particle that can participate in intercalation process

$$\frac{dR_{min}}{dt} = \frac{3|j_r(t)|}{c_{b2} - c_{b1}}. \tag{7}$$

Then we assume that particles with $R \leq R_{min}$ are excluded from the intercalation process (they are fully charged/discharged). The total current redistributes over the interface of residual particles with $R > R_{min}$. Therefore, the galvanostatic mode can be determined by the following equation:

$$j_r(t) \int_{R_{min}}^{\infty} w(R) R^2 dR = j_0 \langle R^2 \rangle, \tag{8}$$

where j_0 is the flux density at the beginning of intercalation process and $w(R)$ is the size distribution function of cathode particles, $\langle R^2 \rangle$ is the squared radius averaged over the whole ensemble of cathode particles.

Combining Equations (7) and (8) leads to the equation for the threshold value R_{min}

$$\frac{dR_{min}}{dt} = \frac{3j_0 \langle R^2 \rangle}{(c_{b2} - c_{b1}) \int_{R_{min}}^{\infty} w(R) R^2 dR}. \tag{9}$$

Let us assume that ensemble of cathode particles is described by the gamma distribution

$$w(R) = \frac{R^{m-1}}{a^m \Gamma(m)} \exp\left(-\frac{R}{a}\right) \tag{10}$$

that is usually employed as PSD for different cathode materials [28,29]. Here a and m are the constant parameters and $\Gamma(m)$ is the gamma function. The expectation value and dispersion of the distribution are $\langle R \rangle = ma$ and $\sigma^2 = ma^2$, respectively.

Integration of Equation (9) with respect to PSD (10) gives the implicit time-dependence of R_{min}

$$\frac{a^3}{\Gamma(m)}\left[\Gamma(m+3) + \frac{R_{min}}{a}\Gamma\left(m+2, \frac{R_{min}}{a}\right) - \Gamma\left(m+3, \frac{R_{min}}{a}\right)\right] = \frac{3j_0 \langle R^2 \rangle t}{(c_{b2} - c_{b1})}. \tag{11}$$

The threshold radius R_{min} alters from zero to infinity, and the charging/discharging time is determined by

$$t_{max} = \frac{a^3(c_{b2} - c_{b1})\Gamma(m+3)}{3j_0 \langle R^2 \rangle \Gamma(m)}.$$

Figure 5 shows the time-dependence of threshold particle radius R_{min} in dimensionless coordinates for different values of parameter m. The results in these coordinates do not depend on parameter a (see Equation (11)).

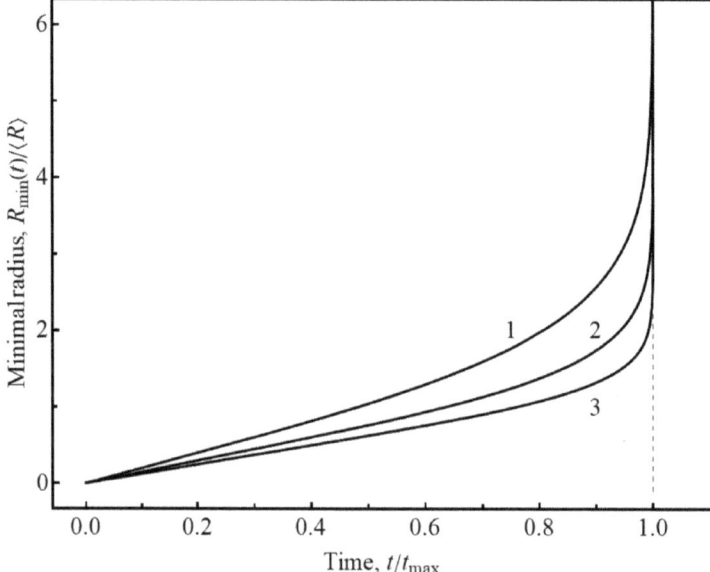

Figure 5. Dynamics of minimal radius R_{min} of active particle in process of intercalation. Lines are the results of analytical solution (11) for different values of parameter m: (1) $m = 2$, (2) $m = 4$, (3) $m = 8$.

Analyzing Figure 5, we conclude that R_{min} depends linearly on time at the beginning of insertion or extraction process that can be described by dependence

$$\frac{R_{min}}{\langle R \rangle} = \frac{\Gamma(m+3)}{m\Gamma(m+2)} \cdot \frac{t}{t_{max}}. \qquad (12)$$

The time interval corresponding to linear dependence is characterized by a constant flux density j_r. After that the flux density j_r of intercalating atoms grows very quickly and goes to infinity when time approaches to t_{max}.

Taking Equation (7) into account, one can transform Equations (5) into

$$\bar{c}(t) = c_0 - (c_{b2} - c_{b1})\frac{R_{min}(t)}{R_0}, \quad 0 \le t < t_s, \qquad (13a)$$

$$\frac{R_0^3 - R_h^3(t)}{R_0^2} = R_{min}(t), \quad t > t_s. \qquad (13b)$$

These equations describe the dynamics of intercalant distribution over the electrode particle at the single-phase and double-phase stage with respect to PSD.

Let us consider the impact of the PSD parameters on the extraction process from the particle ensemble characterized by PSD (10). Here, $\langle R \rangle = 0.1$ μm and $m = 4$. We solve the CH Equation (1) for particle with size of $R_0 = 0.2$ μm and time-dependent interfacial flux density $j_r(t)$ that defines the boundary condition (2).

The time-dependence of flux density $j_r(t)$ can be calculated by Equations (7) and (11). To represent $R_{min}(t)$ given by Equation (11) in explicit form, we choose the following approximation

$$\frac{R_{min}}{\langle R \rangle} = \sum_{m=0}^{n} a_m \left(\frac{t}{t_{max}}\right)^{2m+1} + b \operatorname{arctanh}\left(\frac{t}{t_{max}}\right), \qquad (14)$$

where n is the order of approximation, a_m and b are fitting parameters. Using this approximation, we obtain the explicit dependence of the interfacial flux density for $n = 0$ in the form

$$j_r(t) = j_0\Big(0.512826 + 0.487174/(1-(t/t_{max})^2)\Big).$$

The initial flux density j_0 corresponds to C-rate of 1C for cathode particle with average size $\langle R \rangle$.

The result of simulation of the extraction process from an individual particle of the ensemble is shown in Figure 6. The radius of this particle is chosen equal to $R_0 = 0.2$ µm. The concentration profiles at different times are similar to Figure 2f. Analyzing Figure 6, one can conclude that solution of the CH Equation (1) with time-dependent interfacial flux density agrees well with the general Equation (13b) at the double-phase stage. The size distribution function of the cathode particles causes decrease of the total time of extraction that corresponds to time interval Δt in Figure 6.

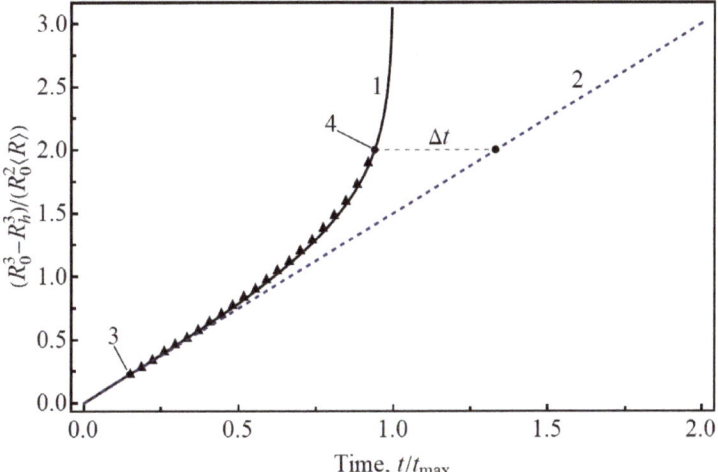

Figure 6. The time-dependence of concentration wave position R_h in dimensionless units for cathode particle with the size of $R_0 = 0.2$ µm at the double-phase stage. The particle is the part of ensemble described by PSD given by Equation (10) with $\langle R \rangle = 0.1$ µm and $m = 4$. Solid line 1 corresponds to solution of Equation (11). Dashed line 2 is the linear dependence corresponding to constancy of the interfacial flux density. Points 3 and 4 correspond to moments of start and finish of double-phase stage for the particle. Triangle points correspond to solution of the CH equation.

The time-dependent flux density is approximately constant at the single-phase stage, therefore the dependence of average concentration $\bar{c}$ at this stage is almost identical to that in Figure 2f. Generally, the time-dependent flux density can influence the average concentration $\bar{c}$ dynamics at the single-phase stage for large particles. The large particles are expected to deviate from the linear dependence that is observed under constant flux density (see Figure 3).

4. Conclusions

We extended the phase-field model based on the Cahn-Hilliard equation to the lithium intercalation dynamics in a cathode with particles of distributed size. Considering the simplified phase-field intercalation in a single submicron spherical particle under galvanostatic condition, we established universal behavior observed for different sizes of electrode particles and a wide range of temperatures and C-rates. The universal rules were formulated for speed of front propagation and evolution of single-phase concentration. Two stages associated with single-phase and double-phase patterns are typical for both charging and discharging processes. The single-phase stage takes approximately 10–15% of the process and plays an important role in the intercalation dynamics.

The particle size distribution causes the time-dependence of the interfacial flux density under galvanostatic conditions. The analytical results for a particle ensemble are presented for the case of the gammadistribution of particle sizes, and can be easily generalized to other PSDs. The universality of the established laws allowed us to formulate the boundary condition with time-dependent flux density for the Cahn-Hilliard equation. Numerical solutions of the corresponding boundary-value problem agree well with the obtained universal relations.

Thus, we obtained thegeneral analytical relations describing the intercalation process in ensemble of submicron electrode particle characterized by the arbitrary PSD. Other approaches based on the diffusion equation for intercalating atoms can consider finite number of particle bins only and have higher computational cost.

The Cahn-Hilliard equation and obtained approximate relations on the phase-field intercalation can be used for modification of the well-known SPM and P2D models of LIBs.

Author Contributions: Conceptualization, P.L. and R.S.; methodology, P.L.; software, P.L.; validation, P.L. and R.S.; formal analysis, P.L.; writing—original draft preparation, P.L.; writing—review and editing, R.S.; visualization, P.L.; supervision, P.L. and R.S.; project administration, R.S. All authors have read and agreed to the published version of the manuscript.

Funding: The authors thank the Russian Science Foundation (project no. 19-71-10063) for financial support.

Conflicts of Interest: The authors declare no conflict of interest.

List of Symbols

a	Parameter of gamma distribution function
a_k	Fitting parameter of time-dependence $R_{\min}(t)$
b	Fitting parameter of time-dependence $R_{\min}(t)$
$c(r,t)$	Concentration of intercalating atom
c_{b1}, c_{b2}	Concentration of intercalating atom in equilibrium states of binary system ($c_{b2} > c_{b1}$)
$\bar{c}$	Average concentration of intercalant in cathode particle at the single-phase stage
c_h	Half-sum of equilibrium compositions of binary system $c_h = (c_{b1} + c_{b2})/2$
D	Diffusion coefficient, m^2/s
$f(c)$	Free energy of mixing of binary system, J/atom
j_r	Interfacial flux density of intercalating atoms, m/s
j_0	Initial flux density of cathode particle ensemble, m/s
M	Mobility of intercalating atoms, m^2/(Js)
m	Parameter of gamma distribution function
t	Time, s
t_s	Time point of the end of single–phase stage, s
t_d	Time point of the end of double-phase stage, s
$t_{\max}$	Duration of charging or discharging process of cathode particle ensemble, s
T	Temperature, K
k_B	Boltzmann constant, J/K
r_0	Intermolecular distance, m
R, R_0	Radius of individual cathode particle, m
$R_{\min}$	Threshold value of radius of active cathode particle ($R > R_{\min}$), m
R_h	Coordinate of concentration wave front at the double-phase stage, m
$w(R)$	Particle size distribution function
$\Gamma(m)$	Gamma function
κ	Gradient energy coefficient, Jm^2
Ω	Interaction parameter of binary system, J
η	Dimensionless coefficient in gradient energy coefficient $\kappa = \eta \Omega r_0^2$
τ_s	Ratio of duration of single-phase and double-phase stages
∇^2	Laplace operator, m^{-2}

References

1. Liu, C.; Cao, G. Fundamentals of Rechargeable Batteries and Electrochemical Potentials of Electrode Materials. In *Nanomaterials for Energy Conversion and Storage*; Wang, D., Cao, G., Eds.; World Scientific (Europe): London, UK, 2018; pp. 394–451.
2. Yuan, L.X.; Wang, Z.H.; Zhang, W.X.; Hu, X.L.; Chen, J.T.; Huang, Y.H.; Goodenough, J.B. Development and challenges of LiFePO$_4$ cathode material for lithium-ion batteries. *Energy Environ. Sci.* **2011**, *4*, 269–284. [CrossRef]
3. Zhang, W.J. Structure and performance of LiFePO$_4$ cathode materials: A review. *J. Power Sources* **2011**, *196*, 2962–2970. [CrossRef]
4. Doyle, M.; Fuller, T.F.; Newman, J. Modeling of galvanostatic charge and discharge of the lithium/polymer/insertion cell. *J. Electrochem. Soc.* **1993**, *140*, 1526–1533. [CrossRef]
5. Hariharan, K.S.; Tagade, P.; Ramachandran, S. *Mathematical Modeling of Lithium Batteries*; Springer International Publishing: Cham, Switzerland, 2018.
6. Sibatov, R.T.; Svetukhin, V.V.; Kitsyuk, E.P.; Pavlov, A.A. Fractional differential generalization of the single particle model of a lithium-ion cell. *Electronics* **2019**, *8*, 650. [CrossRef]
7. Yaroslavtsev, A.B.; Kulova, T.L.; Skundin, A.M. Electrode nanomaterials for lithium-ion batteries. *Rus. Chem. Rev.* **2015**, *84*, 826. [CrossRef]
8. Cogswell, D.A.; Bazant, M.Z. Coherency Strain and the Kinetics of Phase Separation in LiFePO$_4$ Nanoparticles. *ACS Nano* **2012**, *6*, 2215–2225. [CrossRef]
9. Fleck, M.; Federmann, H.; Pogorelov, E. Phase-field modeling of Li-insertion kinetics in single LiFePO$_4$-nano-particles for rechargeable Li-ion battery application. *Comp. Mater. Sci.* **2018**, *153*, 288–296. [CrossRef]
10. Tang, M.; Huang, H.-Y.; Meethong, N.; Kao, Y.-H.; Carter, W.C.; Chiang, Y.-M. Model for the particle size, overpotential, and strain dependence of phase transition pathways in storage electrodes: Application to nanoscaleolivines. *Chem. Mater.* **2009**, *21*, 1557–1571. [CrossRef]
11. Santhanagopalan, S.; Guo, Q.; Ramadass, P.; White, R.E. Review of models for predicting the cycling performance of lithium ion batteries. *J. Power Sources* **2006**, *156*, 620–628. [CrossRef]
12. Han, B.; der Ven, A.V.; Morgan, D.; Ceder, G. Electrochemical modeling of intercalation processes with phase field models. *Electrochim. Acta* **2004**, *49*, 4691–4699. [CrossRef]
13. Zhang, T.; Kamlah, M. Phase-field modeling of the particle size and average concentration dependent miscibility gap in nanoparticles of LiMn$_2$O$_4$, LiFePO$_4$, and NaFePO$_4$ during insertion. *Electrochim. Acta* **2019**, *298*, 31–42. [CrossRef]
14. Huttin, M.; Kamlah, M. Phase-field modeling of stress generation in electrode particles of lithium ion batteries. *Appl. Phys. Lett.* **2012**, *101*, 133902. [CrossRef]
15. Burch, D.; Bazant, M.Z. Size-dependent spinodal and miscibility gaps for intercalation in nanoparticles. *NanoLett.* **2009**, *9*, 3795–3800. [CrossRef] [PubMed]
16. Zelič, K.; Katrašnik, T. Thermodynamically consistent derivation of chemical potential of a battery solid particle from the regular solution theory applied to LiFePO$_4$. *Sci. Rep.* **2019**, *9*, 1–13. [CrossRef] [PubMed]
17. Cahn, J.W.; Hilliard, J.E. Free Energy of a Nonuniform System. I. Interfacial Free Energy. *J. Chem. Phys.* **1958**, *28*, 258–267.
18. Provatas, N.; Elder, K. *Phase-Field Methods in Material Science and Engineering*; John Wiley & Sons: Weinheim, Germany, 2010.
19. L'vov, P.E.; Svetukhin, V.V. Simulation of the first order phase transitions in binary alloys with variable mobility. *Model. Simul. Mater. Sci. Eng.* **2017**, *25*, 75006. [CrossRef]
20. Farkhondeh, M.; Delacourt, C. Mathematical modeling of commercial LiFePO$_4$ electrodes based on variable solid-state diffusivity. *J. Electrochem. Soc.* **2011**, *159*, A177–A192. [CrossRef]
21. Vetter, J.; Novák, P.; Wagner, M.R.; Veit, C.; Möller, K.C.; Besenhard, J.O.; Winter, M.; Wohlfahrt-Mehrens, M.; Vogler, C.; Hammouche, A. Ageing mechanisms in lithium-ion batteries. *J. Power Sources* **2005**, *147*, 269–281. [CrossRef]
22. Capone, I.; Hurlbutt, K.; Naylor, A.J.; Xiao, A.W.; Pasta, M. Effect of the Particle-Size Distribution on the Electrochemical Performance of a Red Phosphorus–Carbon Composite Anode for Sodium-Ion Batteries. *Energy Fuels* **2019**, *33*, 4651–4658. [CrossRef]

23. Röder, F.; Sonntag, S.; Schröder, D.; Krewer, U. Simulating the impact of particle size distribution on the performance of graphite electrodes in lithium-ion batteries. *Energy Technol.* **2016**, *4*, 1588–1597. [CrossRef]
24. Wu, S.; Yu, B.; Wu, Z.; Fang, S.; Shi, B.; Yang, J. Effect of particle size distribution on the electrochemical performance of micro-sized silicon-based negative materials. *RSC Adv.* **2018**, *8*, 8544–8551. [CrossRef]
25. Nagarajan, G.S.; Van Zee, J.W.; Spotnitz, R.M. A mathematical Model for Intercalation Electrode Behaviour. *J. Electrochem. Soc.* **1998**, *145*, 771–779. [CrossRef]
26. Majdabadi, M.M.; Farhad, S.; Farkhondeh, M.; Fraser, R.A.; Fowler, M. Simplified electrochemical multi-particle model for LiFePO$_4$ cathodes in lithium-ion batteries. *J. Power Sources* **2015**, *275*, 633–643. [CrossRef]
27. Biner, S.B. *Programming Phase-Field Modeling*; Springer: Cham, Switzerland, 2017.
28. Westhoff, D.; Feinauer, J.; Kuchler, K.; Mitsch, T.; Manke, I.; Hein, S.; Latz, A.; Schmidt, V. Parametric stochastic 3D model for the microstructure of anodes in lithium-ion power cells. *Comput. Mater. Sci.* **2017**, *126*, 453–467. [CrossRef]
29. Westhoff, D.; Manke, I.; Schmidt, V. Generation of virtual lithium-ion battery electrode microstructures based on spatial stochastic modeling. *Comput. Mater. Sci.* **2018**, *151*, 53–64. [CrossRef]

© 2020 by the authors. Licensee MDPI, Basel, Switzerland. This article is an open access article distributed under the terms and conditions of the Creative Commons Attribution (CC BY) license (http://creativecommons.org/licenses/by/4.0/).

Article

The Physical Manifestation of Side Reactions in the Electrolyte of Lithium-Ion Batteries and Its Impact on the Terminal Voltage Response

Bharat Balagopal * and Mo-Yuen Chow

Department of Electrical & Computer Engineering, North Carolina State University, Raleigh, NC 27606, USA; chow@ncsu.edu
* Correspondence: bbalago@ncsu.edu

Received: 28 August 2020; Accepted: 21 October 2020; Published: 31 October 2020

Abstract: Batteries as a multi-disciplinary field have been analyzed from the electrical, material science and electrochemical engineering perspectives. The first principle-based four-dimensional degradation model (4DM) of the battery is used in the article to connect the interdisciplinary sciences that deal with batteries. The 4DM is utilized to identify the physical manifestation that electrolyte degradation has on the battery and the response observed in the terminal voltage. This paper relates the different kinds of side reactions in the electrolyte and the material properties affected due to these side reactions. It goes on to explain the impact the material property changes has on the electrochemical reactions in the battery. This paper discusses how these electrochemical reactions affect the voltage across the terminals of the battery. We determine the relationship the change in the terminal voltage has due to the change in the design properties of the electrolyte. We also determine the impact the changes in the electrolyte material property have on the terminal voltage. In this paper, the lithium ion concentration and the transference number of the electrolyte are analyzed and the impact of their degradation is studied.

Keywords: sensitivity; electrolyte; lithium ion battery; degradation; 4DM; terminal voltage; side reactions

1. Introduction

Lithium-based batteries are dominating the battery market because of their high energy density and rapidly decreasing manufacturing cost per kWh. While these batteries have many advantages, they also have disadvantages such as safety and recycling. Recycling of lithium ion batteries is a threefold process that involves pyrometallurgy (treatment with heat), hydrometallurgy (treatment with acid/liquid) and recycling through physical processes such as separation by weight. Recycling was mainly performed to recover the rare-earth metals that are hard or expensive to find and mine and hazardous materials that are toxic for the environment [1]. By recycling the used or spent lithium ion batteries, it is possible to recover up to 70% of the cathode material that is made up of rare earth metals [2]. However, the cost of recycling lithium ion batteries is increasing because of the increase in the complexity of lithium ion battery chemistries to ensure stability and improved tolerance to charging rates and temperatures [3]. Safety is a major concern in lithium ion batteries because they are designed to have highly combustible agents (such as organic solvents in electrolytes) and combustion inducing agents (electrochemical reactions that generate heat) in a sealed container. When operating normally, the electrochemical reactions generate very little heat and therefore prevent any kind of combustion or explosion. However, if subjected to extreme operating conditions (e.g., high charging/discharging currents, high temperatures, etc.) these agents can react violently and result in explosions [4]. Inappropriate operation can also lead to dendrite formations that can cause an internal short between the cathode and the anode and result in explosive reactions [5]. The electrolyte is one of the important components in the battery where heat or gas generation can cause problems. This is because most electrolytes in lithium ion batteries are dissolved in organic solvents that are highly flammable [6]. To ensure that these batteries operate as they are designed to, battery management/monitoring systems (BMS) are developed to continuously monitor the states of the battery such as the State of Charge (SOC), State of Health (SOH), Remaining Useful Life (RUL), State of Function (SOF) and temperature of operation. The BMS also monitors the charging and discharging operations of the battery to ensure that the operating currents are within the rated specifications of the battery and that the upper and lower cut-off voltage limits are not exceeded [7]. The temperature of operation of the battery also has a very important role in its operation and performance. When the battery is operated at higher temperatures, the electrolytic resistance decreases initially and then begins to dissociate resulting in an increase in the resistance between the electrodes. Similarly, when the temperature of the battery drops below the operating range, the electrolyte begins to coagulate, resulting in an increase in the resistance to the flow of lithium ions between the electrodes [8]. The C-rate or charging/discharging rate plays a crucial role in the degradation of the battery as well. Using very high C-rates can lead to deposition of lithium ions instead of intercalation. Deposition of the lithium ions will result in loss of active material and lithium inventory and cause the battery to degrade faster [9]. To better understand the operation of lithium ion batteries, a physics-based modeling approach is used to represent the lithium ion battery and its components [9]. Most batteries have four major components—electrodes, electrolytes, separators and current collectors. The electrodes, positive and negative, are the regions where electrochemical reactions take place that generate electrons. The electrolyte acts as a charge transportation medium between the positive and negative electrode and vice versa based on the mode of operation (i.e., charging or discharging) [10]. The electrolyte in lithium ion batteries is often lithium salts such as lithium hexafluorophosphate ($LiPF_6$) dissolved in an organic solvents usually ethylene carbonate (EC) and di-methyl carbonate (DMC) [11]. The ratio of the EC and DMC is determined by the dielectric property and the viscosity requirements of the electrolyte. EC contributes to the dielectric property while DMC makes the electrolyte less viscous [12]. The dielectric property contributes to the charge holding capability and the viscosity determines the resistance to the flow of ions between the electrodes. The separator provides electrical isolation between the electrodes and is doused in the electrolyte to enable movement of ions through the separator. To reduce the effect of self-discharge, the electrolyte is designed to have very high ionic conductivity and minimal electronic conductivity, which means the electrolyte offers low resistance to lithium ion movement and very high resistance to the flow of electrons. This high ionic conductivity and low electronic conductivity is achieved by dissolving the $LiPF_6$ in an organic solvent—the EC and DMC combination. The organic solvent ensures that the electrolyte offers high impedance to electron flow, and the dissolved $LiPF_6$ ensures that it offers a low resistance to lithium ion flow. The electrolyte of the battery in this paper is $LiPF_6$ dissolved in a 2:1 EC:DMC solution [13].

This paper discusses the physical manifestations of side reactions that happen in the electrolyte and the impact these manifestations have on the terminal voltage of the battery. It simulates degradation of the electrolyte through degradation of the salt diffusion coefficient and the transference number and generates the voltage profile when the battery model is subjected to a constant discharge current of 0.4C for a fixed duration of 4500 s or until the lower limit of the terminal voltage (3.5 V) is reached. The degradation of the electrolyte parameters was simulated in intervals of 10% so as to determine the sensitivity of the terminal voltage to the degradation of the parameter in consideration.

This article is organized as follows: Section 2 discusses the side reactions that take place in the battery, Section 3 presents the electrolyte salt diffusion coefficient degradation and the impact it has on the terminal voltage of the battery, Section 4 describes the impact of electrolyte transference number degradation on the voltage across the terminals of the battery and Section 5 concludes the paper and provides a discussion on the future work planned for this research area. The abbreviations, units and initial values of all the parameters used in the simulation are described in Table A1.

2. Side Reactions

The side reactions in a battery are highly reliant on the battery's operating parameters. Based on operating conditions such as temperature and charging and discharging C-rates, there are three areas where side reactions can occur—at the electrode–electrolyte interface [14], at the electrode–current collector interface and in the electrolyte itself.

2.1. Electrode–Electrolyte Interface

At the electrode–electrolyte interface, the side reaction causes an increase in the thickness of a solid electrolyte interface. The solid electrolyte interface (SEI) is a passivation layer that is designed by the manufacturer during the creation of the battery to isolate the electrode from the electrolyte as shown in Figure 1 [15]. If the battery is exposed to temperatures outside it's nominal operating range (higher or lower) or high charging/discharging rates, there is a significant rise in the loss of lithium inventory because of changes in the electrochemical reactions inside the battery. When operated under high charging/discharging rates, the lithium ions react with the SEI and cause a chemical reaction that results in the depletion of the SEI. Upon undergoing this kind of reaction, the anode is exposed to the electrolyte. The anode exposure to the electrolyte causes chemical reactions that produce compounds that inhibit the charge-producing electrochemical reactions [16]. The battery also undergoes different kinds of stress: charging/discharging stress, mechanical stress, temperature-based stress, etc. [17–19]. This paper focuses on the mechanical stress that the battery's electrodes undergo and the resulting side reactions. This kind of stress can cause the battery's electrode to separate from the current collector and cause a barrier to the current flow between the current collector and the electrode. This phenomenon is electrically represented as a growth in the internal resistance of the battery [20]. Exceeding the upper and lower cut-off voltage by forcing the battery to charge or discharge beyond the manufacturer's specification results in a chemical reaction between the current collectors, the electrode and the electrolyte. This corrosive reaction causes a passivation layer to form between the electrode and current collector. The passivation layer hinders the transfer of electrons from the electrode to the current collector and thus causes the resistance of the battery to increase [21].

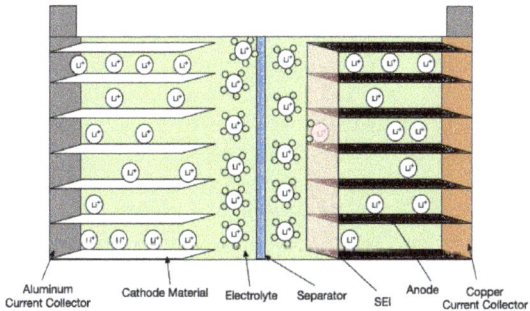

Figure 1. Structure of the battery with manufacturer designed SEI.

2.2. Electrode–Current Collector Interface

The electrode–current collector interface is where the battery is able to accept the electrons from the external circuit to complete the electrochemical reaction. During the design of the battery, the electrode is deposited on the current collector, made of highly conductive metals such as aluminum and copper, to prevent any loss of contact. However, either due to improper design or wear and tear of the electrode or current collector, there may be loss of contact between the current collector and the electrode. This loss of contact between the two surfaces can result in an increase in the internal impedance because of the gap between the surfaces and thus cause energy loss [22].

2.3. Electrolyte

The ideal electrolyte of the battery is a transport medium to enable lithium ions to move without resistance between electrodes. The electrolyte forms a temporary chemical bond with the lithium ions as they move from one electrode to the other. When the lithium ion reaches the other electrode, the SEI acts as a filter, removing the electrolyte surrounding the lithium ion and permitting only the lithium ion to flow through the SEI and into the cathode or anode [23]. This phenomenon is particularly important at the anode because when the lithium ion is bonded with the electrolyte, it has a much bigger size than the expansion capability of the graphite electrode. If the lithium ion with the electrolyte intercalate into the anode, then it results in uneven expansion and cracking of the anode. This cracking of the anode exposes it to the electrolyte and results in a growth of SEI and irreversible compounds. Figure 1 shows the temporary bond that the lithium ion makes with the electrolyte as it moves through the electrolyte and the SEI filtering the electrolyte to enable proper intercalation into the graphite anode. However, when subjected to extreme operating conditions such as high C-rates and temperatures, the electrolyte begins to interact with the lithium ions being transported and forms permanent chemical compounds. Since batteries generate and store energy through electrochemical reactions, the temperature of operation has an important role in the kind of reaction that takes place. When subjected to high temperatures, the ethylene carbonate in the electrolyte solution reacts with lithium ions to form more SEI and ethylene gas as shown in Equation (1) [24]. The ethylene gas causes the expansion of the battery cell because there is no exhaust or outlet for the gas to escape. This expansion applies pressure on the electrode and causes it to crack and results in more electrode material being exposed to the electrolyte. With more electrolyte-electrode exposure, the degradation rate is increased and more electrolyte and electrode material are lost. Equation (1) is an example of one of the side reactions that takes place in the battery that results in SEI formation.

$$\underbrace{(2(CH_2)_2CO_3)}_{\text{ethylene carbonate}} + 2e^- + 2Li^+ \rightarrow \underbrace{((CH_2OCO_2Li)_2)}_{\text{SEI}} + \underbrace{(CH_2 = CH_2 \uparrow)}_{\text{(ethylene gas)}}. \tag{1}$$

Another degradation phenomena that affects the electrolyte is dissociation when subjected to either very high temperatures or potentials. When the electrolyte is subjected to very high temperatures or potential differences across it, the chemical compounds begin to break down. This breakdown of the chemicals results in the inability of the electrolyte to act as a transportation medium between the electrodes.

This paper simulates the degradation of the electrolyte by varying the electrochemical properties of the electrolyte: the salt diffusion coefficient and the transference number using the first principle-based 4 dimensional degradation model (4DM) as shown in Figure 2 [24]. The terminal voltage is studied based on the degradation of these parameters and conclusions are drawn in terms of its sensitivity to the degradation of the electrolyte. The electrolyte salt diffusion coefficient and the transference number are also interdependent. Their interdependence can be observed in Equation (3) where the change in the effective diffusion coefficient is directly proportional to the transference number of the electrolyte. When either of the parameters change, there will be a more severe impact on the rate of change of the terminal voltage of the battery. While the 4DM framework is capable of simulating the interdependencies between the different parameters of the electrolyte, this article focuses on the sensitivity of each individual component of the battery on the performance and voltage response. As a result, the interdependencies are not considered in this article and will be presented in a separate article.

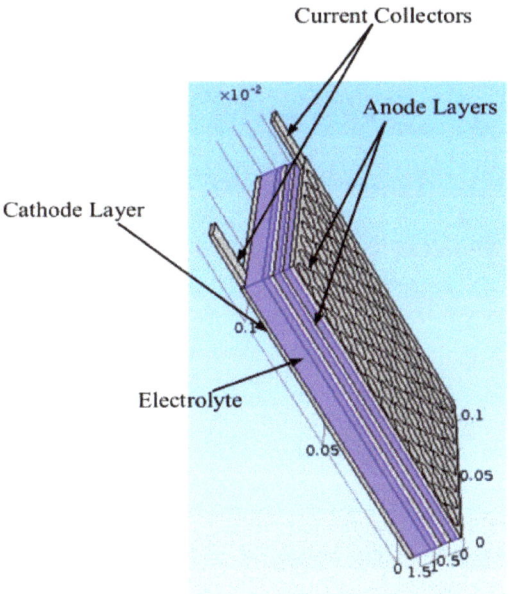

Figure 2. 4DM used to simulate electrolyte degradation.

3. Electrolyte Salt Diffusion Coefficient Degradation

The salt diffusion coefficient of the electrolyte defines the maximum rate of diffusion that is possible in the electrolyte [25]. The diffusion coefficient's relationship with the temperature and electromotive force (EMF) applied is given by the Stokes–Einstein equation [26].

$$D = \mu k_B T. \qquad (2)$$

Using the conservation of mass equation based on Fick's Law when applied to the Pseudo 2D (P2D) model, gives,

$$\frac{\partial(\epsilon_e c_e)}{\partial t} = \frac{\partial}{\partial x}\left(D_e^{eff}\frac{\partial c_e}{\partial x}\right) + \left(\frac{1-t_+^0}{F}\right)j^{Li}. \qquad (3)$$

Given $D_e^{eff} = D_e \epsilon_e^p$, Equation (3) can be written as

$$\frac{\partial(\epsilon_e c_e)}{\partial t} = \left(D_e \epsilon_e^p \frac{\partial^2 c_e}{\partial x^2}\right) + \left(\frac{1-t_+^0}{F}\right)j^{Li}. \qquad (4)$$

For a constant j^{Li}, if ϵ_e^p decreases then $\frac{\partial(\epsilon_e c_e)}{\partial t}$ will decrease. Integrating and solving Equation (4) with respect to time gives:

$$\epsilon_e c_e = D_e \epsilon_e^p \frac{\partial^2 c_e}{\partial x^2} t + \int \left(\frac{1-t_+^0}{F}\right)j^{Li}\partial t. \qquad (5)$$

Integrating on both sides of Equation (5) with respect to x and solving, we get,

$$c_e = \frac{1}{\left(\epsilon_e x^2 - D_e \epsilon_e^p t\right)} \iiint \left(\frac{1-t_+^0}{F}\right)j^{Li}\partial t \partial x \partial x. \qquad (6)$$

Therefore, when ϵ_e^p decreases then $(\epsilon_e x^2 - D_e \epsilon_e^p t)$ increases and c_e decreases. From (3) and (7), for constant j^{Li}, the decrease in c_e will cause an increase in ϕ_e to balance the equation.

$$\frac{\partial}{\partial x}\left(\kappa^{eff}\frac{\partial \phi_e}{\partial x} + \kappa_D^{eff}\frac{\partial \ln c_e}{\partial x}\right) + j^{Li} = 0, \tag{7}$$

$$\kappa^{eff}\frac{\partial^2 \phi_e}{\partial x^2} = -\kappa_D^{eff}\frac{\partial^2 \ln c_e}{\partial x^2} - j^{Li}. \tag{8}$$

Integrating (8) gives:

$$\phi_e = -\frac{\kappa_D^{eff}}{\kappa^{eff}}\ln c_e - \frac{j^{Li} x^2}{\kappa^{eff}}, \tag{9}$$

$$\Delta\phi_e = \frac{\kappa_D^{eff}}{\kappa^{eff}}\ln\frac{c_{e_2}}{c_{e_1}}. \tag{10}$$

The overpotential of the battery can be written as,

$$\eta = \phi_s - \phi_e - U. \tag{11}$$

Since η is constant and the equilibrium potential, U, at any defined concentration is also constant, the only parameter that can vary to compensate for the change in ϕ_e is ϕ_s. Thus,

$$\Delta\eta = \Delta\phi_s - \Delta\phi_e - \Delta U, \tag{12}$$

$$0 = \Delta\phi_s - \Delta\phi_e, \tag{13}$$

$$\Delta\phi_s = \Delta\phi_e. \tag{14}$$

Therefore, when ϕ_e decreases, then ϕ_s will increase to keep η constant. With an increase in ϕ_s, V_t will either increase or decrease based on the battery mode of operation (charging/discharging).

$$V_t = \phi_{s+} - \phi_{s-} - \frac{R_f}{A}I, \tag{15}$$

$$\Delta V_t = \Delta\phi_{s+} - \Delta\phi_{s-}. \tag{16}$$

From Equations (14) and (16),

$$\Delta V_t = f(\Delta\phi_e). \tag{17}$$

Thus, from Equations (10) and (17) we get,

$$\Delta V_t = f\left(\frac{\kappa_D^{eff}}{\kappa^{eff}}\ln\frac{c_{e_2}}{c_{e_1}}\right). \tag{18}$$

From Equation (18), if c_{e_1} is the initial lithium ion concentration in the electrolyte when the battery is designed then the change in terminal voltage to the decrease in lithium ion concentration in the electrolyte follows an exponential decrease.

Figures 3–5 show the terminal voltage response to the change in the electrolyte salt diffusion coefficient degradation. It can be observed that there is a decrease in the terminal voltage when the battery's electrolyte salt diffusion coefficient decreases. This is in correlation with the mathematical derivations obtained in Equations (6) and (18). Figure 6 shows that the response follows an exponential curve. This is obtained using the curve fitting toolbox in the Matlab. The R^2 fit is determined to be 0.9952.

$$\Delta V_t = -0.02406 e^{(-4.001 D_e)}. \tag{19}$$

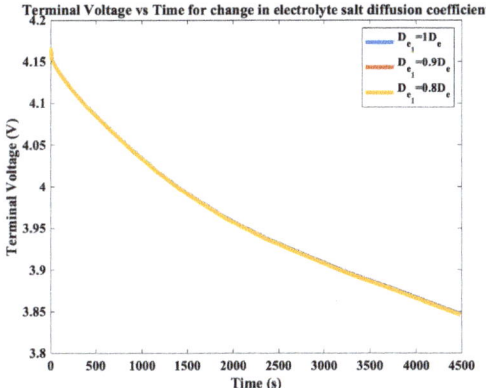

Figure 3. Terminal voltage vs. time for change in electrolyte salt diffusion coefficient from 1.0 to 0.8.

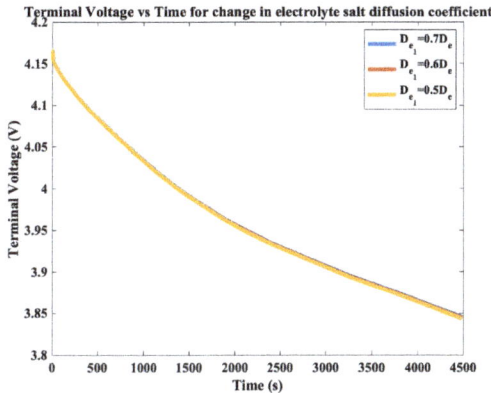

Figure 4. Terminal voltage vs. time for change in electrolyte salt diffusion coefficient from 0.7 to 0.5.

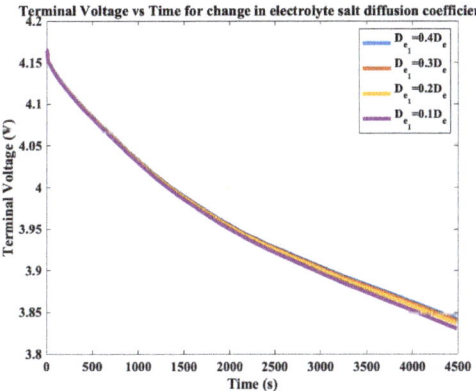

Figure 5. Terminal voltage vs. time for change in electrolyte salt diffusion coefficient from 0.4 to 0.1.

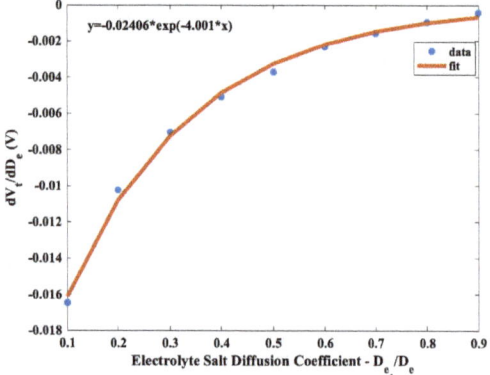

Figure 6. Change of terminal voltage to transference number vs. electrolyte salt diffusion coefficient.

4. Electrolyte Transference Number Degradation

The electrolyte transference number defines the ratio of current carried by ions in an electrolyte to the total current flowing through the electrolyte [27]. Since in any electrolyte there are positive and negative ions, the transference number is usually between 0 and 1 [26]. In lithium ion batteries, the electrolyte ion transference number is around 0.3–0.4.

Using Equation (7) and the conservation of charge per Fick's Law as shown in Equation (21) we have,

$$\frac{\partial}{\partial x}\left(\kappa^{eff}\frac{\partial \phi_e}{\partial x} + \kappa_D^{eff}\frac{\partial \ln c_e}{\partial x}\right) + j^{Li} = 0, \tag{20}$$

$$\kappa_D^{eff} = \frac{2RT\kappa^{eff}}{F}(t_+^0 - 1)\left(1 + \frac{\partial \ln f_\pm}{\partial \ln c_e}\right). \tag{21}$$

For a fixed current being drawn or sent into the battery, j^{Li} is a constant. Assuming that $\frac{\partial \ln f_\pm}{\partial \ln c_e} = 0$ we get,

$$\kappa_D^{eff} = \frac{2RT\kappa^{eff}}{F}(t_+^0 - 1). \tag{22}$$

Differentiating κ_D^{eff} with respect to t_+^0 gives:

$$\frac{\partial \kappa_D^{eff}}{\partial t_+^0} = \frac{2RT\kappa^{eff}}{F}. \tag{23}$$

Therefore, with an increase in t_+^0, there is an increase in κ_D^{eff}. If κ_D^{eff} increases then using Equation (20), the only way to keep the equation equal to 0 for a constant j^{Li} is that $\frac{\partial}{\partial x}\left(\kappa^{eff}\frac{\partial \phi_e}{\partial x}\right)$ must decrease. But κ^{eff} is a constant for any material. Therefore, $\frac{\partial^2 \phi_e}{\partial x^2}$ must decrease. Thus, Equation (20) can be written as,

$$\frac{\partial^2 \phi_e}{\partial x^2} = \frac{\left(-\frac{2RT\kappa^{eff}}{F}(t_+^0 - 1)\frac{\partial^2 \ln c_e}{\partial x^2}\right) - j^{Li}}{\kappa^{eff}}. \tag{24}$$

Integrating (24) gives,

$$\phi_e(x) = -\frac{2RT}{F}(t_+^0 - 1)\ln c_e(x) - \frac{j^{Li} x^2}{\kappa^{eff}}. \tag{25}$$

Therefore, from Equations (17) and (25), a conclusion that the variation in the terminal voltage is directly proportional to the change in the electrolyte potential can be drawn.

$$\Delta \phi_e(x) = -\frac{2RT}{F}(\Delta t_+^0) \ln c_e(x). \tag{26}$$

From Equations (17) and (26), there is a direct relationship between the change in the transference number and the change in the terminal voltage. However, the relationship also has a negative slope.

$$\Delta V_t = f\left(-\frac{2RT}{F}(\Delta t_+^0) \ln c_e(x)\right). \tag{27}$$

From Equation (6), we know that c_e is a function of t_+^0. Therefore, Equation (27) can be rewritten as,

$$\Delta V_t = f(\Delta t_+^0). \tag{28}$$

From the curve fitting toolbox in Matlab, the smallest order of the polynomial that fits the function is a second order polynomial with an R^2 fit of 0.9982.

$$\Delta V_t = -0.003808 {\Delta t_+^0}^2 + 0.02002 \Delta t_+^0 - 0.02652. \tag{29}$$

The transference number varies from 0.04 to 1.12. Figures 7–12 show the transference number varies from 0.1 to 2.8 because the base transference number is set to 0.4 and is being scaled between 0.04 and 1.12. With a decrease in the transference number, there is an increased impedance to lithium ion flow across the electrolyte. This increase in the resistance causes an increased potential drop across the electrolyte and results in the battery reaching its cut-off voltage sooner. Figure 13 shows the relationship between the change in terminal voltage to the change in transference number vs. the change in the transference number. This figure is consistent with the results obtained from (28).

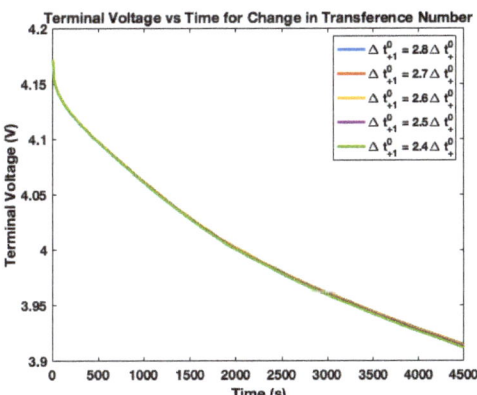

Figure 7. Terminal voltage vs. time for change in transference number from 2.8 to 2.4.

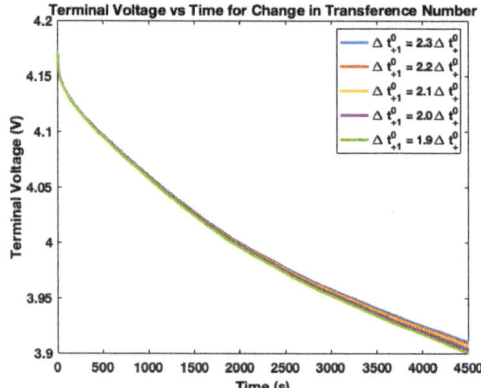

Figure 8. Terminal voltage vs. time for change in transference number from 2.3 to 1.9.

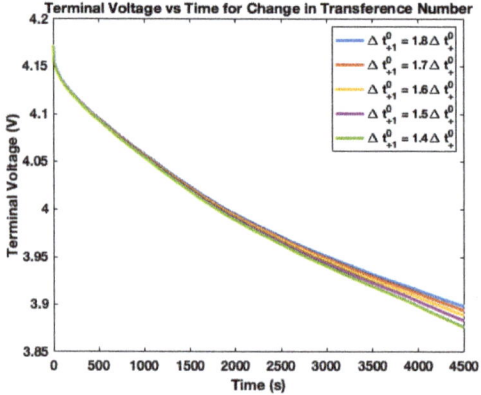

Figure 9. Terminal voltage vs. time for change in transference number from 1.8 to 1.4.

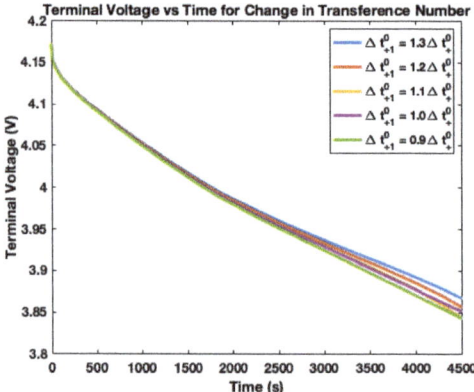

Figure 10. Terminal voltage vs. time for change in transference number from 1.3 to 0.9.

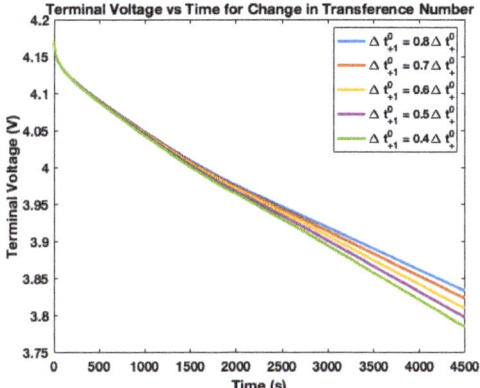

Figure 11. Terminal voltage vs. time for change in transference number from 0.8 to 0.4.

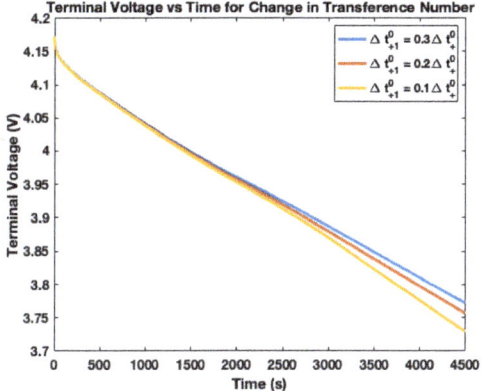

Figure 12. Terminal voltage vs. time for change in transference number from 0.3 to 0.1.

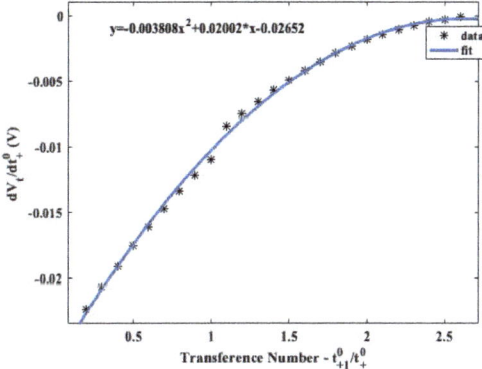

Figure 13. Change of terminal voltage to transference number vs. transference number.

5. Conclusions and Future Work

There are three major types of side reactions, and they occur at the solid–electrolyte interface, the current collector–electrode interface and in the electrolyte. This article highlights the impact of electrolyte degradation on the performance of the battery and the physical manifestation that this degradation phenomenon has on the terminal voltage of the battery. There are two ways to represent the degradation of the electrolyte—loss of electrolyte salt concentration and change in transference number. The loss of electrolyte concentration increases the resistance to the flow of lithium ions across the electrolyte. The electrolyte loses lithium ions when the battery is subjected to harsh operating conditions such as high temperatures or charging/discharging rates. The lithium ions react with the organic solvents in the electrolyte to form irreversible chemical compounds as shown in Equation (1). The decrease in the transference number of the electrolyte symbolizes the loss of charge-carrying lithium ions or the increase in the electrons that are carrying charge across the electrolyte (i.e., self-discharge). From Equations (18) and (27), it can be seen that the terminal voltage has an exponential decrease for decrease in the concentration of lithium ions in the electrolyte while it follows a quadratic function with the transference number. This implies that the change in the terminal voltage is more sensitive to the decrease in lithium ion concentration in the electrolyte than to the variation in transference number. Depending on the rate of change of the voltage, it is possible to determine the kind of side reaction that is dominant in the electrolyte.

Future work involves performing more sensitivity analysis on the degradation of different physical battery components and determining the sensitivity of the change in terminal voltage and capacity to the change in component degradation.

Author Contributions: Conceptualization: M.-Y.C. and B.B.; Methodology: B.B.; Software: B.B.; Validation: B.B.; Formal Analysis: Bharat Balagpopal; Investigation: B.B.; Data Curation: B.B.; Writing Original Draft preparation: B.B.; Writing-review and editing: M.-Y.C.; Visualization: B.B.; Supervision: M.-Y.C.; Project Administration: M.-Y.C.; Funding Acquisition: M.-Y.C. and B.B. All authors have read and agreed to the published version of the manuscript.

Funding: This research was partially supported by National Science Foundation Award Number IIP—1500208.

Conflicts of Interest: The authors declare no conflict of interest.

Appendix A

Table A1. Table of Symbols.

Symbol	Definition	Units
D	Diffusion Coefficient	m^2/s
μ	Particle Mobility	m/Vs
V	Applied Electromotive Force	V
k_B	Boltzmann Constant	J/K
T	Temperature	K
ϵ_e	Electrolyte Phase Volume Fraction	Unitless
c_e	Electrolyte Lithium Ion Concentration	mol/m^3
D_e^{eff}	Effective Electrolyte Diffusion Coefficient	m^2/s
x	Variable to define position along the length of the battery	m
t	Time	s
t_+^0	Positive Ion (Li$^+$) Transference Number	Unitless
F	Faraday's Constant	C/mol
j^{Li}	Volumetric Electrochemical Reaction Rate at the Surface of the Electrode	A/m^3
D_e	Electrolyte Diffusion Coefficient	m^2/s
κ^{eff}	Effective Electrolyte Ionic Conductivity	S/m

Table A1. *Cont.*

Symbol	Definition	Units
ϕ_e	Electrolyte Phase Potential	V
$c_{s,e}$	Lithium Ion Concentration at the Solid—Electrolyte Interface	mol/m^3
κ_D^{eff}	Effective Diffusion Conductivity	S/m
R	Universal Gas Constant	J/mol K
$f_\pm$	Electrolyte Activity Coefficient	Unitless
ϕ_s	Solid Phase Potential	V
U	Thermodynamic Equilibrium Potential	V
η	Over Potential	V
V_t	Terminal Voltage	V
I	Input Current	A
R_f	Current Collector—Electrode Contact Resistance	Ω

References

1. Zheng, X.; Gao, W.; Zhang, X.; He, M.; Lin, X.; Cao, H.; Zhang, Y.; Sun, Z. Spent lithium-ion battery recycling—Reductive ammonia leaching of metals from cathode scrap by sodium sulphite. *Waste Manag.* **2017**, *60*, 680–688. [CrossRef] [PubMed]
2. Gaines, L. Lithium-ion battery recycling processes: Research towards a sustainable course. *Sustain. Mater. Technol.* **2018**, *17*, e00068. [CrossRef]
3. Velázquez-Martínez, O.; Valio, J.; Santasalo-Aarnio, A.; Reuter, M.; Serna-Guerrero, R. A critical review of lithium-ion battery recycling processes from a circular economy perspective. *Batteries* **2019**, *5*, 68. [CrossRef]
4. Doughty, D.; Roth, E.P. A general discussion of Li Ion battery safety. *Electrochem. Soc. Interface* **2012**, *21*, 37–44. [CrossRef]
5. Liu, B.; Jia, Y.; Li, J.; Yin, S.; Yuan, C.; Hu, Z.; Wang, L.; Li, Y.; Xu, J. Safety issues caused by internal short circuits in lithium-ion batteries. *J. Mater. Chem. A* **2018**, *6*, 21475–21484. [CrossRef]
6. Wang, Q.; Jiang, L.; Yu, Y.; Sun, J. Progress of enhancing the safety of lithium ion battery from the electrolyte aspect. *Nano Energy* **2019**, *55*, 93–114. [CrossRef]
7. Rahimi-Eichi, H.; Chow, M. Adaptive parameter identification and State-of-Charge estimation of lithium-ion batteries. In Proceedings of the IECON 2012—38th Annual Conference of the IEEE Industrial Electronics Society, Montreal, QC, Canada, 25–28 October 2012; pp. 4012–4017.
8. Balagopal, B.; Chow, M.Y. The state of the art approaches to estimate the state of health (SOH) and state of function (SOF) of lithium Ion batteries. In Proceedings of the 2015 IEEE International Conference on Industrial Informatics (INDIN 2015), Cambridge, UK, 22–24 July 2015; pp. 1302–1307. [CrossRef]
9. Balagopal, B.; Chow, M.Y. Effect of anode conductivity degradation on the Thevenin Circuit Model of lithium ion batteries. In Proceedings of the IECON 2016—42nd Annual Conference of the IEEE Industrial Electronics Society, Florence, Italy, 24–27 October 2016; pp. 2028–2033. [CrossRef]
10. Wu, Y. *Lithium-ion Batteries: Fundamentals and Applications*; CRC Press: Boca Raton, FL, USA, 2015; p. 1.
11. Jow, T.R.; Delp, S.A.; Allen, J.L.; Jones, J.P.; Smart, M.C. Factors Limiting Li + Charge Transfer Kinetics in Li-Ion Batteries. *J. Electrochem. Soc.* **2018**, *165*, A361–A367. [CrossRef]
12. Logan, E.R.; Tonita, E.M.; Beaulieu, L.Y.; Ma, X.; Li, J.; Dahn, J.R.; Gering, K.L. A Study of the Physical Properties of Li-Ion Battery Electrolytes Containing Esters. *J. Electrochem. Soc.* **2018**, *165*, A21–A30. [CrossRef]
13. Balagopal, B.; Huang, C.S.; Chow, M.Y. Effect of Calendar Aging on Li Ion Battery Degradation and SOH. In Proceedings of the IECON 2017—43rd Annual Conference of the IEEE, Beijing, China, 29 October–1 November 2017; pp. 7647–7652.
14. Guan, P.; Liu, L. Lithium-ion diffusion in solid electrolyte interface (SEI) predicted by phase field model. *Mater. Res. Soc. Symp. Proc.* **2015**, *1753*, 31–37. [CrossRef]
15. Lee, J.T.; Nitta, N.; Benson, J.; Magasinski, A.; Fuller, T.F.; Yushin, G. Comparative study of the solid electrolyte interphase on graphite in full Li-ion battery cells using X-ray photoelectron spectroscopy, secondary ion mass spectrometry, and electron microscopy. *Carbon* **2013**, *52*, 388–397. [CrossRef]

16. Laresgoiti, I.; Kabitz, S.; Ecker, M.; Sauer, D.U. Modeling mechanical degradation in lithium ion batteries during cycling: Solid electrolyte interphase fracture. *J. Power Sources* **2015**, *300*, 112–122. [CrossRef]
17. Gao, Y.; Jiang, J.; Zhang, C.; Zhang, W.; Ma, Z.; Jiang, Y. Lithium-ion battery aging mechanisms and life model under different charging stresses. *J. Power Sources* **2017**, *356*, 103–114. [CrossRef]
18. Cannarella, J.; Arnold, C.B. Stress evolution and capacity fade in constrained lithium-ion pouch cells. *J. Power Sources* **2014**, *245*, 745–751. [CrossRef]
19. Cannarella, J.; Arnold, C.B. State of health and charge measurements in lithium-ion batteries using mechanical stress. *J. Power Sources* **2014**, *269*, 7–14. [CrossRef]
20. Pinson, M.B.; Bazant, M.Z. Theory of SEI Formation in Rechargeable Batteries: Capacity Fade, Accelerated Aging and Lifetime Prediction. *J. Electrochem. Soc.* **2012**, *160*, A243–A250. [CrossRef]
21. Lawder, M.T.; Northrop, P.W.C.; Subramanian, V.R. Model-Based SEI Layer Growth and Capacity Fade Analysis for EV and PHEV Batteries and Drive Cycles. *J. Electrochem. Soc.* **2014**, *161*, A2099–A2108. [CrossRef]
22. Williard, N.; He, W.; Osterman, M.; Pecht, M. Reliability and failure analysis of Lithium Ion batteries for electronic systems. In Proceedings of the 2012 13th International Conference on Electronic Packaging Technology & High Density Packaging, Guilin, China, 13–16 August 2012; pp. 1051–1055. [CrossRef]
23. Valoen, L.O.; Reimers, J.N. Transport Properties of LiPF$_6$-Based Li-Ion Battery Electrolytes. *J. Electrochem. Soc.* **2005**, *152*, A882. [CrossRef]
24. Balagopal, B.; Huang, C.S.; Chow, M.Y. Effect of Calendar Ageing on SEI growth and its Impact on Electrical Circuit Model Parameters in Lithium Ion Batteries. In Proceedings of the IEEE International Conference on Industrial Electronics for Sustainable Energy Systems, Hamilton, New Zealand, 31 January–2 February 2018.
25. Hayamizu, K. Direct relations between ion diffusion constants and ionic conductivity for lithium electrolyte solutions. *Electrochim. Acta* **2017**, *254*, 101–111. [CrossRef]
26. Chintapalli, M.; Timachova, K.; Olson, K.R.; Mecham, S.J.; Devaux, D.; Desimone, J.M.; Balsara, N.P. Relationship between Conductivity, Ion Diffusion, and Transference Number in Perfluoropolyether Electrolytes. *Macromolecules* **2016**, *49*, 3508–3515. [CrossRef]
27. Diederichsen, K.M.; McShane, E.J.; McCloskey, B.D. Promising Routes to a High Li+ Transference Number Electrolyte for Lithium Ion Batteries. *ACS Energy Lett.* **2017**, *2*, 2563–2575. [CrossRef]

Publisher's Note: MDPI stays neutral with regard to jurisdictional claims in published maps and institutional affiliations.

© 2020 by the authors. Licensee MDPI, Basel, Switzerland. This article is an open access article distributed under the terms and conditions of the Creative Commons Attribution (CC BY) license (http://creativecommons.org/licenses/by/4.0/).

Article

Unification of Internal Resistance Estimation Methods for Li-Ion Batteries Using Hysteresis-Free Equivalent Circuit Models

S M Rakiul Islam [1,*], Sung-Yeul Park [1] and Balakumar Balasingam [2]

1. Electrical and Computer Engineering Department, University of Connecticut, Storrs, CT 06269, USA; sung_yeul.park@uconn.edu
2. Electrical and Computer Engineering Department, University of Windsor, 401 Sunset Avenue, Office 3051, Windsor, ON N9B3P4, Canada; singam@uwindsor.ca
* Correspondence: s.islam@uconn.edu; Tel.: +1-860-486-0915

Received: 10 April 2020; Accepted: 26 May 2020; Published: 3 June 2020

Abstract: Internal resistance is one of the important parameters in the Li-Ion battery. This paper identifies it using two different methods: electrochemical impedance spectroscopy (EIS) and parameter estimation based on equivalent circuit model (ECM). Comparing internal resistance, the conventional parameter estimation method yields a different value than EIS. Therefore, a hysteresis-free parameter identification method based on ECM is proposed. The proposed technique separates hysteresis resistance from the effective resistance. It precisely estimated actual internal resistance, which matches the internal resistance obtained from EIS. In addition, state of charge, open circuit voltage, and different internal equivalent circuit components were identified. The least square method was used to identify the parameters based on ECM. A parameter extraction algorithm to interpret impedance spectrum obtained from the EIS. The algorithm is based on the properties of Nyquist plot, phasor algebra, and resonances. Experiments were conducted using a cellphone pouch battery and a cylindrical 18650 battery.

Keywords: internal resistance; battery parameters; equivalent circuit model; electrochemical impedance spectroscopy

1. Introduction

It is essential to know the different parameters of a battery to track, control, and forecast its dynamics [1–3]. Efficiency, charging control, safety, and the lifespan of a battery can be enhanced by utilizing these parameters. Terminal voltage, current, and temperature are useful parameters to operate a battery within safety limits [4]. Cycle life and calendar life are useful parameters to determine the aging and health condition of a battery [5]. State of charge (SOC), depth of discharge, capacity, and open circuit voltage (OCV) are useful for fuel gauging a battery [6]. State of health (SOH) and remaining useful life (RUL) are good indicators of performance degradation [7,8]. The internal resistance is one of the battery parameters which is relevant to determine the aging mechanism, SOH, and RUL [9,10]. Internal equivalent impedance of a battery is useful for improving charging efficiency [11]. Internal resistance and other equivalent circuit parameters can be correlated to the conduction loss, ions loss, and active material loss in a Li-Ion battery [12].

Different methods have been developed to identify specific battery parameters. Some parameters (Voltage, Current, and Temperature) are identified using direct measurements. Other parameters are calculated, estimated, and predicted based on the initial measurements [1–3,13–19]. Hybrid pulse power characterization (HPPC), Galvanostatic intermittent titration technique (GITT), Equivalent circuit model (ECM) estimation and electrochemical impedance spectroscopy (EIS) are renowned methods for parameter identification. ECM estimation and the EIS based method are used. It is expected that the same battery parameter obtained using different identification methods should be equal. However, experimental results show that internal resistance obtained using the conventional estimation method for ECM does not match with the results obtained using EIS. This paper proposes a solution by introducing a hysteresis-free parameter identification method based on ECM to reduce the mismatch.

The relationship between SOC and OCV is obtained from experiments considering an equivalent circuit model (ECM). For parameter estimation based on ECM, a battery is charged and discharged fully by a very small amount of current (C/26-rate). Terminal voltage and current are measured and analyzed [2]. Recent studies show that the ECM based parameter estimation method can also determine the internal DC resistance, RUL, and SOH [15,20,21]. Parameter estimation based on ECM is easy to implement and reliable for SOC determination. However, the test for parameter estimation based on ECM is prone to hysteresis [22]. Hysteresis affects the internal resistance estimation. Different models have been used to characterize hysteresis [23–25]. Hysteresis voltage in a battery depends on the SOC level, temperature, charging/discharging current rate, and aging [22,26,27]. Based on the recent literature, there is no universally defined model for hysteresis in a battery [25]. Considering the effects of hysteresis, a hysteresis-free parameter estimation method is developed in this paper.

Impedance spectrum (impedances corresponding to specific frequencies) of a battery is measured with the EIS method. It can accurately explain the internal electrochemical characteristics of a battery. A change of SOH is due to the aging and can be determined using the impedance spectrum [12]. The chemistry of aging, frequency sweep, and EIS have been discussed in [28–31]. EIS uses either voltage or current perturbation to get the battery response. The voltage and current measurements are used to obtain the impedance spectrum by frequency domain analysis [32]. Despite the expensive equipment, EIS provides the opportunity to obtain highly accurate values for internal equivalent circuit parameters. Unlike ECM estimation by charging–discharging test, a battery is not affected by hysteresis during the EIS test [25]. Conventionally, the curve fitting method is used to characterize and track the SOC with internal equivalent circuit parameters from EIS test data [17]. The curve fitting method uses regression analysis and iterative processes. A curve fitting based parameter extraction algorithm is introduced in [10], which can be implemented offline and require high computing power. Therefore, the curve fitting method is inconvenient for fast interpretation of the impedance spectrum. Fast interpretation of impedance spectrum is necessary for online EIS application. To resolve this issue, an internal equivalent circuit parameter extraction algorithm is proposed in [1]. The algorithm looks through the properties of the impedance spectrum and do not have iterative processes for regression analysis. Therefore, it provided the scope to interpret impedance spectrum with less computing power and can be implemented online. In addition to [1], impedance spectrum for different SOC has been interpreted using the proposed algorithm for two different types of Li-Ion batteries in this paper.

Although parameter identification based on ECM and EIS methods uses voltage and current measurements, their intended objectives, procedures, and analyses are different. As a result, each method identifies internal resistances with different values. This happens because of the role of hysteresis in a conventional parameter estimation method based on ECM. The conventional method determines effective resistance as internal resistance. On the other hand, the aforementioned parameter extraction algorithm gives Ohmic resistance as internal resistance which was introduced [1]. The algorithm is applied to the results obtained using EIS tests in this paper. To unify the internal resistance from both methods, the following contributions are made in this paper:

- The tests for ECM and EIS are conducted on two different types of Li-Ion batteries.
- Parameter extraction algorithm is proposed and validated theoretically and experimentally.
- The identified parameters from two methods are compared.
- Difference is found for internal resistance obtained from two methods.
- To unify the values of internal resistance, a hysteresis-free ECM based parameter estimation method is proposed.

- The values from EIS tests matched with the results obtained from the tests for ECM using the proposed hysteresis-free method.

To unify the identification methods, tests have been conducted on a Samsung-B600BC pouch cellphone battery (Seoul, South Korea) and a Samsung-18650 cylindrical Li-Ion battery. Theoretical explanation of ECM based parameter estimation, EIS, and the parameter extraction algorithm are discussed in Sections 2–4. The ECM based parameter estimation method is explained by using optimization theories and equivalent circuit models. EIS and the parameter extraction algorithm are explained by Nyquist theory, resonance, and phasor algebra. In addition to the development of theories and experimentation, analyses have been performed for validity in Section 5. Analytical results obtained from the tests for ECM and EIS are compared to justify the proposed method. The analysis shows how hysteresis affects resistance identification for the conventional ECM based parameter estimation. Results finally show that the estimated internal resistance obtained from the proposed ECM based parameter estimation matches with that obtained from EIS.

2. ECM Based Parameter Estimation

The ECM based parameter identification method can be implemented using steady state response of a battery. The details of conventional ECM based parameter estimation techniques are discussed in [2]. The basic structure of the ECM, battery response and test set up, conventional, and proposed parameter estimation technique are discussed as follows.

2.1. Equivalent Circuit Model

The response of a battery in a steady state condition can be explained by an equivalent circuit model as shown in Figure 1. The circuit consists of a DC OCV source, V_o, DC equivalent resistance, R_0, and hysteresis voltage source, h. The open circuit voltage is considered as a function of SOC, s. The terminal voltage can be expressed by Equation (1):

$$V[k] = V_o(s[k]) + h[k] + i[k]R_0 + n_v[k] \tag{1}$$

where terminal voltage of kth measurement is $V[k]$, current through the battery is $i[k]$, and measurement noise is n_v.

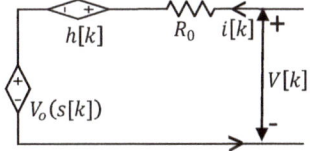

Figure 1. Equivalent circuit model of a battery for estimation.

2.2. Response of a Battery in a Charging–Discharging Test

The block diagram and actual experimental set up for ECM based parameter estimation are shown in Figure 2. The test set up consists of a bench power supply, power resistor, programmable load, data logging system, etc. To estimate ECM parameters, a Samsung-B600BC cellphone's 2600 mAh 3.8 V 9.88 Wh Li-Ion battery (LCO) was used. To validate the idea further, the test was also conducted on a Samsung-18650 Li-Ion battery (NCA) which has 2550 mAh nominal capacity. The batteries were charged and discharged slowly with constant 100 mA current, considering the effect of hysteresis, rated capacity, and safety limits.The maximum discharging/charging voltage limits used for the tests for B600BC and 18,650 batteries are 3.3 V/4.4 V and 3.0 V/4.2 V, respectively. As time passes, the terminal voltage and SOC level of the battery change. The values of terminal voltage and current have been measured and recorded using a scopecorder (DLP750P) for analysis. The responses for the charging–discharging test for batteries are shown in Figure 3. Up and down arrows denote charging voltage, V_{ch}, and discharging voltage, V_{dch}, respectively. SOC has been calculated by the Coulomb counting method using current integration (trapezoidal). The actual OCV for a specific SOC level has been calculated by averaging the charging and discharging voltages. The measured voltage and actual OCV with respect to SOC are shown in Figure 4. The actual OCV is calculated by Equation (2):

$$V_o(s) = \frac{V_{ch}(s) + V_{dch}(s)}{2} \tag{2}$$

where V_{ch} and V_{dch} denote the measured voltage at specific SOC level s, during charging and discharging. The effective voltage drop V_{0h} which is combined with hysteresis voltage and Ohmic drop is determined by the following equations:

$$V_{0h}(s) = V_{ch}(s) - V_o(s) = V_o(s) - V_{dch}(s) \tag{3}$$

$$V_{0h}(s) = \frac{V_{ch}(s) - V_{dch}(s)}{2} \tag{4}$$

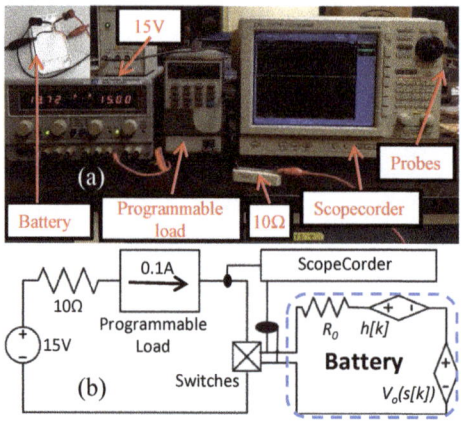

Figure 2. Equivalent circuit model (ECM) experiment: (**a**) test bench; (**b**) block diagram.

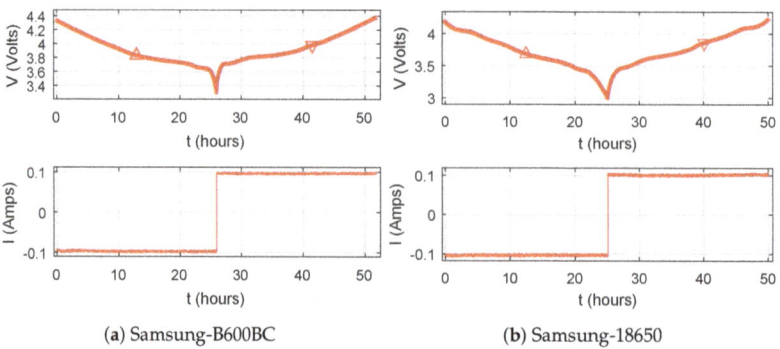

Figure 3. Response of batteries for the charging–discharging test.

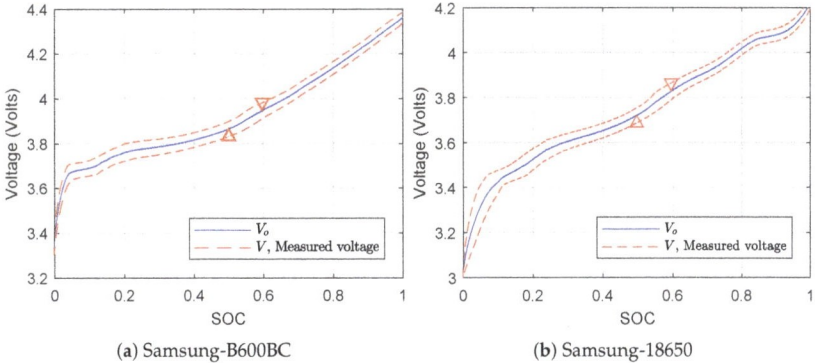

(a) Samsung-B600BC (b) Samsung-18650

Figure 4. Battery voltage profiles for charging–discharging tests.

2.3. Theory of Conventional Parameter Estimation

The SOC-OCV relationship is estimated from the voltage and current measurements, $V[k]$ and $i[k]$. Battery current, $i[k]$, is integrated to get the present capacity, total capacity, and SOC levels for every data logging instant from the measurements. By considering equal voltage drop during charging and discharging for the same SOC levels, the average OCV is calculated from the measured terminal voltage as in Figure 4. Estimation (1) can be re-written as in Equation (5), which is adopted from [2].

$$V[k] = V_o(s[k]) + i[k]R_h + i[k]R_0 + n_v[k] \tag{5}$$

Due to a small value of $i[k]$ and dependency on hysteresis resistance, R_h, hysteresis voltage $h[k]$ is replaced by $i[k]R_h$ in Equation (5). By combining R_h and R_0 to the effective resistance, R_{0h}, Equation (5) is simplified in Equation (6).

$$V[k] = V_o(s[k]) + i[k]R_{0h} + n_v[k] \tag{6}$$

where

$$R_{0h} = R_0 + R_h \tag{7}$$

The ECM parameters are estimated from $V[k]$, $i[k]$ and s. Thus, Equation (6) is rearranged to matrix form, as shown in Equation (8).

$$V[k] = \begin{bmatrix} p_o(s[k])^T & i[k] \end{bmatrix} \begin{bmatrix} \mathbf{k_o} \\ R_{0h} \end{bmatrix} \tag{8}$$

The first portion of Equation (8) forms the matrix $p[k]$ as shown in Equation (9).

$$p[k] = \begin{bmatrix} p_o(s[k])^T & i[k] \end{bmatrix} \tag{9}$$

The function $p_0(s[k])$ provides the estimation equation for the battery. p_o can be expressed using various models. The combined model [2] for p_o is shown in Equation (10).

$$p_o(s)^T = \begin{bmatrix} 1 & s & \frac{1}{s} & \ln(s) & \ln(1-s) \end{bmatrix} \tag{10}$$

For the Combined+3 model [2], p_o is expressed by Equation (11).

$$p_o(s)^T = \begin{bmatrix} 1 & s & \frac{1}{s} & \frac{1}{s^2} & \frac{1}{s^3} & \frac{1}{s^4} & ln(s) & ln(1-s) \end{bmatrix} \quad (11)$$

The second portion of Equation (8) has coefficient vector, $\mathbf{k_o}$, and the effective resistance, R_{0h}. The size of vector $\mathbf{k_o}$ depends on the model p_o. If p_o has M number of terms in its model, then $\mathbf{k_o} = [k_0, k_1, k_2, \ldots, K_M]^T$.

Now, for N number of measurements, the equation for voltage can be written as

$$v = P\mathbf{k} + n \quad (12)$$

where

$$V[k] = \begin{bmatrix} V[1] & V[2] & \cdots & V[N] \end{bmatrix}^T \quad (13)$$

$$P = \begin{bmatrix} p[1] & p[2] & \cdots & p[N] \end{bmatrix}^T \quad (14)$$

$$n = \begin{bmatrix} n[1] & n[2] & \cdots & n[N] \end{bmatrix}^T \quad (15)$$

$$\mathbf{k} = \begin{bmatrix} \mathbf{k_o} \\ R_{0h} \end{bmatrix} \quad (16)$$

The parameter vector, $\mathbf{k}$, is determined by using a parameter estimation technique. For the least square estimation technique, the estimated parameter vector, $\hat{\mathbf{k}}$, is calculated by Equation (17).

$$\hat{\mathbf{k}} = (P^T P)^{-1} P^T V \quad (17)$$

Coefficient vector, $\hat{\mathbf{k}_o}$, and the effective resistance, $\hat{R}_{0h}$, are obtained from the estimated parameter vector, $\hat{\mathbf{k}}$. From the estimated coefficient vector, $\hat{\mathbf{k}_o}$, the SOC-OCV relationship is determined by Equation (18):

$$\hat{V}_o[k] = p_o(s[k])^T \hat{\mathbf{k}_o} \quad (18)$$

The actual and estimated SOC-OCV curves are shown in the result section. The estimated values for $\hat{\mathbf{k}_o}$, and $\hat{R}_{0h}$ are also presented in the results section.

2.4. Proposed Hysteresis-Free Estimation

The effective resistance, $\hat{R}_{0h}$, of a battery was considered to be constant for different SOC levels in conventional ECM based parameter estimation technique [2]. The consideration is usable for SOC-OCV relation identification but not valid for resistance identification. $h[k]$ changes with SOC [22–24,33–35]. It indicates that the hysteresis resistance R_h also changes with SOC. Recent study shows that hysteresis voltage depends on SOC levels, temperature, charging/discharging current, and aging [23,27,36]. Our test was conducted at room temperature with a new battery. Hysteresis voltage is represented by the only multiplication of current and constant hysteresis resistance in Equation (5). In the proposed method, terminal voltage will be expressed by Equation (19) instead of by Equation (1).

$$V[k] = V_o(s[k]) + h(s[k]) + i[k]R_0 + n_v[k] \quad (19)$$

Hysteresis voltage of battery can be represented by Equation (20).

$$h(s[k]) = i[k]R_h(s[k]) \quad (20)$$

As a result, the effective resistance, $\hat{R}_{0h}$, model will also be changed, and is expressed in Equation (21).

$$R_{0h}(s[k]) = R_0 + R_h(s[k]) \tag{21}$$

Hysteresis voltage has been modeled using differential equations, four states model, variable dependent model and Tackas model, and so on [23,36–40]. Authors of [2,25,36] claim that hysteresis in a battery is not a one to one relation, impractical to model perfectly and has no absolute model. A simplified computationally effective quadratic model for hysteresis resistance is proposed in this paper for simplicity. SOC dependency on hysteresis resistance, R_h, is modeled in quadratic form in the following equations:

$$R_h(s[k]) = \begin{bmatrix} 1 & s & s^2 \end{bmatrix} \mathbf{k_h} \tag{22}$$

$$\mathbf{k_h} = \begin{bmatrix} k_{h1} & k_{h2} & k_{h3} \end{bmatrix}^T \tag{23}$$

where $\mathbf{k_h}$ is the coefficient vector for the hysteresis resistance model. Considering the proposed model, the effective resistance can be expressed by Equation (24).

$$R_{0h}(s[k]) = R_0 + \begin{bmatrix} 1 & s & s^2 \end{bmatrix} \mathbf{k_h} \tag{24}$$

Since the first term of hysteresis resistance is constant, Equation (24) could be reduced to Equation (25).

$$R_{0h}(s[k]) = \begin{bmatrix} 1 & s & s^2 \end{bmatrix} \mathbf{k_{0h}} \tag{25}$$

where $\mathbf{k_{0h}}$ is the coefficient vector for the effective resistance model:

$$\mathbf{k_{0h}} = \begin{bmatrix} R_0 + k_{h1} & k_{h2} & k_{h3} \end{bmatrix}^T \tag{26}$$

By using this hysteresis model, measured voltage can be expressed by Equation (27) instead of by Equation (8):

$$V[k] = \begin{bmatrix} p_o(s[k])^T & i(s[k])^T \end{bmatrix} \begin{bmatrix} \mathbf{k_o} \\ \mathbf{k_{0h}} \end{bmatrix} \tag{27}$$

where

$$i(s[k])^T = \begin{bmatrix} i[k] & i[k]s[k] & i[k](s[k])^2 \end{bmatrix} \tag{28}$$

To estimate the SOC-OCV parameters, the p is formed by Equation (29) instead of by Equation (9):

$$p[k] = \begin{bmatrix} p_o(s[k])^T & i(s[k])^T \end{bmatrix} \tag{29}$$

The matrix P is formed considering Equations (29) and (14). The modified version of $\mathbf{k}$ is expressed by

$$\mathbf{k} = \begin{bmatrix} \mathbf{k_o} \\ \mathbf{k_{0h}} \end{bmatrix} \tag{30}$$

Once $\hat{\mathbf{k}}$ is estimated from Equation (17), $R_{0h}(s[k])$ can be calculated from Equation (25). At 100% SOC level, current should not be injected into the battery. Experimental observation shows that, at 100% SOC, the battery has minimum hysteresis resistance. The hysteresis resistance is very small compared to effective resistance. Therefore, only the internal resistance mainly exists in the effective resistance in this case. This special case can be expressed by Equation (31):

$$R_0 \approx R_{0h}(1) \tag{31}$$

At 100% SOC level, effective resistance can be expressed from Equation (25) and written as in Equation (32):

$$R_{0h}(1) = \begin{bmatrix} 1 & 1 & 1 \end{bmatrix} \mathbf{k_{0h}} \qquad (32)$$

From Equations (31) and (32), internal resistance could be calculated from the obtained parameter using Equation (33).

$$R_0 = \sum_{m=1}^{3} \mathbf{k_{0h}}(m) \qquad (33)$$

where m is the element number of vector $\mathbf{k_{0h}}$. The estimated internal resistance, R_0, is free from the hysteresis.

3. Electrochemical Impedance Spectroscopy

EIS measures perturbation response in a battery. Voltage and current measurements are converted from the time domain to the frequency domain to get the internal impedance spectrum. From the impedance spectrum, internal resistance can be obtained. The impedance spectrum represents internal circuit parameters of a battery, which give a good insight into electrochemical behavior. EIS is inherently hysteresis free because of its AC perturbation signal [25]. To understand EIS, it is necessary to explain the Adaptive Randles Equivalent Circuit Model (AR-ECM) [12].

3.1. Adaptive Randles Equivalent Circuit Model

The AR-ECM for a battery as shown in Figure 5 consists of a voltage source, E_{cell}, stray inductance, L, Ohmic resistance, R_Ω, solid electrolytic interface impedance, Z_{SEI}, Faradaic impedance (double layer capacitance, C_{DL} and charge transfer resistance, R_{CT}), and Warburg impedance, Z_W.

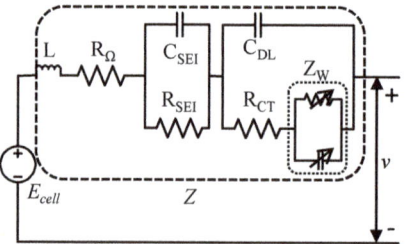

Figure 5. AR-ECM for a Li-ion battery.

3.1.1. Output Voltage Model

The cell potential is analogous to OCV, and can be determined by the Nernst Equation (34):

$$E_{cell} = E^0 - \frac{RT}{nF} \ln Q \qquad (34)$$

where E_{cell} is the cell potential, E^0 is the standard cell potential, R is the gas constant, T is the absolute temperature, n is the number of electron transfers in the reaction, F is the Faraday constant, and Q is the reaction quotient.

3.1.2. Impedance Model

Li-Ion battery battery impedance, Z, is comprised of different contributing factors as shown in Figure 5. Those factors are discussed below.

Inductance

A negligible amount of inductance, L, occurs due to parasitics in the wire and metal. Inductive impedance, Z_L, is a frequency-dependent component and expressed by Equation (35).

$$Z_L(j\omega) = j\omega L \qquad (35)$$

where $j = \sqrt{-1}$, and ω is the angular frequency, $\omega = 2\pi f$.

Ohmic Resistance

Ohmic resistance, R_Ω, is the resistance of bulk material in a battery. R_Ω is comprised of electrolyte resistance, current collector resistance, electrode resistance, and binder resistance. Electrolyte resistance is dominant here and depends on ionic concentration, geometry of the cell, temperature and SOC. R_Ω is independent of frequency, which is shown in Equation (36).

$$R_\Omega(j\omega) = R_\Omega \qquad (36)$$

Solid Electrolyte Interface impedance

Solid Electrolyte Interface impedance, Z_{SEI}, occurs because of mass transfer and polarization. Frequency response of Z_{SEI} is expressed using resitance and capacitance in Equation (37).

$$Z_{SEI}(j\omega) = \frac{1}{\frac{1}{R_{SEI}} + \frac{1}{\frac{1}{j\omega C_{SEI}}}} \qquad (37)$$

Faradaic impedance

Faradaic impedance is comprised of double layer capacitance, C_{DL}, and charge transfer resistance, R_{CT}. C_{DL} depends on the porosity and tortuosity of the electrodes.

Warburg Impedance

Warburg impedance, Z_W, is related to the diffusion of particles [41]. Resistive and capacitive elements are equal for Z_W. The value of this impedance is significant at low frequencies. Z_W is expressed by Equations (38) and (39).

$$Z_w(j\omega) = (1-j)\frac{\sigma}{\sqrt{\omega}} = \sigma\sqrt{\frac{2}{j\omega}} \qquad (38)$$

$$\sigma = \frac{RT}{n^2 F^2 A \sqrt{2}} \left(\frac{1}{C_O^b \sqrt{D_O}} + \frac{1}{C_R^b \sqrt{D_R}} \right) \qquad (39)$$

where σ is the Warburg Coefficient, A is the surface area of the electrode, C is the concentration, D is the diffusion coefficient, $_O$ is the oxidant, $_R$ is the reductant, and b represents the bulk material. The models of Faradaic and Warburg impedance are combined in Equation (40).

$$Z_{CT+DL+W}(j\omega) = \frac{1}{\frac{1}{R_{CT}+Z_w(j\omega)} + \frac{1}{\frac{1}{j\omega C_{DL}}}} \qquad (40)$$

The overall impedance, Z, of a battery in the frequency domain is expressed by Equation (41).

$$Z(j\omega) = j\omega L + R_\Omega + \frac{1}{\frac{1}{R_{SEI}} + \frac{1}{\frac{1}{j\omega C_{SEI}}}} + \frac{1}{\frac{1}{R_{CT}+Z_w(j\omega)} + \frac{1}{\frac{1}{j\omega C_{DL}}}} \qquad (41)$$

3.2. Response of a Battery in EIS Test

The EIS test has been conducted for the same batteries that were used in the ECM based parameter estimation. The experimental set up and block diagram for the EIS test is shown in Figure 6. The Solartron 1455 cell test system has been used to conduct this test. AC perturbation current 70 mA on top of 200 mA DC current was applied to a Samsung-B600BC battery. The frequency has been swept from 0.1 Hz to 10 kHz using a chirp perturbation signal. For Samsung-18650 battery, 100 mA AC perturbation was applied from 0.05 Hz to 5 kHz. The celltest system can measure the few milivolt voltage response caused by current perturbation. It also performs the analysis, and gives V, I, Z_{im}, Z_r, f and ω. An explanatory impedance plot (inverted Nyquist plot) of the battery is shown in Figure 7. The impedance plot is used to analyze electrochemical behavior of the battery.

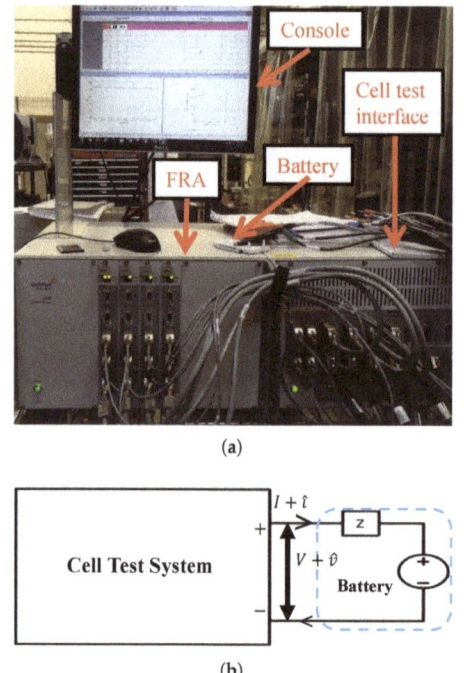

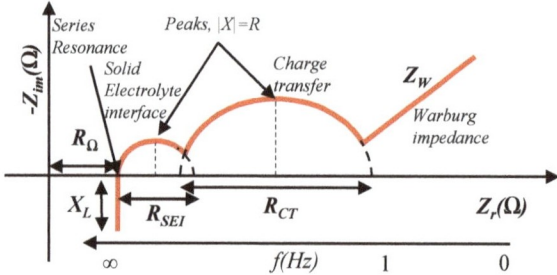

Figure 6. EIS experimental set up: (**a**) test bench; (**b**) block diagram.

Figure 7. Properties of an impedance plot for a battery [1].

3.3. Computation Technique in EIS

Frequency response includes perturbation and measurement of voltage and current over a range of frequencies. The time domain voltage and current measurement is converted to frequency domain by Fast Fourier Transformation (FFT) as expressed by Equation (42). After performing the FFT, impedance is calculated in the complex plane. The complex impedance in frequency domain provides the Nyquist plot and Bode plot.

$$\begin{bmatrix} V_1 & I_1 & t_1 \\ V_2 & I_2 & t_2 \\ \vdots & \vdots & \vdots \\ V_n & I_n & t_n \end{bmatrix} \rightarrow \begin{bmatrix} v(j\omega_1) & i(j\omega_1) & \omega_1 \\ v(j\omega_2) & i(j\omega_2) & \omega_2 \\ \vdots & \vdots & \vdots \\ v(j\omega_N) & i(j\omega_N) & \omega_N \end{bmatrix} \rightarrow \begin{bmatrix} Z_r(j\omega_1) & Z_{im}(j\omega_1) & \omega_1 \\ Z_r(j\omega_2) & Z_{im}(j\omega_2) & \omega_2 \\ \vdots & \vdots & \vdots \\ Z_r(j\omega_N) & Z_{im}(j\omega_N) & \omega_N \end{bmatrix} \quad (42)$$

4. Proposed Parameter Extraction Algorithm

Recently, EIS is implemented in conjunction with a charger, which is known as online EIS [42–44]. Online EIS provides the opportunity to track the internal condition of a battery while it is in operation. Online EIS needs a fast interpretation method for impedance spectrum. Parameter estimation and curve fitting methods are available to interpret impedance spectrum [10,17]. These methods are based on iterative regression analysis and required high computing power. For fast interpretation of impedence spectrum, an algorithm shown in Figure 8 is proposed.

The proposed algorithm interprets the properties of impedance spectrum rather than iterative regression analysis. It is based on Nyquist plot, peak detection, resonance, and phasor algebra. The proposed algorithm is able to get the AR-ECM parameters from an impedance spectrum.

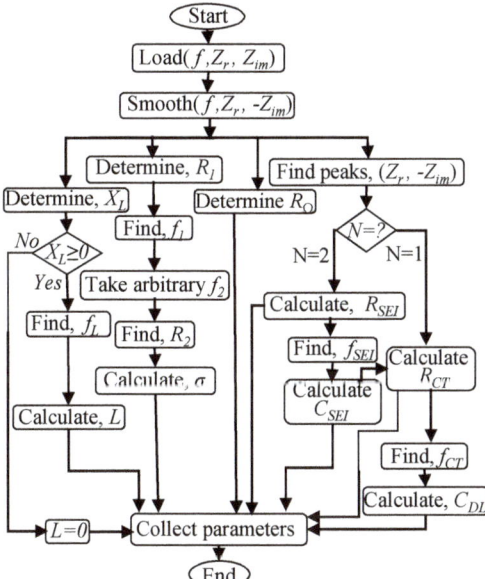

Figure 8. Parameter extraction algorithm [1].

First, frequency and corresponding impedances are loaded to analyze. Then, the impedance plot is smoothed. Finally, circuit parameters are extracted from the smoothed impedance plot. At very high frequencies, capacitive reactance is almost zero and only resistive and inductive parts exist. Thus, the total impedance at high frequencies is expressed by Equation (43).

$$Z_{Total,f_{high}} \approx jX_L + R_\Omega \qquad (43)$$

For the imaginary part, the inductive reactance X_L becomes dominant. Equation (43) can therefore be written as Equation (44).

$$Z_{im,f_{high}} \approx X_L \qquad (44)$$

Now, to get X_L from the impedance plot, Equation (45) can be used.

$$X_L = |min(-Z_{im})| \qquad (45)$$

From the value of X_L, L is easily determined by Equation (46).

$$L = \frac{X_L}{2\pi f} \qquad (46)$$

For sub-Hertz frequencies, Warburg impedance is dominant. It causes a 45-degree slope in the impedance plot, and the resistive part $R_{w,f}$ can be expressed by Equation (47).

$$R_{w,f} = \frac{\sigma}{\sqrt{2\pi f}} \qquad (47)$$

Two sub-Hertz level frequencies are chosen to determine the Warburg coefficient, σ. First, the maximum value of resistance, R_1, ($R_1 = max(Z_r)$) and the corresponding minimum frequency, f_1, are selected. Then, an arbitrary frequency f_2 is chosen such that $f1 < f_2 < 1Hz$. R_2 is the corresponding resistance to f_2, which is found from the impedance spectrum. Now, Equation (48) is used to calculate σ.

$$\sigma = \frac{R_1 - R_2}{\sqrt{2\pi f_1} - \sqrt{2\pi f_2}} \qquad (48)$$

For frequencies close to series resonance, battery impedance can be represented by Equation (49).

$$Z_{Total,f_{close}} \approx jX_L + R_\Omega - jX_C \qquad (49)$$

The value of ohmic resistance, R_Ω, is found at the series resonance frequency where it crosses the x-axis. In this case, X_L and X_C cancel each other out. As a result, only the resistive part R_Ω exists, which is found using Equation (50) at the minimum resistance condition from the impedance plot.

$$Z_{Total,f_{series}} = Z_{r,f_{series}} = min(Z_r) = R_\Omega \qquad (50)$$

To find the R_{SEI}, C_{SEI}, R_{CT}, and C_{DL}, the peaks in the impedance plot are used. For the peaks, the absolute value for specific resistance and reactance are equal, as in Equation (51).

$$\frac{1}{2}R = \frac{1}{2}|X| = |Z_{im}| \qquad (51)$$

For the first peak corresponding to the lowest frequency, R_{CT} and C_{DL} are calculated utilizing Equations (52) and (53).

$$R_{CT} = 2|-Z_{im}(CT_{peak})| \qquad (52)$$

$$C_{DL} = \frac{1}{2\pi f_{(CT_{peak})} R_{CT}} \qquad (53)$$

If another peak exists at a higher frequency, then from that peak R_{SEI} and C_{SEI} are calculated utilizing Equations (54) and (55).

$$R_{SEI} = 2|-Z_{im}(SEI_{peak})| \tag{54}$$

$$C_{SEI} = \frac{1}{2\pi f_{(SEI_{peak})} R_{SEI}} \tag{55}$$

5. Results and Analysis

EIS and ECM based charging/discharging tests are conducted for two Li-Ion batteries. Internal resistance and other parameters obtained from both methods are analyzed and compared as follows.

5.1. Parameter Estimation Based on ECM

The actual test voltage, current, SOC, and OCV are presented in Figures 3 and 4. The estimated SOC-OCV relationship has been presented in Figure 9 which are obtained by a conventional method using two models: the combined model (10) and combined+3 model (11). Figure 9 also shows the SOC-OCV parameter estimation using a proposed hysteresis-free method. The error (ΔV) of the estimation is shown in Figure 10. The estimated parameters are shown in Tables 1 and 2. The parameters are relevant to (8), (10), (11), (27), and (33). The estimated parameters of B600BC battery are different than the 18650 battery.

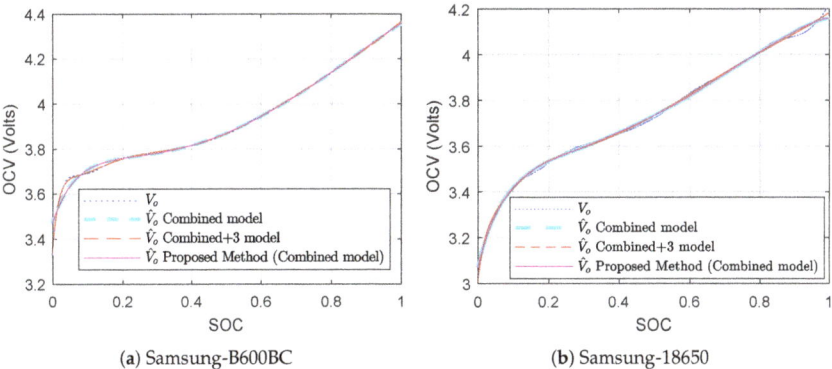

Figure 9. SOC vs. OCV estimation in ECM.

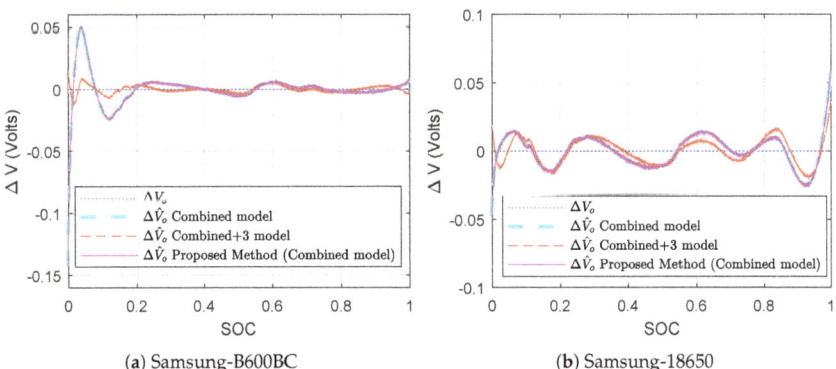

Figure 10. OCV estimation error in ECM.

The estimated DC internal resistance, $\hat{R}_0$, from the proposed method versus the actual value of effective resistance, R_{0h}, and estimated effective resistance, $\hat{R}_{0h}$, from different methods are shown in Figure 11. R_{0h} is considered constant with respect to SOC in the conventional estimation technique. The proposed method considers R_{0h} as a varying quantity.

The actual R_{0h} has been calculated from Δv (measured voltage-actual OCV) and current. The pattern of calculated R_{0h} matches earlier research of hysteresis [24]. Figure 11 shows that, in the conventional method, $\hat{R}_{0h}$ is the average of actual R_{0h}. The proposed method fits the curve to actual R_{0h} in quadratic form. The estimated value of $\hat{R}_{0h}$ from the proposed method gives the value of $\hat{R}_0$ at 100% SOC. The estimated $\hat{R}_0$ from the proposed method is hysteresis-free and can be compared to the measurement from the EIS test. The proposed $\hat{R}_0$ differs significantly from the conventional $\hat{R}_{0h}$. ECM experimentation and the proposed hysteresis-free method have been validated by repeated tests and comparing results with [2,23–25].

Table 1. Estimated parameters: Samsung-B600BC.

Conventional Methods					Proposed Method	
Combined Model (10)		Combined+3 Model (11)			Combined Model (10)	
$\hat{k}_0$	−0.1804	$\hat{k}_0$	−0.9964		$\hat{k}_0$	−0.1804
$\hat{k}_1$	−0.5298	$\hat{k}_1$	40.1725		$\hat{k}_1$	−0.5298
$\hat{k}_2$	5.8554	$\hat{k}_2$	−6.6619		$\hat{k}_2$	5.8555
$\hat{k}_3$	−3.3401	$\hat{k}_3$	0.6954		$\hat{k}_3$	−3.3401
$\hat{k}_4$	0.1904	$\hat{k}_4$	−0.0311		$\hat{k}_4$	0.1904
$\hat{R}_{0h}$	342.3625 mΩ	$\hat{k}_5$	−29.7565		$\hat{k}_{0h}(1)$	0.4238
		$\hat{k}_6$	56.6907		$\hat{k}_{0h}(2)$	−0.1291
		$\hat{k}_7$	−0.3978		$\hat{k}_{0h}(3)$	−0.0582
		$\hat{R}_{0h}$	342.3572 mΩ		$\hat{R}_0$	236.5139 mΩ

Table 2. Estimated parameters: Samsung-18650.

Conventional Methods					Proposed Method	
Combined Model (10)		Combined+3 Model (11)			Combined Model (10)	
$\hat{k}_0$	−0.3502	$\hat{k}_0$	4.2904		$\hat{k}_0$	−0.3502
$\hat{k}_1$	−0.5807	$\hat{k}_1$	34.7860		$\hat{k}_1$	−0.5807
$\hat{k}_2$	6.3877	$\hat{k}_2$	−5.3094		$\hat{k}_2$	6.3877
$\hat{k}_3$	−3.3733	$\hat{k}_3$	0.5128		$\hat{k}_3$	−3.3733
$\hat{k}_4$	0.4131	$\hat{k}_4$	−0.0215		$\hat{k}_4$	0.4131
$\hat{R}_{0h}$	37.7168 mΩ	$\hat{k}_5$	−31.3209		$\hat{k}_{0h}(1)$	0.0641
		$\hat{k}_6$	53.8363		$\hat{k}_{0h}(2)$	−0.0527
		$\hat{k}_7$	−0.4718		$\hat{k}_{0h}(3)$	0
		$\hat{R}_{0h}$	37.7168 mΩ		$\hat{R}_0$	11.4161 mΩ

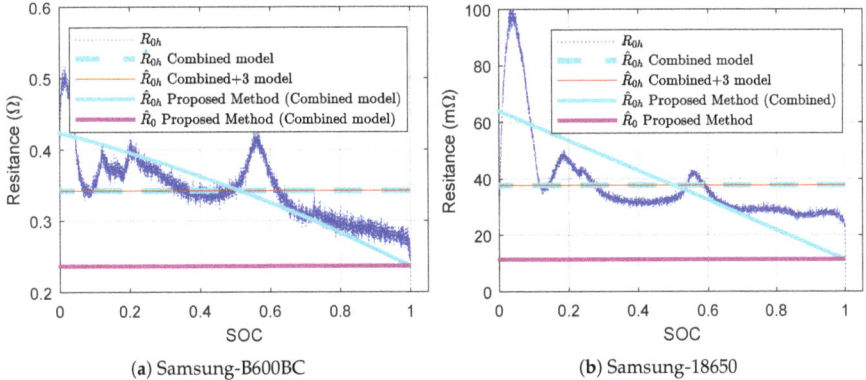

Figure 11. SOC vs. resistance estimation in ECM.

5.2. Parameter Extraction from EIS Tests

The EIS test has been conducted for the same battery used in the ECM based parameter estimation. The test has been conducted for different SOC levels varying from 0% to 100% in a discharging cycle as shown in Figure 12.

Impedance spectra have been obtained using Solartron 1455, and stored in Excel files. The Excel file has been used later in the parameter extraction algorithm. The extracted parameters for different SOC levels are shown in Table 3. R_Ω is the main contributing factor of battery impedance and its value is $\approx$ 230 mΩ for B600BC battery and $\approx$ 13 mΩ for 18,650 battery. The value of R_{CT} is almost 1/10th of R_Ω for B600BC battery and is almost 1/3rd for 18,650 battery. The value of C_{DL} is a few Farads for the B600BC battery and a fraction of a Farad for a 18650 battery. R_{SEI} and C_{SEI} are not included in the Table 3 because their values are negligible. There was only one peak for each impedance plots shown in Figure 12. This indicates that the value of R_{SEI} and C_{SEI} are negligible compared to other portions of the impedance. The values are negligible because a new battery is used for the EIS tests. The extracted parameters can form the AR-ECM, as described in Section 3.

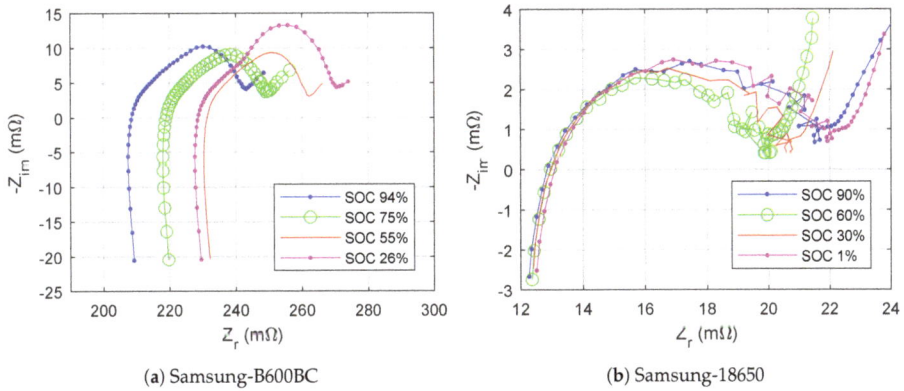

Figure 12. Impedance plots from EIS tests.

Table 3. Extracted parameters from EIS.

SOC (%)	L (µH)	R_Ω (mΩ)	R_{CT} (mΩ)	C_{DL} (F)	$\sigma \times 10^3$
(a) Samsung-B600BC					
100	0.329	205	22.2	1.8	3.98
94	0.327	207	20.5	1.55	11.3
88	0.327	209	19.1	1.32	11.49
82	0.326	213	18.6	1.36	12.58
75	0.325	218	18.1	1.39	12.95
65	0.325	221	18.4	1.37	11.98
55	0.323	230	18.7	1.35	8.17
45	0.323	237	19.8	1.61	8.13
40	0.322	229	22.5	1.78	8.01
34	0.325	226	24.8	2.03	9.44
26	0.325	228	26.5	2.39	9.62
20	0.324	227	29.6	2.7	9.8
13	0.323	228	37.4	2.69	17.8
6	0.322	228	38.3	2.62	75.26
(b) Samsung-18650					
100	0.140	13.00	5.36	0.23	2.09
90	0.141	12.88	5.00	0.15	1.82
80	0.142	12.91	5.04	0.25	1.53
70	0.145	12.93	4.59	0.21	1.04
60	0.145	12.95	4.57	0.17	0.59
50	0.143	12.94	4.61	0.21	0.37
40	0.129	13.04	5.18	0.19	1.30
30	0.130	13.05	4.92	0.16	0.86
20	0.134	13.04	5.13	0.24	0.47
10	0.131	13.04	4.85	0.16	0.54
5	0.131	13.13	5.31	0.23	0.79
1	0.133	13.16	5.48	0.23	0.94

5.3. Unification of Methods for Resistance Identification

The results from ECM based parameter estimation have been compared with the EIS test results for the same battery in Figure 13. The charging discharging tests have been conducted consecutively after the EIS tests. Thus, the change in R_Ω is insignificant. At the end of the EIS tests, R_Ω was 227.75 mΩ for B600BC battery and R_Ω was 13.16 mΩ for 18650 battery. The resistance varies between 18650 battery and B600BC battery because of different chemistry and shape. $\hat{R}_{0h}$ identified from conventional ECM based parameter estimation methods are ≈342 mΩ for B600BC battery and ≈ 37 mΩ for 18,650 battery. These values differ 50% from R_Ω for B600BC battery. For 18,650 battery, the difference is 172%. Therefore, we conclude that $\hat{R}_{0h}$ does not represent the internal resistance properly. $\hat{R}_{0h}$ can be useful for fuel gauging but useless to determine internal electrochemical behavior of a battery. The effective resistance found by conventional ECM is $R_{0h} = R_0 + R_h$, while effective resistance found by EIS testing is $R = R_\Omega + R_{CT} + R_{SEI} + R_W$. The proposed method estimates $\hat{R}_0$, which is 236.51 mΩ for B660BC battery and $\hat{R}_0$ is only 3.5% different from the actual internal resistance obtained from EIS. For 18,650 battery, the estimated value of $\hat{R}_0$ by the proposed method is 11.41 mΩ and differs only 1.76 mΩ from the value of R_Ω. Therefore, it can be concluded that $\hat{R}_0 \approx R_\Omega$. This unified understanding of internal resistance will be useful for estimating internal electrochemical behavior of a battery.

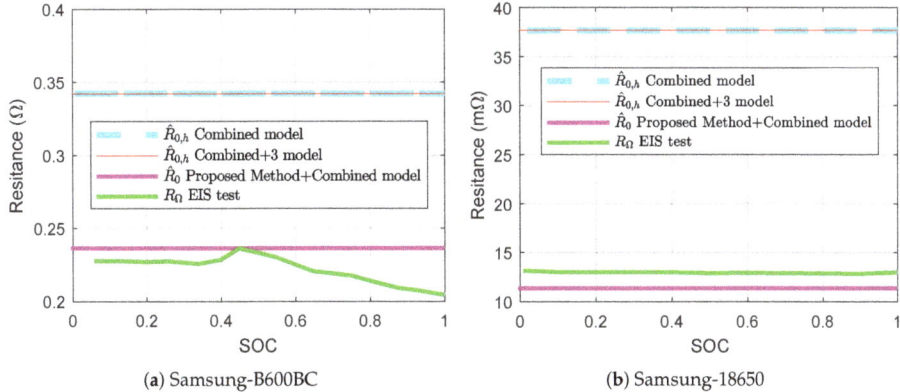

Figure 13. Identified internal resistances from different methods.

6. Conclusions

Parameter identification methods (based on ECM and EIS) for a Li-Ion battery are compared considering their similarities and differences. Hysteresis is the major difference between the two methods for resistance identification. To compare the methods, the following theories are developed:

- A parameter extraction algorithm to interpret the properties of impedance spectrum.
- A hysteresis-free ECM based parameter estimation method.
- Internal resistance obtained from EIS test is approximately equal to hysteresis-free resistance from ECM based parameter estimation

These theories are validated by experimental results. The parameter extraction algorithm was applied for impedance spectra of two different batteries. The proposed hysteresis-free resistance identification method is used to analyze data obtained from a charging–discharging test. Finally, a match is found for the value of internal resistance from both methods with an insignificant tolerance. The proposed parameter extraction will be useful for the online EIS where fast interpretation of impedance spectrum is necessary. The quadratic model of hysteresis resistance separation is computationally effective and will be useful for the modeling of a battery. The proposed methods will be supplementary to track electrochemical parameters properly which can improve the accuracy of SOH prediction.

Author Contributions: All the authors contributed substantially to the manuscript.Contributions of each author are as follows: conceptualization, S.M.R.I., S.-Y.P., and B.B.; methodology, S.M.R.I.,and B.B.; software, S.M.R.I.; validation, S.M.R.I. and S.-Y.P.; formal analysis, S.M.R.I.; investigation, S.M.R.I.; resources, S.I. and S.-Y.P.; data curation, S.M.R.I.; writing—original draft preparation, S.M.R.I.; writing—review and editing, S.M.R.I. and S.-Y.P.; visualization, S.M.R.I.; supervision, S.-Y.P. and B.B.; project administration, S.-Y.P.; funding acquisition, S.-Y.P. All authors have read and agreed to the published version of the manuscript.

Funding: This research was funded by the National Science Foundation of under Grant No. 1454578.

Acknowledgments: This work was supported by the National Science Foundation under Grant No. 1454578. However, any opinions, findings, conclusions, or recommendations expressed herein are those of the authors and do not necessarily reflect the views of the National Science Foundation.

Conflicts of Interest: The authors declare no conflict of interest.The funders had no role in the design of the study; in the collection, analyses, or interpretation of data; in the writing of the manuscript, or in the decision to publish the results.

References

1. Islam, S.M.R.; Park, S.Y.; Balasingam, B. Circuit parameters extraction algorithm for a lithium-ion battery charging system incorporated with electrochemical impedance spectroscopy. In Proceedings of the IEEE Applied Power Electronics Conference and Exposition (APEC), San Antonio, TX, USA, 4–8 March 2018; pp. 3353–3358. [CrossRef]
2. Pattipati, B.; Balasingam, B.; Avvari, G.; Pattipati, K.; Bar-Shalom, Y. Open circuit voltage characterization of lithium-ion batteries. *J. Power Sources* **2014**, *269*, 317–333. [CrossRef]
3. Jossen, A. Fundamentals of battery dynamics. *J. Power Sources* **2006**, *154*, 530–538. [CrossRef]
4. Brand, M.; Gläser, S.; Geder, J.; Menacher, S.; Obpacher, S.; Jossen, A.; Quinger, D. Electrical safety of commercial Li-ion cells based on NMC and NCA technology compared to LFP technology. In Proceedings of the World Electric Vehicle Symposium and Exhibition (EVS27), Barcelona, Spain, 17–20 November 2013; pp. 1–9. [CrossRef]
5. Ecker, M.; Nieto, N.; Käbitz, S.; Schmalstieg, J.; Blanke, H.; Warnecke, A.; Sauer, D.U. Calendar and cycle life study of Li(NiMnCo)O2-based 18650 lithium-ion batteries. *J. Power Sources* **2014**, *248*, 839–851. [CrossRef]
6. Balasingam, B.; Avvari, G.; Pattipati, B.; Pattipati, K.; Bar-Shalom, Y. A robust approach to battery fuel gauging, part I: Real time model identification. *J. Power Sources* **2014**, *272*, 1142–1153. [CrossRef]
7. Pastor-Fernández, C.; Widanage, W.D.; Chouchelamane, G.H.; Marco, J. A SoH diagnosis and prognosis method to identify and quantify degradation modes in Li-ion batteries using the IC/DV technique. In Proceedings of the 6th Hybrid and Electric Vehicles Conference (HEVC 2016), London, UK, 2–3 November 2016; pp. 1–6. [CrossRef]
8. Zhang, D.; Dey, S.; Perez, H.E.; Moura, S.J. Remaining useful life estimation of Lithium-ion batteries based on thermal dynamics. In Proceedings of the American Control Conference (ACC), Seattle, WA, USA, 24–26 May 2017; pp. 4042–4047. [CrossRef]
9. Andre, D.; Appel, C.; Soczka-Guth, T.; Sauer, D.U. Advanced mathematical methods of SOC and SOH estimation for lithium-ion batteries. *J. Power Sources* **2013**, *224*, 20–27. [CrossRef]
10. Tröltzsch, U.; Kanoun, O.; Tränkler, H.R. Characterizing aging effects of lithium ion batteries by impedance spectroscopy. *Electrochim. Acta* **2006**, *51*, 1664–1672. [CrossRef]
11. Lee, Y.D.; Park, S.Y. Electrochemical State-Based Sinusoidal Ripple Current Charging Control. *IEEE Trans. Power Electron.* **2015**, *30*, 4232–4243. [CrossRef]
12. Pastor-Fernández, C.; Widanage, W.D.; Marco, J.; Gama-Valdez, M.Á.; Chouchelamane, G.H. Identification and quantification of ageing mechanisms in Lithium-ion batteries using the EIS technique. In Proceedings of the IEEE Transportation Electrification Conference and Expo (ITEC), Dearborn, MI, USA, 27–29 June 2016; pp. 1–6. [CrossRef]
13. Cuadras, A.; Kanoun, O. SoC Li-ion battery monitoring with impedance spectroscopy. In Proceedings of the 6th International Multi-Conference on Systems, Signals and Devices, Djerba, Tunisia, 23–26 March 2009; pp. 1–5. [CrossRef]
14. Lotfi, N.; Landers, R.G.; Li, J.; Park, J. Reduced-Order Electrochemical Model-Based SOC Observer With Output Model Uncertainty Estimation. *IEEE Trans. Control Syst. Technol.* **2017**, *25*, 1217–1230. [CrossRef]
15. Schweiger, H.G.; Obeidi, O.; Komesker, O.; Raschke, A.; Schiemann, M.; Zehner, C.; Gehnen, M.; Keller, M.; Birke, P. Comparison of Several Methods for Determining the Internal Resistance of Lithium Ion Cells. *Sensors* **2010**, *10*, 5604–5625. [CrossRef]
16. Nagaoka, N.; Ametani, A. An estimation method of Li-ion battery impedance using z-transform. In Proceedings of the IEEE 13th Workshop on Control and Modeling for Power Electronics (COMPEL), Kyoto, Japan, 10–13 June 2012; pp. 1–6. [CrossRef]
17. Wang, Q.; He, Y.; Shen, J.; Hu, X.; Ma, Z. State of Charge-Dependent Polynomial Equivalent Circuit Modeling for Electrochemical Impedance Spectroscopy of Lithium-Ion Batteries. *IEEE Trans. Power Electron.* **2018**, *33*, 8449–8460. [CrossRef]
18. Lai, X.; Gao, W.; Zheng, Y.; Ouyang, M.; Li, J.; Han, X.; Zhou, L. A comparative study of global optimization methods for parameter identification of different equivalent circuit models for Li-ion batteries. *Electrochim. Acta* **2019**, *295*, 1057–1066. [CrossRef]

19. Chu, Z.; Jobman, R.; Rodríguez, A.; Plett, G.L.; Trimboli, M.S.; Feng, X.; Ouyang, M. A control-oriented electrochemical model for lithium-ion battery. Part II: Parameter identification based on reference electrode. *J. Energy Storage* **2020**, *27*, 101101. [CrossRef]
20. Gould, C.R.; Bingham, C.M.; Stone, D.A.; Bentley, P. New Battery Model and State-of-Health Determination Through Subspace Parameter Estimation and State-Observer Techniques. *IEEE TTrans. Veh. Technol.* **2009**, *58*, 3905–3916. [CrossRef]
21. Zhang, J.; Lee, J. A review on prognostics and health monitoring of Li-ion battery. *J. Power Sources* **2011**, *196*, 6007–6014. [CrossRef]
22. Somakettarin, N.; Funaki, T. Study on Factors for Accurate Open Circuit Voltage Characterizations in Mn-Type Li-Ion Batteries. *Batteries* **2017**, *3*, 8. [CrossRef]
23. Roscher, M.A.; Bohlen, O.; Vetter, J. OCV hysteresis in Li-ion batteries including two-phase transition materials. *Int. J. Electrochem.* **2011**, *2011*, 984320. [CrossRef]
24. Baronti, F.; Femia, N.; Saletti, R.; Zamboni, W. Comparing open-circuit voltage hysteresis models for lithium-iron-phosphate batteries. In Proceedings of the IECON 2014—40th Annual Conference of the IEEE Industrial Electronics Society, Dallas, TX, USA, 29 October–1 November 2014; pp. 5635–5640. [CrossRef]
25. Plett, G.L. *Battery Management Systems, Volume I: Battery Modeling, Chapter 2*; Artech House: Norwood, MA, USA, 2015; Chapter 2, pp. 29–63.
26. Rahimi Eichi, H.; Chow, M. Modeling and analysis of battery hysteresis effects. In Proceedings of the IEEE Energy Conversion Congress and Exposition (ECCE), Raleigh, NC, USA, 15–20 September 2012; pp. 4479–4486. [CrossRef]
27. Marongiu, A.; Nußbaum, F.G.W.; Waag, W.; Garmendia, M.; Sauer, D.U. Comprehensive study of the influence of aging on the hysteresis behavior of a lithium iron phosphate cathode-based lithium ion battery—An experimental investigation of the hysteresis. *Appl. Energy* **2016**, *171*, 629–645. [CrossRef]
28. Legrand, N.; Raël, S.; Knosp, B.; Hinaje, M.; Desprez, P.; Lapicque, F. Including double-layer capacitance in lithium-ion battery mathematical models. *J. Power Sources* **2014**, *251*, 370–378. [CrossRef]
29. Erol, S. Electrochemical Impedance Spectroscopy Analysis and Modeling of Lithium Cobalt Oxide/Carbon Batteries. Ph.D. Thesis, University of Florida, Gainesville, FL, USA, 2015.
30. Müller, S.; Massarani, P. Transfer-function measurement with sweeps. *J. Audio Eng. Soc.* **2001**, *49*, 443–471.
31. Instruments, G. *Introduction to Electrochemical Impedance Spectroscopy*; Application Note: Warminster, PA, USA, 2015.
32. Orazem, M.E.; Tribollet, B. *Electrochemical Impedance Spectroscopy*; John Wiley & Sons: Pennington, NJ, USA, 2011; Volume 48.
33. Liu, C.; Neale, Z.G.; Cao, G. Understanding electrochemical potentials of cathode materials in rechargeable batteries. *Mater. Today* **2016**, *19*, 109–123. [CrossRef]
34. Lin, S.H.; Zhao, H.; Burke, A. *A First-Order Transient Response Model for Lithium-ion Batteries of Various Chemistries: Test Data and Model Validation*; Technical Report; Institute of Transportation Studies, University of California: Davis, CA, USA, 2012.
35. Fuller, T.F.; Doyle, M.; Newman, J. Relaxation Phenomena in Lithium-Ion-Insertion Cells. *J. Electrochem. Soc.* **1994**, *141*, 982–990. [CrossRef]
36. Zhang, H.; Chow, M. On-line PHEV battery hysteresis effect dynamics modeling. In Proceedings of the IECON 2010—36th Annual Conference on IEEE Industrial Electronics Society, Glendale, AZ, USA, 7–10 November 2010; pp. 1844–1849. [CrossRef]
37. Zhao, X.; de Callafon, R.A. Modeling of battery dynamics and hysteresis for power delivery prediction and SOC estimation. *Appl. Energy* **2016**, *180*, 823–833. [CrossRef]
38. Windarko, N.A.; Choi, J. Hysteresis modeling for estimation of State-of-Charge in NiMH battery based on improved Takacs model. In Proceedings of the INTELEC 2009—31st International Telecommunications Energy Conference, Incheon, Korea, 18–22 October 2009; pp. 1–6. [CrossRef]
39. Hussein, A.A.; Kutkut, N.; Batarseh, I. A hysteresis model for a Lithium battery cell with improved transient response. In Proceedings of the Twenty-Sixth Annual IEEE Applied Power Electronics Conference and Exposition (APEC), Fort Worth, TX, USA, 6–11 March 2011; pp. 1790–1794. [CrossRef]
40. Sasaki, T.; Ukyo, Y.; Novák, P. Memory effect in a lithium-ion battery. *Nat. Mater.* **2013**, *12*, 569. [CrossRef] [PubMed]

41. Taylor, S.; Gileadi, E. Physical interpretation of the Warburg impedance. *Corrosion* **1995**, *51*, 664–671. [CrossRef]
42. Qahouq, J.A.A.; Xia, Z. Single-Perturbation-Cycle Online Battery Impedance Spectrum Measurement Method With Closed-Loop Control of Power Converter. *IEEE Trans. Ind. Electron.* **2017**, *64*, 7019–7029. [CrossRef]
43. Varnosfaderani, M.A.; Strickland, D. A Comparison of Online Electrochemical Spectroscopy Impedance Estimation of Batteries. *IEEE Access* **2018**, *6*, 23668–23677. [CrossRef]
44. Din, E.; Schaef, C.; Moffat, K.; Stauth, J.T. A Scalable Active Battery Management System With Embedded Real-Time Electrochemical Impedance Spectroscopy. *IEEE Trans. Power Electron.* **2017**, *32*, 5688–5698. [CrossRef]

© 2020 by the authors. Licensee MDPI, Basel, Switzerland. This article is an open access article distributed under the terms and conditions of the Creative Commons Attribution (CC BY) license (http://creativecommons.org/licenses/by/4.0/).

Article

Effect of Current Rate and Prior Cycling on the Coulombic Efficiency of a Lithium-Ion Battery

Seyed Saeed Madani *, Erik Schaltz and Søren Knudsen Kær

Department of Energy Technology, Aalborg University, DK-9220 Aalborg, Denmark
* Correspondence: ssm@et.aau.dk; Tel.: +45-93-56-22-49

Received: 4 June 2019; Accepted: 13 August 2019; Published: 16 August 2019

Abstract: The determination of coulombic efficiency of the lithium-ion batteries can contribute to comprehend better their degradation behavior. In this research, the coulombic efficiency and capacity loss of three lithium-ion batteries at different current rates (C) were investigated. Two new battery cells were discharged and charged at 0.4 C and 0.8 C for twenty times to monitor the variations in the aging and coulombic efficiency of the battery cell. In addition, prior cycling was applied to the third battery cell which consist of charging and discharging with 0.2 C, 0.4 C, 0.6 C, and 0.8 C current rates and each of them twenty times. The coulombic efficiency of the new battery cells was compared with the cycled one. The experiments demonstrated that approximately all the charge that was stored in the battery cell was extracted out of the battery cell, even at the bigger charging and discharging currents. The average capacity loss rates for discharge and charge during 0.8 C were approximately 0.44% and 0.45% per cycle, correspondingly.

Keywords: lithium-ion batteries; coulombic efficiency; capacity loss

1. Introduction

Air pollution and global climate change are fundamental issues for today's society. New technological innovations are necessary to overcome these problems. Considerable eco-friendly changes have to be made for principal way of transport, which is mostly based on the internal combustion engine. One of the possibilities to have cleaner environment is the electrification of buses, cars, and trucks.

The applications of lithium-ion batteries are increasing in different sectors, such as space and automotive industries and consumer electronics to meet the power and energy requirements [1]. Notwithstanding, understanding a battery's rate of useful life or capacity loss in these applications is necessary, especially in automotive and space industry. In addition, determining the durability and performance of the lithium-ion batteries are critical [1].

Li-ion batteries have so many applications in different sectors. One of the problems related to these batteries is their lifetime. Their lifetime is not limitless, and they have a restricted lifetime due to some limitations in technology.

It is possible to expand their market by increasing their cycle life. In the past few years, substantial efforts have been accomplished for model development and to anticipate capacity fade in lithium ion batteries [1–3]. Notwithstanding, experimental data are necessary for the investigation of the capacity fading mechanisms and the aging processes of a battery system [1].

A factor influencing the rechargeable capacity of a lithium-ion battery cell was described [4]. It was seen that diminution resistive electrolytes and oxidation are essential to improve the discharge and charge coulombic efficiencies of both the negative and positive electrodes [4]. It was concluded that electrochemical investigations on the diminution of electrolytes and oxidation, accompanied with the chemical investigation of reaction products, would assistance anticipation in safety and advance cycle life for a lithium-ion cell [4].

Testing life cycle under many cycles such as fifteen thousand cycles and undergoing situation that simulate real application is significantly problematic due to the testing time, which is an extraordinarily long time. The capacity retention plays an important role in the lifetime of Li-ion batteries.

According to the data that were assembled from the cycle life experiments of two kinds of commercial lithium-ion battery cells containing NMC battery cells and LFP battery cells, which were experimentally studied the long-term coulombic efficiency development and its correspondence with the battery cell degradation. The findings demonstrate that NMC and LFP cells display two different aging behaviors [5].

A semi-empirical model that was obtained from the correspondence among battery degradation and coulombic efficiency was suggested to seize the capacity degradation behavior of several cylindrical lithium-ion batteries [6]. The suggested model seizes the convexity of the degradation arc competently, exhibits a superior goodness-of-fit than the generally employed square-root-of time model. In addition, it introduces an extreme robustness versus simulated data, with dissimilar aging shapes [6].

Coulombic efficiency and continuous-time energy efficiency of several lithium titanate batteries were investigated according to dissimilar discharge current rates and state of charge sections. The experimental outcomes demonstrated the coulombic efficiency and energy efficiency discrepancy in dissimilar state of charge sections and changing discharge rates [7].

Different investigations regarding the coulombic efficiency of lithium-ion batteries have been done. Notwithstanding, effect of different prior cycling and current rates on the coulombic efficiency of lithium-ion batteries were not precisely and comprehensively studied. Therefore, the main objective of this investigation was to determine the impact of important parameters such as current rate and prior cycling on the coulombic efficiency of the battery cell by accomplishing different experiments.

Although many investigations about discharge and charge processes on lithium-ion batteries have been accomplished [8,9] most of them were accomplished by applying discharging and charging for different current rates. In this investigation, a comprehensive investigation of discharge and charge parameters of a lithium-ion battery was demonstrated. The coulombic efficiency of the lithium-ion battery at different current rates was determined. In addition, dependence and impact of the discharging and charging intensity, on the coulombic efficiency of the battery cell was studied.

2. Experimental

The CT0550 was used to test the battery cells. The CT0550 contains eighty channel cell tester, which is ideal for evaluating and testing battery cells. In addition, it is used for big volume testing. Two commercial lithium-ion battery cells were taken from suppliers to accomplish these experiments. The charging and nominal voltage of the battery is 4.2 V and 3.6 V, correspondingly. The battery cells were cycled between 2.6 and 4.2 V. Each of battery cell was cycled by using applied currents of 0.2 C, 0.4 C, 0.6 C, and 0.8 C at 25 °C.

The other experiments were done by using CT0550, which includes eighty independent 5 V/50 A channels per rack and 1 microprocessor per five channels. There is liquid cooling with central heat exchanger for stability and accuracy of high power and ultra-fast increase, decrease, and switching time between the charging and discharging modes. Three identical and new Li-ion batteries were employed for this experiment. Maximum discharge current of the batteries for continuous discharge is 8000 mA and for not continuous discharge is 13,000 mA.

3. Results and Discussion

3.1. Effect of Current Rate

The battery cells were cyclically discharged and charged at different current rates and between a lower and upper voltage limit. Figure 1 illustrates the current, voltage, and temperature profiles from the applied experiment for 0.4 C and 0.8 C.

When performing the test, each of the battery cells were initially rested for 24 h, and then being charged at a constant current rate equal to 0.4 C and 0.8 C. Following each of the charging processes, immediately, the battery cells were being discharged at a constant current rate equal to 0.4 C and 0.8 C, correspondingly. Lower and upper voltage limits were assigned as 2.65 V and 4.2 V to fulfill the lesser and uppermost voltage limit, correspondingly.

To automate the experiments, safety procedures were applied in the battery cycler to stop the experiment in the case special events are triggered. Each part of the experiment is finished if the measured voltage attains some limits, for example, 2.65 V and 4.2 V during discharging and charging, correspondingly.

In addition, another constraint was implemented, which was restricting the charging and discharging time. For example, in case the current rate is C/5, the battery cell needs maximum 5 h to attain each voltage limitation. This time was selected in such a way that the battery cells reach the lower and upper voltage thresholds. Because, stopping each discharge and charge cycle before reaching the threshold voltage leads to forcing the battery cells to settle to a dissimilar relaxed voltage. Two different loading profiles were applied to the batteries. One of them (number 1) was charged and then discharged with 0.4 C for 20 times and another cell (number 2) was charged and discharged with 0.8 C with the same amount of cycle number.

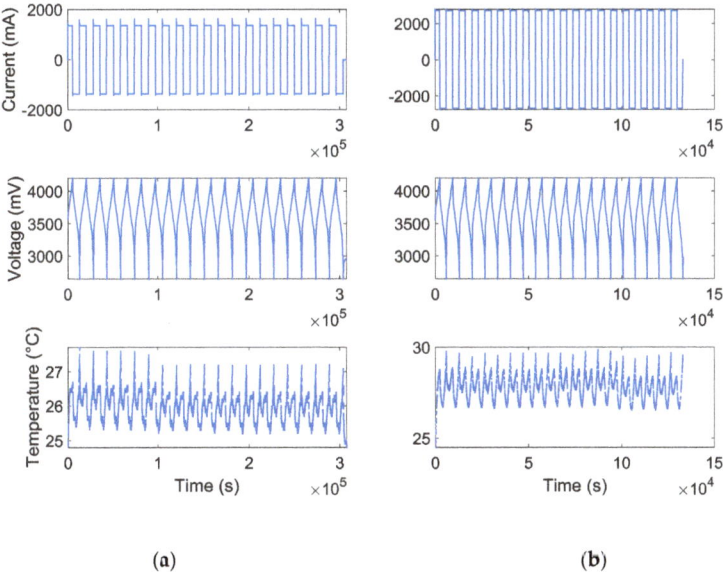

Figure 1. The load profile that was applied to the battery. (**a**) 0.4 C; (**b**) 0.8 C.

Charge End Capacity (CEC) and Discharge End Capacity (DEC) of the battery cells are illustrated in Figures 2 and 3, correspondingly. All of the end points of charge and discharge capacities continually decrease as lithium-ion cells are cycled and this could be a conventional feature of all lithium-ion battery cells.

The battery cell which was cycled at bigger C rates lose capacity quicker than another battery which was cycled at lower C rates. Discharge end capacity is less than charge end capacity for 0.4 C in all cycles. Notwithstanding, the cycling type less affects 0.8 C. In other words, the charge end capacity and discharge end capacity are almost the same for 0.8 C. The average capacity loss rates for discharge and charge during 0.4 C were approximately 0.076% and 0.09% per cycle, correspondingly. This was calculated over the 19 cycles.

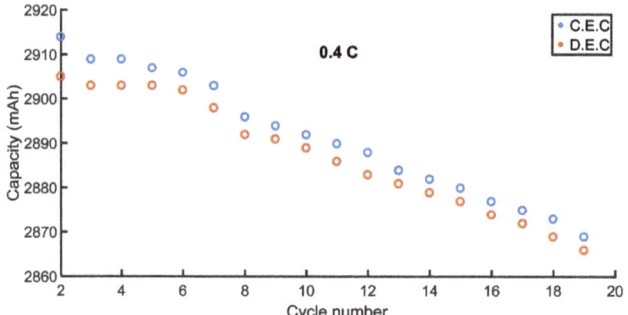

Figure 2. End Capacities for 0.4 C.

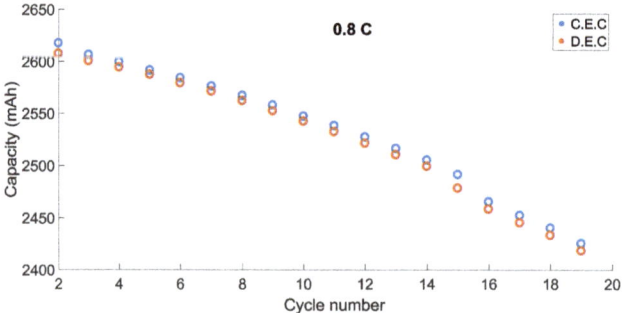

Figure 3. End Capacities for 0.8 C.

The life cycle of a lithium-ion battery cell is not unlimited, because of smart parts of battery cell ingredients that are utilized by parasitic reactions throughout the time of each cycle likely constructing electrolyte oxidation and capacity fade [10]. The quantity of these parasitic reactions could be displayed by accurate measurements of coulombic efficiency [10]:

Coulombic Efficiency = charge out/charge in

Figure 4 illustrates the coulombic efficiency vs cycle number. The presence of these reactions could be distinguished by a coulombic efficiency fewer than 1.000. The coulombic inefficiency vs cycle number is shown in Figure 5. Coulombic inefficiency divided by time of each discharge and charge cycle vs time is illustrated in Figure 3.

The quantity of parasitic reactions that happen for a specified cycle is straightforwardly the time of each cycle multiplied by the parasitic reaction rate. This causes a conventional expression for the coulombic inefficiency for any particular cycle [10]:

$$(1 - CE) = (\text{time of one cycle}) \times k(T,t)$$

where t: Calendar time; $k(T,t)$: The parasitic reaction; T: The cell temperature.

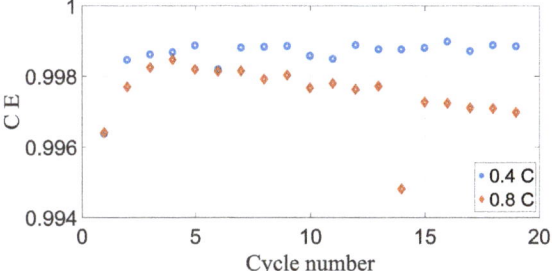

Figure 4. Coulombic efficiency vs cycle number.

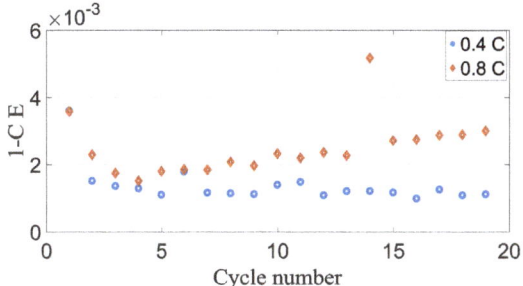

Figure 5. Coulombic inefficiency vs cycle number.

Reaction processes that use electrolyte ingredients or active lithium in lithium-ion battery cells are frequently considered as parasitic reactions [11]. As was mentioned before, parasitic reactions that occur in the battery cell and $k(T,t)$ are related to this parasitic reaction rate, which is as a function of the battery cell calendar time and temperature. As could be seen from Figure 6, these parasitic reactions are higher for 0.8 C when compared to 0.4 C.

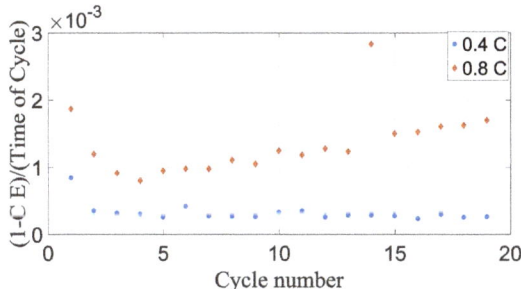

Figure 6. Coulombic inefficiency divided by time of each cycle vs cycle number.

The intensity of these parasitic reactions decreases as the battery cells age for the reason that coulombic inefficiency comes within reach of unification [10]. This is unquestionably attributable to the growing of interface layers between the electrolyte and electrodes [10].

As could be seen from the Figure 4, the battery cells demonstrated well capacity retention throughout the beginning twenty cycles. However, there are clear differences in the coulombic efficiency for both cases, which would cause differences in the capacity retentions. The coulombic efficiency outcomes demonstrated that all the battery cells that were discharged and charged have coulombic efficiency less than 1000.

The rate of battery capacity loss is proximate to the divergence of the coulombic efficiency from 1000. This correspondence is to be assumed in the application which Li absent at the negative electrode by becoming more concentrated of solid electrolyte interphase that the derivation of the coulombic inefficiency is [12]. It is essential to contemplate the experimental factors that should be controlled to measure the coulombic efficiency accurately during a constant current discharge and charge among fixed voltage limits. There are several factors that need to be contemplated, such as accuracy of voltage, currents, battery cell temperature, and time among voltage measurements [12].

3.2. Effect of Prior Cycling

The investigation of rechargeable batteries, particularly lithium-ion battery cells, in the present circumstances, is of great technological and scientific attentiveness. Additionally, experimental investigations targeting at engineer more accomplished batteries. A considerable quantity of modeling has been attempted to comprehend the electrochemical processes that happen throughout battery application.

To satisfy the demands for some applications, it is needed to prolong the lifetime of Li-ion batteries. For instance, solar and wind energy storage systems have more demanding lifetime requirements. The capacity of the Li-ion batteries decreases during cycling. In an automotive application, this lessening in Li-ion battery capacity demonstrates a lessening in the uttermost driving scope of an electric vehicle. Li-ion battery cell capacity, accordingly, is an appropriate metric for characterizing the state of health of a Li-ion battery cell [13–15].

The lithium-ion batteries are distinguished rechargeable batteries. In these batteries, lithium ions are commuted internally between two electrodes, during which electrons are carried by the external circuit and perform the electrical function. The electrodes are generally inserting porous electrodes that, in a perfect instance, reversibly keep lithium in their construction. The electrode that is at the greater electrochemical potential is considered the positive electrode and another that at the lesser potential is considered the negative electrode.

An electrolyte is employed as surroundings of transmission for the lithium ions among the electrodes. A separator that permits ion transportation is employed to stop physical contact among the electrodes.

From beginning to end of charging, lithium ions are transferred from the positive electrode to the negative electrode by the separator and electrolyte. Electrons relocate in the corresponding direction by the exterior circuit. The opposite process happens throughout discharging.

The effectiveness of lithium-ion batteries worsens over time, even if they are used or not. Ageing without and with use are called calendar ageing and cycle ageing, correspondingly.

The two principal outcomes of ageing are power and energy fade. In an electric vehicle utilization, for example, the power specifies the utmost acceleration the vehicle could gain, and the energy specifies the utmost distance that the vehicle could travel through a single charge.

Energy declining could be induced through a diminution of battery capacity or in the increase of the impedance. Diminution of counterbalancing of the active electrode material or the cyclable lithium is the principal origins of capacity fade.

The increase in impedance is attributable to the physical or chemical conversion of the diverse interfaces and materials. An increase in impedance consequently results in a power fade moreover to an energy fade. Typically, both power and energy fades happen contemporary and their comparative importance relies on the specific application. For example, power fade is less critical than energy fade in an electric vehicle.

Lithium-ion batteries have different classifications of ageing mechanism. They could be either mechanical or chemical in character. The mechanisms are dissimilar on the negative electrode side and on the positive electrode side. The most essential ageing mechanism on the negative electrode side is the development of a solid electrolyte interphase, which utilizes the cyclable lithium.

This interphase layer between the electrolyte and graphite is produced due to the fact that the functioning graphite potential on the surface is greater than the stability range of the mostly utilized carbonate electrolytes.

Generally, configuration cycling is accomplished in a lithium-ion battery after battery cell structure where the commencing solid electrolyte interphase is made. Notwithstanding, continuing cycling induces the graphite particles to thicken and construct cracks in the solid electrolyte interphase layer. This revealing novel surface is responsible for supplementary solid electrolyte interphase expansion.

The solid electrolyte interphase enlargement declines at rate that is accompanied by time. Nevertheless, it proceeds over the length and breadth of the lifetime of the battery cell and it uses the cyclable lithium [16].

As was mentioned before, there are different types of ageing mechanism. There is ageing mechanisms within the confines of the graphite that comprise gas development, lithium plating, graphite depilation, and current collector erosion. A considerably slim impervious layer of the electrolyte oxidation production establishes on the electrode surface that brings about the increase in the battery cell impedance.

Life cycle is essentially necessary in implementations of rechargeable batteries. Nevertheless, lifetime prognostication is predominantly based upon empirical trends, instead of mathematical models. In practicable lithium-ion batteries, capacity fade happens over a large amount of cycles, which is restricted by sluggish electrochemical processes, for instance, the creation of a solid-electrolyte interphase in the negative electrode.

Throughout the discharge and charge of a lithium-ion battery cell, the active lithium-ion in the battery cell is inserted out of and into the negative electrode, correspondingly. For the duration of each cycle a tiny quantity of that active lithium-ion reacts with the intention of thickening a passive layer on the surface of the electrode. This is identified under the name of solid electrolyte interphase.

The life cycle of a lithium-ion battery cell is not boundless because little fractions of battery cell ingredients are used up by parasitic reactions throughout each cycle. These undesirable reactions could appear by several different processes, such as solid electrolyte interphase repair and growth, electrolyte oxidation, progression metal ions from out of the positive electrode, and destruction of the positive electrode. Each of these processes could have different reasons for instance solid electrolyte interphase growth and repair is because of lithium-ion loss at the negative electrode [17].

The significance of the coulombic efficiency was acknowledged in a thoughtful research paper on factors that influence capacity retention of lithium ion cells. In the mentioned research paper, it was declared that matched coulombic efficiencies for the negative and positive electrodes, notwithstanding could result in outstanding life cycle for full Li-ion battery cells [4].

It was shown that accuracy measurements of coulombic efficiency are achievable and could lead to bigger comprehension of the degradation processes to be accomplished at the electrodes of Li-ion battery cells [4].

As mentioned before, coulombic efficiencies for the Li-ion battery cells were calculated as the ratio of the capacity of the discharge instantaneously following the previous charge capacity. Consequently, for the Li-ion battery cells:

$$CE = Q_d/Q_c$$

where: Q_d: Discharge capacity; Q_c: Charge capacity

Three new Li ion battery cells were selected for the experiments. Two of the Li-ion battery cells were discharged and then charged at 25 °C by using currents corresponding to 0.4 C and 0.8 C. Another Li-ion battery cell was discharged and charged at the same temperature, but with the currents corresponding to 0.2 C, 0.4 C, 0.6 C, and 0.8 C.

Figures 7 and 8 demonstrate the coulombic efficiency of the new and cycled commercial battery cells plotted vs cycle number. Figure 9 demonstrates the result of coulombic efficiency measurements of Li-ion battery cells for different current rates. It can be seen from Figures 7 and 8 that cases coulombic efficiencies of the new cell is approximately bigger than the cycled battery cell for both 0.4 C and 0.8 C.

Another observation from the figures is an almost similar pattern of coulombic efficiencies for both 0.4 C and 0.8 C cases.

Figure 9 shows a comparison between coulombic efficiencies for different current rates from 0.2 C and 0.8 C. It is clear from the figure that the coulombic efficiency for 0.8 C is lesser than other cases and, in addition, it follows an almost different pattern as compared to other C rates.

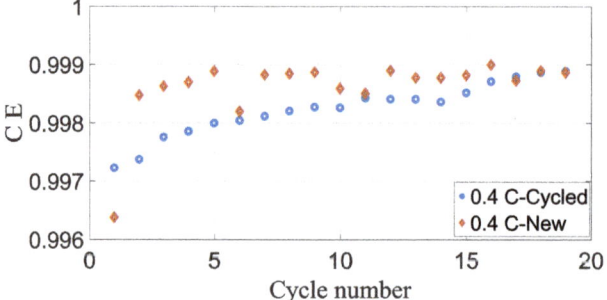

Figure 7. A comparison between coulombic efficiencies for 0.4 C.

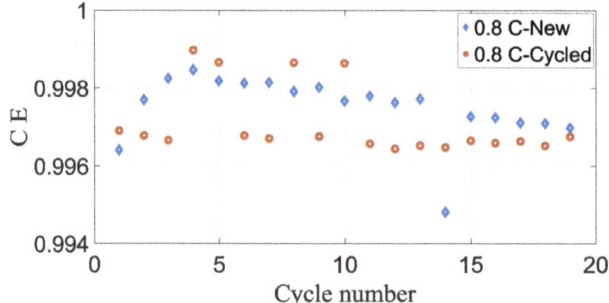

Figure 8. A comparison between coulombic efficiencies for 0.8 C.

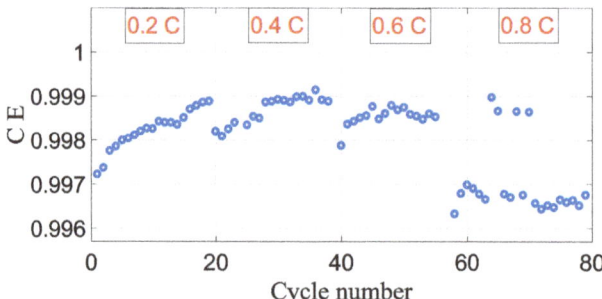

Figure 9. A comparison between coulombic efficiencies for different current rates.

4. Conclusions

The effects of the current rates on the coulombic efficiency of the lithium-ion batteries were studied. The battery cells experienced continuous discharge and charge cycles under constant discharging and charging currents. Three different load profiles were applied to the battery cells. The achieved results demonstrated an approximately identical capacity fade vs cycle number for the dissimilar current rates at the same temperature. The discharge and charge end point capacities decreased together with cycle number for 0.4 C. However, at a considerably sluggish rate when compared to the discharge and charge end point capacities for 0.8 C. The battery cell that was cycled with 0.8 C demonstrated a considerable and obvious capacity loss in relation to cycle number as compared to 0.4 C. It was concluded that parasitic reactions of the battery cells moved away more greatly from unification as the cycling rates were increased.

Author Contributions: S.S.M. proposed the idea of the paper; S.S.M. wrote the paper; E.S. provided suggestions on the content and structure of the paper; S.K.K. and E.S. has been reviewing the draft manuscripts.

Funding: This research received no external funding.

Conflicts of Interest: The authors declare no conflict of interest.

References

1. Wang, J.; Liu, P.; Hicks-Garner, J.; Sherman, E.; Soukiazian, S.; Verbrugge, M.; Tataria, H.; Musser, J.; Finamore, P. Cycle-life model for graphite-LiFePO$_4$ cells. *J. Power Sources* **2010**, *196*, 3942–3948. [CrossRef]
2. Cheng, Y.; Tao, L.; Yang, C. Lithium-Ion Battery Capacity Estimation: A Method Based on Visual Cognition. *Complexity* **2017**, *2017*. [CrossRef]
3. Santhanagopalan, S.; Guo, Q.; Ramadass, P.; White, R.E. Review of models for predicting the cycling performance of lithium ion batteries. *J. Power Sources* **2006**, *156*, 620–628. [CrossRef]
4. Ohzuku, T.; Ueda, A.; Yamamoto, N.; Iwakoshi, Y. Factor affecting the capacity retention of lithium-ion cells. *J. Power Sources* **1995**, *54*, 99–102. [CrossRef]
5. Yang, F.; Wang, D.; Zhao, Y.; Tsui, K.L.; Bae, S.J. A study of the relationship between coulombic efficiency and capacity degradation of commercial lithium-ion batteries. *Energy* **2018**, *145*, 486–495. [CrossRef]
6. Yang, F.; Song, X.; Dong, G.; Tsui, K.L. A coulombic efficiency-based model for prognostics and health estimation of lithium-ion batteries. *Energy* **2019**, *171*, 1173–1182. [CrossRef]
7. Liu, Y.; Zhang, L.; Jiang, J.; Wei, S.; Liu, S.; Zhang, W. A data-driven learning-based continuous-time estimation and simulation method for energy efficiency and coulombic efficiency of lithium ion batteries. *Energies* **2017**, *10*, 597. [CrossRef]
8. Kane, S.N.; Mishra, A.; Dutta, A.K. Preface: International Conference on Recent Trends in Physics (ICRTP 2016). *J. Phys. Conf. Ser.* **2016**, *755*. [CrossRef]
9. Silva, S.P.D.; Silva, P.R.C.D.; Urbano, A.; Scarminio, J. Analysis of a commercial portable lithium-ion battery under low current charge discharge cycles. *Quim. Nova* **2016**, *39*, 901–905. [CrossRef]
10. Smith, A.J.; Burns, J.C.; Dahn, J.R. A High Precision Study of the Coulombic Efficiency of Li-Ion Batteries. *Electrochem. Solid-State Lett.* **2010**, *13*, 177–179. [CrossRef]
11. Glazier, S.L.; Odom, S.A.; Kaur, A.P.; Dahn, J.R. Determining Parasitic Reaction Enthalpies in Lithium-Ion Cells Using Isothermal Microcalorimetry. *J. Electrochem. Soc.* **2018**, *165*, A3449–A3458. [CrossRef]
12. Smith, A.J.; Burns, J.C.; Trussler, S.; Dahn, J.R. Precision Measurements of the Coulombic Efficiency of Lithium-Ion Batteries and of Electrode Materials for Lithium-Ion Batteries. *J. Electrochem. Soc.* **2010**, *157*, 196–202. [CrossRef]
13. Han, X.; Ouyang, M.; Lu, L.; Li, J.; Zheng, Y.; Li, Z. A comparative study of com- mercial lithium ion battery cycle life in electrical vehicle: Aging mechanism identification. *J. Power Sources* **2014**, *251*, 38–54. [CrossRef]
14. Zhang, Y.; Wang, C.Y.; Tang, X. Cycling degradation of an automotive LiFePO$_4$ lithium-ion battery. *J. Power Sources* **2011**, *196*, 1513–1520. [CrossRef]
15. Kassem, M.; Bernard, J.; Revel, R.; Pelissier, S.; Duclaud, F.; Delacourt, C. Calendar aging of a graphite/LiFePO$_4$ cell. *J. Power Sources* **2012**, *208*, 296–305. [CrossRef]
16. Pinson, M.B.; Bazant, M.Z. Theory of SEI Formation in Rechargeable Batteries: Capacity Fade, Accelerated Aging and Lifetime Prediction. *J. Electrochem. Soc.* **2012**, *160*, A243–A250. [CrossRef]
17. Smith, A.J.; Burns, J.C.; Xiong, D.; Dahn, J.R. Interpreting High Precision Coulometry Results on Li-ion Cells. *J. Electrochem. Soc.* **2011**, *158*, A1136–A1142. [CrossRef]

© 2019 by the authors. Licensee MDPI, Basel, Switzerland. This article is an open access article distributed under the terms and conditions of the Creative Commons Attribution (CC BY) license (http://creativecommons.org/licenses/by/4.0/).

Article

In-Operando Impedance Spectroscopy and Ultrasonic Measurements during High-Temperature Abuse Experiments on Lithium-Ion Batteries

Hendrik Zappen [1,2,3,*], **Georg Fuchs** [1,2,3] **and Alexander Gitis** [1,2,3] **and Dirk Uwe Sauer** [1,2,4,5]

1. Electrochemical Energy Conversion and Storage Systems Group, Institute for Power Electronics and Electrical Drives (ISEA), RWTH Aachen University, Jaegerstr. 17/19, 52066 Aachen, Germany; georg.fuchs@isea.rwth-aachen.de (G.F.); alexander.gitis@isea.rwth-aachen.de (A.G.); sr@isea.rwth-aachen.de (D.U.S.)
2. Jülich Aachen Research Alliance, JARA-Energy, Forschungszentrum Jülich GmbH, 52425 Jülich, Germany
3. Safion GmbH, Hüttenstr. 7, 52068 Aachen, Germany
4. Institute for Power Generation and Storage Systems (PGS), E.ON ERC, RWTH Aachen University, Mathieustraße 10, 52074 Aachen, Germany
5. Helmholtz Institute Münster (HI MS), IEK-12, Forschungszentrum Jülich GmbH, 52425 Jülich, Germany
* Correspondence: hendrik.zappen@isea.rwth-aachen.de; Tel.: +49-241-80-49795

Received: 9 March 2020; Accepted: 18 April 2020; Published: 22 April 2020

Abstract: Lithium-Ion batteries are used in ever more demanding applications regarding operating range and safety requirements. This work presents a series of high-temperature abuse experiments on a nickel-manganese-cobalt oxide (NMC)/graphite lithium-ion battery cell, using advanced in-operando measurement techniques like fast impedance spectroscopy and ultrasonic waves, as well as strain-gauges. the presented results show, that by using these methods degradation effects at elevated temperature can be observed in real-time. These methods have the potential to be integrated into a battery management system in the future. Therefore they make it possible to achieve higher battery safety even under the most demanding operating conditions.

Keywords: EIS; electrochemical impedance spectroscopy; lithium-ion batteries; characterization; diagnostics; abuse test; high temperature; degradation; safety; ultrasonics; ultrasound; strain; gassing; gas evolution

---

1. Introduction

Lithium-Ion batteries can nowadays be found in many applications, ranging from mobile computing devices over electric mobility up to multi-megawatt battery storage systems. Often, the battery cells are used under demanding operating conditions, like high charge and discharge current rates and high cyclic depth. the ever increasing requirements on energy and power density of battery systems leads to the development of improved electrode materials. On the anode side, alloy materials containing silicon are promising a greatly increased specific capacity compared to pure graphite. However, the large volume change during Li alloying and de-alloying processes is problematic for stability of the anodic solid electrolyte interphase (SEI) and can lead to reduced cycle life of the battery cell [1]. On the cathode side, nickel rich *NMC811* (Nickel/Manganese/Cobalt) materials have a specific capacity of up to 250 mAh/g [2], which is over 60 % increase compared to the conventional $LiCoO_2$ cathode material. the resulting significant increase in energy density comes at the price of reduced thermal stability: the onset of exothermal decomposition of *NMC811* is as low as 135 °C and can lead to a very strong oxygen release [3]. This reduced thermal stability consequently means a significantly reduced safety margin, compared to more stable materials with lower Ni content.

Operating conditions such as charging at high current rates, or charging at low temperatures, can lead to lithium-plating, which is the deposition of metallic lithium at the anode surface [4].

Besides the possibility of the forming of lithium-metal dendrites, which can lead to internal short circuits,this process also leads to localized exothermal decomposition of the electrolyte [4]. the resulting temperature rise can be especially dangerous for a cell with reduced thermal safety margin, and can ultimately lead to a thermal runaway and destruction of the cell.

From the perspective of battery system technology, the question arises as to how to deal with this risk and which measures can be taken to reduce it. the temperature at which exothermic processes can develop, or a risk of strong aging exists, is not much higher than the maximum temperatures occurring during operation. This condition often makes it necessary for applications with high specific power requirements either to use a generously designed cooling system or to have the available charge- or discharge power reduced by the battery management system (BMS) even at moderate temperatures. But even if this is the case, there are error mechanisms, such as internal short circuits or overcharging, which can ultimately lead to a cell failure. Battery cell manufacturers generally specify a maximum operating temperature in the cell data sheet. However, it is not possible to deduce directly from this information alone whether safety risks are imminent even if the temperature is exceeded for a short period of time, or whether the maximum temperature specified serves above all to limit the ageing that occurs more strongly for prolonged operation at high temperatures.

This poses an interesting question—if the onset of these degradation effects can be reliably detected during operation, would it be conceivable to allow the usual maximum temperature to be exceeded for a short time under certain circumstances, if this operating condition only rarely occurs? Also, there are technical applications in which high ambient temperatures regularly occur, for example in mining or space travel. In these applications, active temperature control for lithium-ion cells is used. However, irregularities in the cooling system must also be detected within a short time. the online detection and forecast of critical battery states, before hazardous processes such as decomposition and internal gas evolution take place is still one of the main challenges in developing safety algorithms for the battery management system.

The goal of this work is therefore to assess the suitability of different *in-operando* measurement methods for detecting degradation phenomena at high temperatures. In the following chapters, a series of high-temperature abuse experiments on a pouch-type lithium ion cell is presented. A variety of instrumentation techniques, such as fast impedance spectroscopy, strain gauges and acoustically guided ultrasonic waves are used to observe degradation phenomena in a commercial cell in real time.

1.1. Behaviour of Lithium-Ion Batteries under High-Temperature Conditions

The optimum operating temperature for lithium-ion batteries is usually between 20 °C to 40 °C [5]. In this temperature range, the electrode materials and electrolyte systems commonly used achieve a high performance with acceptable lifetimes. the results of calendar and cyclic ageing studies suggest that higher temperatures are generally expected to lead to stronger ageing. Capacity loss and internal resistance increase often show an Arrhenius behaviour [6], which is based on an approximate doubling of the reaction speed for a temperature increase of 10 K [7].

The increase in the internal resistance of lithium-ion batteries with graphite-based anode during aging is mainly due to growth in solid electrolyte interphase (SEI). the SEI growth is caused by decomposition of the electrolyte and deposition of components of the conductive salt [8]. Another ageing effect is the anode-side deposition of transition metals such as manganese, which are components of the cathode [8,9]. This causes a loss of the cathode-side active material, which leads to a capacity reduction.

Towards even higher temperatures above 60 °C these processes run at higher speeds according to the Arrhenius law. This accelerated reaction leads to a very rapid aging, combined with other observed effects such as gas formation [10]. With regard to practical operation, such a temperature range is usually avoided by a suitable design of the thermal management. Especially when using large cells in the battery pack for applications with high current rates it is likely that such temperatures can be reached locally. the same applies if the cooling system is insufficiently designed or if other misuse scenarios arise.

High temperatures become safety-critical through the onset of gas generation and cell-internal, exothermic processes. Gas formation leads to an increase in the internal cell pressure and, especially in pouch cells, to a strong deformation of the cell housing. In extreme cases, this can damage the cell housing and lead to electrolyte leakage. One mechanism for cell internal gassing is the formation of CO_2 by lithium metal oxide cathode materials [11,12]. the likely cause is interactions between the electrolyte and surface films on the cathode, consisting of Li_2CO_3 or LiOH [11,13]. This mechanism is particularly promoted by high charge states and the corresponding strong delithiation of the cathode.

Another potential cause of gas formation at high temperatures is the decomposition processes of the solid electrolyte interphase [14] on the anode and also of the electrolyte: If the SEI is decomposed, the liithiated graphite may come into direct contact with the electrolyte [15]. This interaction of the graphite with the electrolyte leads to exothermic reactions, which generate further heat. The onset of exothermic processes can be investigated with so-called Accelerated Rate Calometry (ARC) measurements. Material samples or solid cells are heated gradually, followed by a resting phase in which the self-heating of the sample is measured. the results of such investigations show that from approximately 80–90 °C exothermic reactions may occur in cells [15,16]. However, other literature sources also give higher values in the range of approximately 110 °C [14]. The exact temperature thresholds depend very much on the exact material combinations and concentrations and can hardly be predicted due to the large number of processes involved. However, as soon as significant exothermic processes begin, the so-called *thermal runaway* (TR) can occur. Self-reinforcing exothermic processes lead to an increase in temperature until the cell is completely destroyed.

According to the literature, the basic sequence of events at rising temperature, which can ultimately lead to thermal runaway of a lithium-ion battery can be summarized as follows (also summarized in Figure 1).

- 85–90 °C: Exothermic decomposition of the SEI at the anode begins [15].
- 90–110 °C: Commonly used electrolyte solvents, like dimethyl carbonate (DMC) and ethyl methyl, carbonate (EMC) reach their boiling points [17], which leads to gas generation [18].
- Above 110 °C: Formation of a secondary SEI film and successive decomposition [19]. This SEI film contains polymers and has significant electrical conductivity. This leads to further exothermic reactions of the solvent in the electrolyte with the anode material [20].
- Above 130–160 °C: Melting point of polyethylene (PE) separator (Polypropylene-containing separators have a slightly higher melting temperature) [21] Formation of local internal short circuits and thus further heat generation.
- Above 140 °C: Evaporation of the solvent in the electrolyte. In the presence of free oxygen ignition possible. This oxygen can be produced, for example, by decomposition reactions in the cathode [15].
- 150–250 °C: Decomposition of the cathode material with oxygen formation [15]. $LiFePO_4$ has the highest thermal stability of all common cathode materials. With this cathode material, the probability of ignition and explosion is therefore lower, since the formation of oxygen only begins at a significantly higher temperature. As already discussed above, Nickel-rich cathode materials suffer have a significantly lower thermal stability.

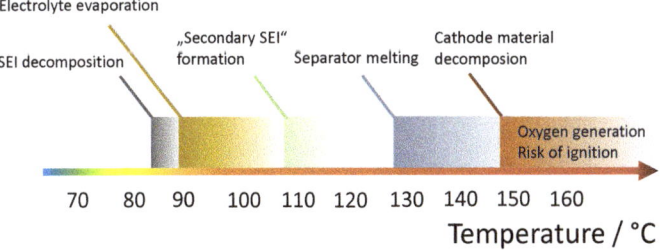

Figure 1. Degeneration effects of lithium-ion batteries at high temperatures.

2. Experimental

2.1. Fast Impedance Spectroscopy Measurements

Electrochemical Impedance Spectroscopy (EIS) measurements capture the complex impedance of a battery cell over a predefined frequency range, typically in the range of below 1 Hz up to several Kilohertz. the most often used measurement type is a frequency sweep method, in which by exciting the battery cell with subsequent sinusoidal currents of different frequencies and measuring the voltage response of the battery cell for each frequency point is measured. By using discrete fourier transform (DFT), the impedance can then be calculated in the frequency domain by dividing the complex values of voltage and current. A typical measurements takes up to several minutes. Conventional EIS techniques are therefore often not sufficient to acquire the impedance of a battery during abuse experiments. the non-stationarity and fast changes of the battery state throughout such an experiment would lead to the violation of the stationarity criterion for valid EIS measurements [22,23]. To overcome this, a multi-sine EIS technique, described in detail in reference [24], is used for all EIS measurements in this work: Using a specifically crafted excitation signal with superposed sine waves of different frequencies from 1 kHz–6 kHz down to 1 Hz allows to acquire an impedance spectrum every second. Figure 2 depicts the procedure—during continuous excitation of a multi-frequency excitation signal with a base period of T_{base}, a sliding window over a period $3 \cdot T_{base}$ is used to calculate a new impedance spectrum from the sampled voltage and current signals for each signal period. the used frequency range covers the typical ranges for the anodic and cathodic charge transfer reaction, as well as interfacial or SEI effects at higher frequencies. the solid state diffusion processes in the active materials, which are typically visible in the very low frequency range below 0.1 Hz cannot be acquired with the used technique under dynamic operating conditions. However, this drawback is negligible for the performed measurements in this work, as the main processes of interest are changes in interfacial properties due to the decomposition of SEI and electrolyte solvents at high temperatures.

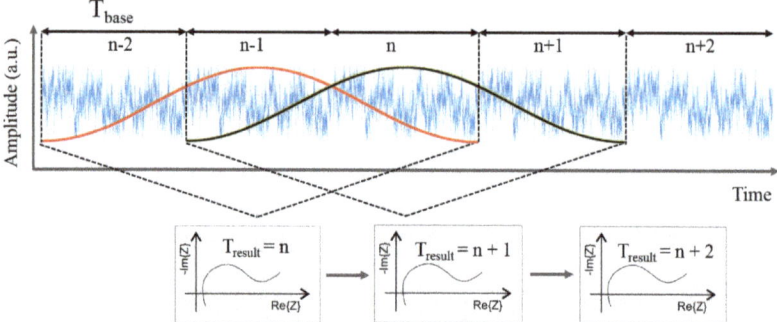

Figure 2. Schematic illustration of the used time-resolved Electrochemical Impedance Spectroscopy (EIS) technique (adapted from [24]).

2.2. Strain Gauge Measurements

To measure the expansion of the cell's pouch foil due to internal gas buildup, a strain gauge strip (Hottinger Baldwin Messtechnik 6/120A LY11) was glued onto the cell with an cyanoacrylate adhesive (Loctite 454) as an additional measuring instrument for one of the performed experiments. Measurement values were recorded by means of a bridge circuit via a National Instruments NI9235 measurement system. Figure 3 shows the assembled strain gauge sensor glued to the battery cell under test.

The used strain gauge is rated for a temperature of up to 200 °C [25]. As all strain gauges, it has a temperature drift, which can only be corrected if the thermal expansion coefficient of the substrate material on which it is applied is exactly known. Additionally, the used adhesive shows a reduction in strength at elevated temperatures [26]. This necessitates a careful interpretation of the acquired data, as a superposition of temperature related phenomena can be observed.

Figure 3. Strain gauge sensor applied to pouch cell.

2.3. Ultrasonic Acoustic Guided Wave Measurements

Ultrasonic methods have been used for non-destructive testing of materials for decades. Recently, measurements with ultrasonic acoustic waves on battery cells have gained attention as a non-destructive tool to evaluate the mechanical characteristics of lithium-ion batteries [27,28], and to use the acquired parameters for state of health (SOH) and state of charge (SOC) estimation [29,30]. Using piezoelectric transducers, which can either be glued or pressed onto the cell casing, a voltage pulse excites a mechanical compressional wave. This wave travels through the cell materials and can be picked up by a second transducer. Different parameters, like time-of-flight, maximum amplitude and intensity can be evaluated.

For the course of this work, an experimental Ultrasound measurement system (Safion US100, prototype) is used to detect changes in the mechanical structure of the tested battery cell due to electrolyte and SEI dissolution. the measurement system consists of a pulse generator circuit for the generation of an excitation of up to 100 V and a low noise amplifier for response signal amplification, as well as filter- and matching circuits. the signal is digitized by a 16.25 MSPS analog to digital converter with a resolution of 14 bit. the battery cell is equipped with three piezoelectric disk transducers (PI Ceramic), glued to the cell surface. Two (generator and receiver A) are placed on the top side of the cell casing, the third one (receiver B) on the bottom side. An overview of the measurement procedure and sensor arrangement is shown in Figure 4. One of the sensors on the top side is used as a generator. This sensor configuration means that two acoustic signal paths exist: the path from the generator to receiver A, hereafter referred to as *in-plane* path captures the wave component along the surface of the pouch foil. the path from generator to receiver B (*trough-plane*) captures the component of the wave which is transferred through the individual layers of the cell acting as an waveguide.

A preliminary test was undertaken to assess the influence of elevated temperatures to the piezoelectric transducers, as well as the stability of the used adhesive—two transducers were glued to an aluminum sheet. While taking ultrasound measurements, the assembly was heated to 100 °C for a period of 2 h and cooled down to 25 °C thereafter. At 100 °C, a reduction in receive signal amplitude of 25 % was observed, mainly caused by the temperature induced change in resonance frequency and coupling factor of the piezoelectric material. After cool-down, no change compared to the initial state was visible in the received signals. This proves the stability of the used joining process. the reduction of signal amplitude at elevated temperatures has to be taken into account when interpreting the obtained results during high-temperature experiments.

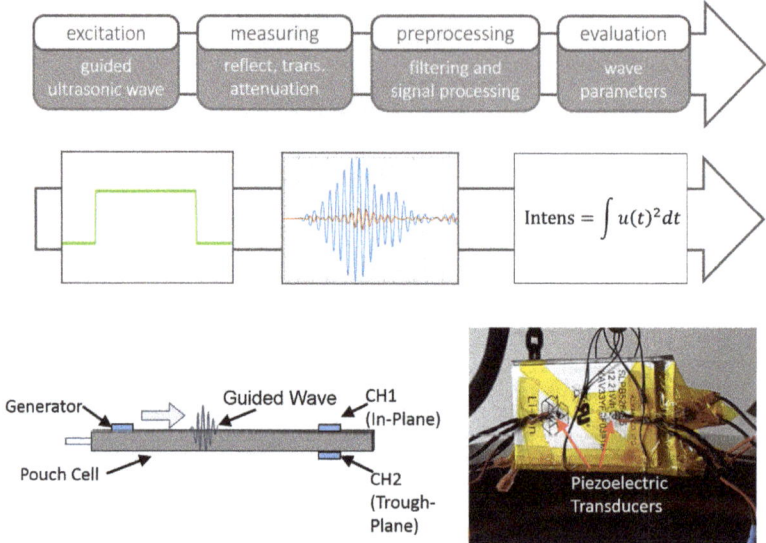

Figure 4. Top: Measurement concept of the used Ultrasonic technique. Bottom: Schematic overview of applied sensors and photograph of the prepared cell.

2.4. High-Temperature Abuse Experiments

In order to find out whether it is possible to detect safety critical degradation processes occurring at high temperatures using the methods introduced in the last sections, investigations were carried out in an adiabatic HEL BTC500 calorimeter (see Figure 5). This device has a pressure-resistant test chamber that can safely absorb any out-gassing of the cell that may occur, even during a thermal runaway. Integrated heating elements and temperature sensors allow an analysis of the thermal behaviour. the cell temperature can be measured using two K type thermocouples.

A total of three high-temperature abuse experiments were carried out, denoted as HT1–HT3 in the following sections. An overview of the experimental conditions, as well as the used instrumentation for each experiment can be found in Table 1. the temperature over time for all three experiments is shown in Figure 6. In all experiments, a *Kokam SLPB526495* lithium-ion battery cell was used. the pouch-type cell with a nominal capacity of 3.3 Ah is composed of NMC/graphite as active materials and an electrolyte consisting of Ethylene Carbonate(EC)/Ethymethyl Carbonate(EMC) as solvent with lithium hexafluorophosphate $LiPF_6$ as conductive salt.

In the first experiment HT1, a cell with at 50% state of charge was heated in the adiabatic calorimeter with a continuous temperature ramp of a set-point value of 2 K/min. the cell had already undergone initial and cycle tests at the time of the experiment. Figure 7 shows the cell before and after the experiment.

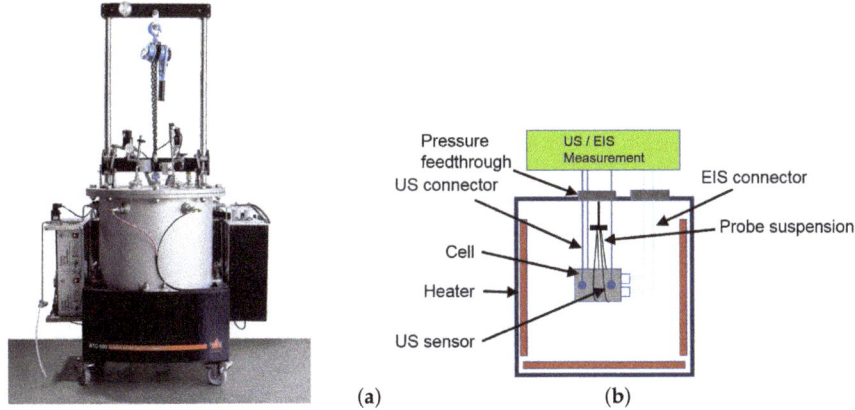

Figure 5. Left (**a**): Used HEL BTC500 type adiabatic calorimeter. Right (**b**): Schematic depiction of cell assembly inside of the calorimeter. Temperature and strain gauge sensors not shown.

Table 1. Performed experiments and used instrumentation (marked with 'x').

Experiment No.	HT1	HT2	HT3
Cell Type		Kokam SLPB526495	
SOC	50%	50%	100%
SOH (Initial Capacity)	95%	96%	100%
EIS Minimum Frequency f_{min}	1 Hz	1 Hz	1 Hz
EIS Maximum Frequency f_{max}	1 kHz	6 kHz	6 kHz
EIS Amplitude Control	constant	manual	manual
Cell Temperature Sensor	x	x	x
Tank Gas Pressure Sensor	x	x	x
Strain Gauge Sensor	-	x	-
Ultrasonic Sensors	-	-	x
Heating Procedure	Ramp 2 K/min until thermal runaway	Ramp with pauses until venting	Ramp with pauses until 110 °C and cool-down

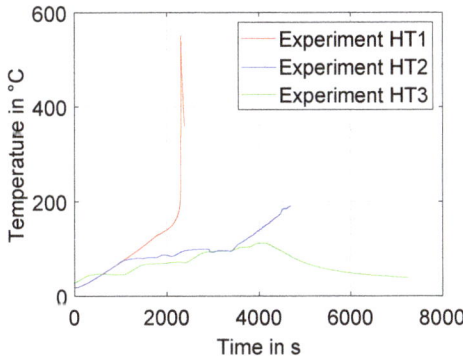

Figure 6. Course of temperature of the three performed high-temperature abuse experiments HT1–HT3.

During the heating process continuous EIS measurements with a galvanostatic excitation with constant maximum amplitude in the range from 1 Hz to 1 kHz were performed. the amplitude was set so that the voltage response at room temperature is limited to 10 mV.

The second experiment, HT2, expanded the frequency range of the EIS measurement to an upper limit of 6 kHz. Also, the excitation amplitude of the galvanostatic EIS measurement was adjusted manually throughout the experiment, to limit the voltage response of the cell to 10 mV over the whole temperature range. Additionally, a strain gauge sensor, shown in Figure 3, was placed on the cell. the used cell had undergone initial cyclic tests, and had a remaining capacity of 96 percent compared to its initial value at the time of the experiment. The state of charge was set to 50 %.

Figure 7. Experiment HT1: Left: Cell with applied temperature sensors before the experiment. Right: Cell after the experiment.

For the third experiment, HT3, a pristine cell at a SOC of 100 % was used. the cell was fitted with three ultrasonic sensors, as described in Section 2.4. the EIS measurements were performed similar to the previous experiment. In this test, the cell was heated to up to 110 °C. After switching off heaters, the cell was allowed to cool off to ambient temperature.

3. Results and Discussion

3.1. Experiment HT1

Figure 8 (left) shows the time-dependent course of the measured cell temperature for experiment HT1. the right part of the figure shows the time derivative of the cell temperature over the absolute cell temperature. the rate of increase is up to a temperature of approx. 150 °C below the set-point change rate of the heaters of 2 K/min. From then on, distinct exothermic processes begin, which eventually lead to a thermal passage of the cell. Peak temperatures of over 550 °C are measured. In Figure 7 (right) accordingly strong burn marks can be seen on the cell after the experiment. Interesting is the clear drop of the temperature rise rate between about 128 °C and 147 °C, which indicates an endothermic process taking place there. This may be the melting of the polyethylene (PE) portion of the separator, as the melting point of common PE separators is very close to this value [31].

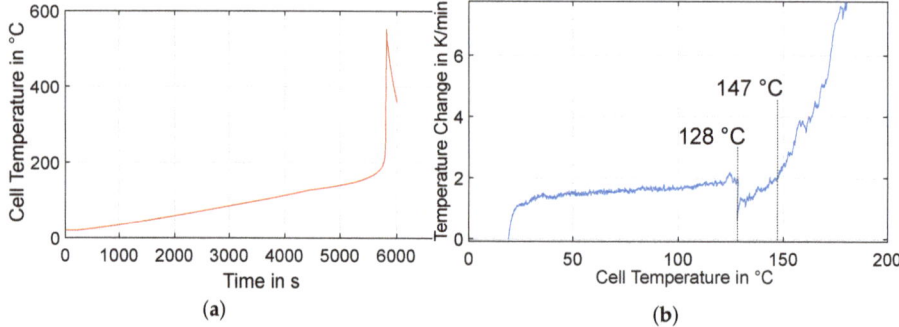

Figure 8. Experiment HT1: Left (**a**): Behavior of the cell temperature over time. Right (**b**): Rate of temperature change as a function of cell temperature.

Figure 9 shows the course of cell voltage as a function of cell temperature during the experiment. the values shown are the moving average of the voltage over one second. Since this corresponds exactly to the base period of the average-free EIS excitation signal, the influence of the excitation current on the cell voltage is filtered out. Thus this value corresponds approximately to the open circuit voltage of the cell.

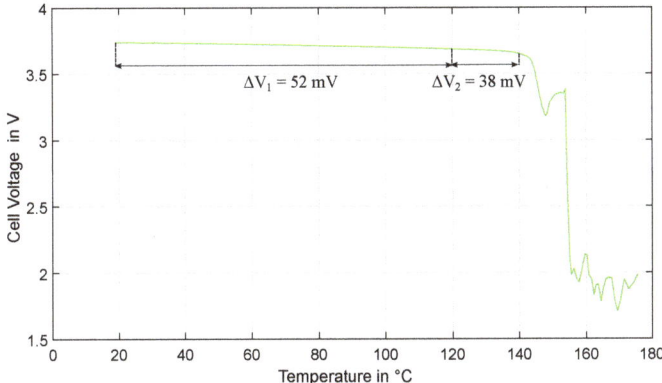

Figure 9. Experiment HT1: Cell voltage curve as a function of cell temperature.

It can be clearly seen that the cell voltage gradually decreases as the temperature increases. the cell voltage decreases between 20 °C and 120 °C by about 52 mV. It is theoretically conceivable that the superimposed EIS signal has an undesired DC offset. However, the offset current of the EIS measurement device is in the range of a maximum of a few milliamperes. the resulting change in charge state during the measurement period is far too small to cause such a voltage change: Even at an offset current of 10 mA, the current in one hour (the time for the temperature rise from 20 °C to 100 °C) would lead to only 10 mAh charge conversion. This would correspond to a SOC drift of only 0.3 %. Considering the open circuit voltage (OCV) to SOC relationship of the cell, this would cause a voltage change of less than 1 mV. It is therefore highly probable that the voltage change is not caused by the EIS superposition.

The variation of the battery cell OCV over temperature and SOC is governed by the Nernst equation [32], which describes the half cell potential for each electrode:

$$\Phi_e = \Phi_0 + \frac{RT}{nF} \ln \frac{C_O}{C_R}. \tag{1}$$

Here, Φ_0 is the lithiation dependent equilibrium potential, R the universal gas constant, T the absolute temperature (in Kelvin), n is the number of electrons involved in the reaction (equals to 1 for lithium-ion batteries), F the Faraday constant and C_O and C_R are the concentrations of oxidized and reduced species at the electrode surface, respectively. the open circuit voltage is the difference of both half cell voltages:

$$V_{OCV} = \Phi_{cathode} - \Phi_{anode}. \tag{2}$$

As can be seen from Equation (1), the temperature dependency of the open circuit voltage is a function of the ratio of the concentrations C_O and C_R on each electrode surface. This temperature dependency of the half cell potentials corresponds to a change in entropy [33]:

$$\Delta S = nF \frac{\delta \Phi_e}{\delta T}. \tag{3}$$

Viswanathan et al. [33] shows that these material-dependent change of entropy for the anode and cathode materials depends also on the state of charge, and may result in positive or negative change of the OCV over temperature change.

The reduction of cell voltage at high temperatures was also observed in other similar experiments: Feng et al. ([16,34]) performed Accelerating Rate Calorimetry (ARC) measurements in combination with pulse resistance measurements. Besides the strong increase of the pulse resistance by melting of the separator at temperatures above 140 °C, a decrease of the voltage already at temperatures of 80 °C can be observed in their work. the change of the cell voltage remains even during cooling, which indicates that it is not only a temperature dependence of the voltage value, but a degradation process causes the voltage change. Ishikawa et al. [35] performed storage tests in which cells of the 18,650 format were stored at different states of charge at a temperature of 100 °C. There, too, a gradual drop in the rest voltages could be observed. In addition, all cells (except those stored at 0 % SOC) showed a sudden drop in voltage after a certain time. the authors attribute this to gas formation, which leads to an internal pressure increase and a triggering of the CID (current-interrupt device).

The cell considered in this experiment does not have a CID. the strong decrease in cell voltage observed from 140 °C is therefore probably due to the melting of the separator, which causes the cell to lose its function. the comparison with Figure 8 (right) shows that strongly exothermic processes occur in this temperature range. This could be due to the formation of internal short circuits due to separator failure.

Figure 10 shows the impedance spectra of the cell in the range from 1 Hz to 1 kHz at different temperatures. In the right part of the figure, the calculated residuals of the spectra in regard to the Kramers-Kronig criteria is shown. the values are calculated using the *LinKK* software tool, developed by Karlsruhe Institute of Technology [36–38]. These are consistently very low and thus indicate valid impedance spectra in compliance with the linearity and time invariance (LTI) criteria.

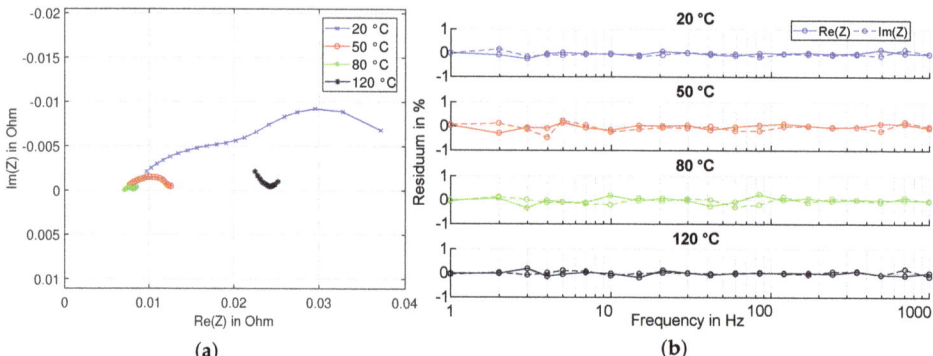

Figure 10. Left (**a**): Experiment HT1 - Nyquist plot of the complex impedance in the range from 1 Hz to 1 kHz of the cell heated at 50 % state of charge in the calorimeter. Right (**b**): Calculated Kramers-Kronig residuals of impedance spectra, showing that stationarity conditions are fulfilled.

As can also be expected, the impedance spectrum changes very strongly with increasing temperature—with increasing temperature, the series resistance (strictly speaking, the true series resistance, which corresponds to the value of the real part at an imaginary part of zero is not visible in the frequency range considered here for the measurements carried out at lower temperatures, since the imaginary part is still negative at 1 kHz. Further experiments, shown later in this section, confirm however that the series resistance at first decreases with increasing temperature.), and also the amount of the negative imaginary part decreases significantly due to the decreasing time constants of the electrochemical processes. the impedance spectrum recorded at 120 °C, however, shows an unusual shape—first of all, compared to the spectrum recorded at 80 °C, there is a clear shift towards larger values on the real axis. On the other hand, a partial capacitive semicircle is formed at high frequencies (towards the measuring points with decreasing real part). Such unusual behaviour is not observed under normal operating conditions and indicates a process with very short time constants. Due to the maximum excitation frequency limited to 1 kHz at the top, this cannot be recognized here as a typical semicircle in the Nyquist diagram.

Therefore, the question arises as to at which temperature this processes begins to get recognizable in the impedance spectrum. To answer this question, it is not helpful to look at Nyquist diagrams at individual selected temperatures. Rather, the continuous course of the impedance must be considered. This is shown in Figure 11 for the different excitation frequencies separately for real and imaginary part. Since the excitation amplitude was not readjusted in this experiment, the signal-to-noise ratio of the voltage response is relatively poor, especially at higher temperatures. This results in an increased variance of the individual impedance values, especially for lower frequencies. For this reason, the time course of the spectra was smoothed using a median filter. Nevertheless, especially with the imaginary part in the range from 1 Hz to 5 Hz, slight noise influences are still recognizable.

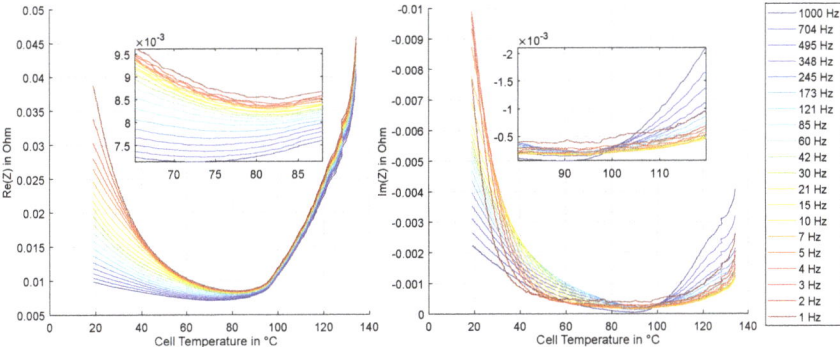

Figure 11. Experiment HT1: Real and imaginary part of impedance at different frequencies over temperature.

The course of the real part over the rising temperature first shows the expected, decreasing behavior for all frequencies. In the range from 20 °C to 75 °C, for example, the real part of the impedance at 1 Hz is reduced by about 75 %. As in this frequency range, the charge transfer process is typically dominant in the impedance spectrum, this is most likely caused by a reduction of the charge transfer resistance of one or both electrodes, which typically shows an exponential temperature dependence according to the Arrhenius law [39,40]. In contrast, the temperature dependence of the real part at higher frequencies, which are no longer dominated by the charge transfer process, is significantly lower. From about 75 °C to 80 °C there is a clear change in the trend: the real part begins to rise again as the temperature continues to rise. the corresponding area is shown enlarged on the left within Figure 11. the temperature at which the change in trend begins is slightly different: at higher frequencies, the onset of the increase can be observed somewhat earlier. For 1 kHz, already from approx. 75 °C an increase is observable, whereas at 1 Hz, the increase begins at approximately 83 °C. From about 90 °C this increase accelerates strongly, so that at 100 °C the real part for all frequencies has already increased significantly. This corresponds to the reported temperature range, at which strong SEI dissolution starts [15]. At approx. 130 °C a jump in the real part is recognizable. This coincides with the endothermic process already visible in the temperature curve in Figure 8.

The consideration of the imaginary part shows up to 90 °C with a strong decrease first a basically similar course. This is not surprising either, since the real and imaginary parts are of course coupled via the Kramers-Kronig relationships. An increase of the imaginary part becomes clearly recognizable from 90 °C upwards. Here, however, the course is strongly frequency-dependent, with a significantly stronger increase in the range of higher frequencies.

The temperature at which this process is getting visible in the impedance spectrum coincides with the temperature range at which a decomposition of the SEI is described in the literature [14–16,41,42]. the deduction that it is this process that shows up in the high frequency range of the impedance spectrum is obvious. However, this cannot be proven directly with the research methodology used.

3.2. Experiment HT2

In order to investigate the effect of the high-frequency process in the impedance spectrum more closely, the experiment was repeated in a slightly modified form, as already described in Section 2.4. This experiment, in the following denoted as "HT2", results in a better signal-to-noise ratio in temperature ranges with a very low impedance, and also ensures that the measurement takes place in the linear range even with a sharp increase in impedance.

Figure 12 shows the course of the real and imaginary parts of the impedance. In contrast to Figure 11, the values are shown over time instead of temperature, since there was no constant temperature rise in this experiment. For reasons of clarity, only the first 10,000 s are plotted here, since the impedance then assumes very high values, also due to venting, and therefore the previous behavior would no longer be recognizable. First a temperature of 80 °C was reached, then the temperature was increased in several steps up to 100 °C. Afterwards the temperature was increased again, until the cell was destroyed. The graph of the cell temperature during the experiment is shown in Figure 13. In contrast to the previous experiment, no thermal runaway of the cell was observed.

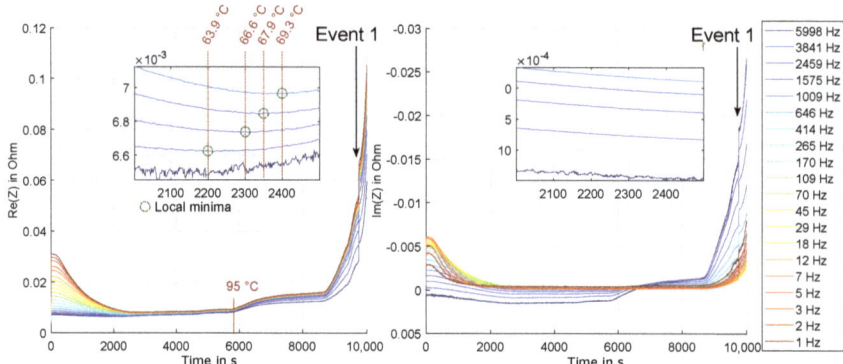

Figure 12. Experiment HT2: Real and imaginary parts of the impedance over the first 10,000 s of the experiment. In the zoomed in area, the trend change of the real part is shown for frequencies between 1 kHz–6 kHz. In the same temperature range, the imaginary part does not show any distinct change.

The basic development of the impedance is similar to the behaviour observed in the previous experiment, even if the initial impedance values are slightly different due to the influence of contact resistance, cables and the different ageing state.

Of special interest in this experiment is again the trend development in the range from approximately 60 °C. In Figure 12 this area is shown enlarged. Compared to the previous experiment, it can be observed that the trend change (local minima) towards an increase in the real part of the impedance can be observed much earlier at higher frequencies: While the signal-to-noise ratio at 6 kHz makes it difficult to find a distinct local minimum, the real part at 3841 Hz shows a clear local minimum at about 64 °C. With 1 kHz, this can only be observed at about 69 °C.

This hints at first degeneration processes already happening at these temperatures. Typically, this is considered a temperature range at which accelerated calendric aging of cells is occurring. This seems to happen at a rate, which makes it observable in real time with EIS measurements. Upwards of the temperature of the local minimum of the real part of impedance for a given frequency, the impedance increase due to occurring degeneration and aging processes outweigh the acceleration (and therefore impedance reduction) of charge transfer processes.

In Appendix A, results from equivalent circuit model (ECM) fitting to the impedance spectra of experiment HT2 are shown. the results of the ECM fitting are in agreement with the findings presented previously, and show a different view of the data.

In Figure 13, the trend of cell temperature and relative expansion of the strain gauge over time is shown. the expansion of the cell is, as expected, a function of the temperature due to the thermal expansion of the various materials. Especially at temperatures above 100 °C the loss of strength the used adhesive will influence the measured values, so that the accuracy ob the absolute value is unclear here. At high temperatures, however, two very noticeable, sudden increases in expansion can be observed. the first, named *Event 1* in Figure 13, occurs at about the same temperature of 130 °C as was seen in the previous experiment HT1. Again, this is also visible in the impedance spectrum, plotted in Figure 12 over time. The second peak value (*Event 2*) can be associated with a sudden rise in temperature and an increase in vessel pressure. This most likely means that an outgassing (venting) took place at this time. Since this process takes place within fractions of a second and is therefore not stationary during the measurement time window, the recorded impedance values do not meet the stationarity criteria.

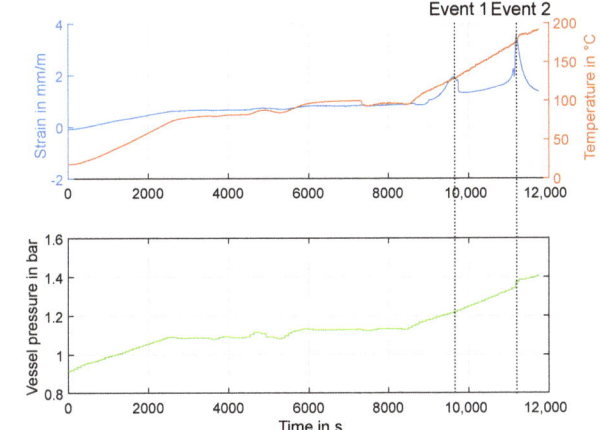

Figure 13. Experiment HT2: Behavior of cell temperature and strain, as well as air pressure in calorimeter.

3.3. Experiment HT3

The question arises to what extent the changes in impedance spectra observed here are completely irreversible or not if the experiment is discontinued and the battery cell cools down. Therefore, one goal of the third experiment HT3 is to investigate this behaviour. the overall procedure is similar to the previous experiments, but the heating is stopped at a cell surface temperature of 110 °C and the cell is left to cool again to its initial temperature.

Figure 14 shows the impedance spectra in the range from 40 °C to 110 °C in steps of 10 °C.

The behavior is almost identical to the previous experiments. the two additionally shown impedance spectra during the cool-down phase at 80 °C and 40 °C show only minor changes compared to the one at 110 °C, and differ very clearly from the spectra recorded at the same temperature during the heating period. This proves that irreversible damage to the cell happens here.

As mentioned in Section 2.3, for this experiment the cell was also outfitted with three piezoelectric transducers, which are able to excite and measure an in-plane and trough-plane path of an ultrasonic guided wave. the received, amplified and digitally bandpass-filtered signals are shown in Figure 15 for a temperature of 50 °C and 100 °C. These show the distinct and different shape of both waveforms at lower temperature. At 100 °C, the in-plane signal is still visible, albeit with significantly reduced amplitude. the through-plane signal is not visible above the noise floor of the measurement any more. To interpret the raw time-series signals, different metrics can be used. These should have the property of a high noise immunity, but also a high sensitivity to changes in the wanted signal. For the course of this work, two metrics have been evaluated: the first is the *intensity* I of the signal, which is defined as the integral of the squared amplitude values, and is closely related to the signal energy. For a discrete time signal v_{us} with N samples, the intensity can be calculated as follows:

$$I = \sum_{k=1}^{N} v_{us}[k]^2. \qquad (4)$$

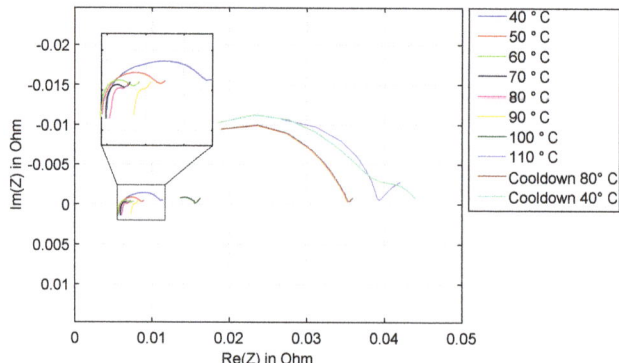

Figure 14. Experiment HT3: Nyquist plot of the complex impedance of a Kokam 3.3 Ah cell in the range from 1 Hz to 6 kHz at different temperatures, including cool-down.

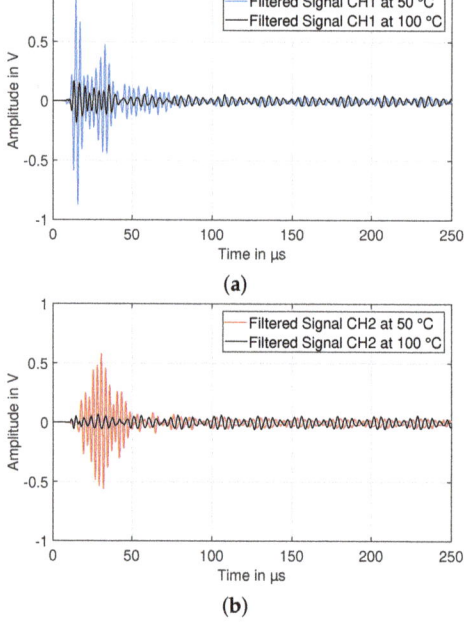

Figure 15. Received signal (amplified and filtered) of in-plane (CH1, (**a**)) and through-plane (CH2, (**b**)) ultrasonic wave for two different temperature values.

Using an integral quantity of a signal has the advantage of a high rejection of zero-average noise. However, as already noted in Section 2.3, the used transducers and adhesive also have the property of a reduced amplitude at higher temperatures, which could potentially mask a battery-cell related change in the signal. Therefore, a second signal quantity, which is more independent of the absolute amplitude should also be considered. One such quantity is the so-called *center of gravity* C of a pulse signal, which relates to the time value at which the signal energy is centered. As such, it has a dependency not only on the amplitude of the excited wave, but also on the speed of sound in the media, which influences the time of flight of the ultrasonic wave. It is defined as follows:

$$C = \frac{\sum_{k=1}^{N} k \cdot v_{us}[k]}{\sum_{k=1}^{N} v_{us}[k]}. \tag{5}$$

In Figure 16, the time course of temperature, as well as signal intensity and center of gravity is shown. Due to a power supply malfunction of the calorimeter, the temperature control was interrupted in the time interval of 5200–6000 s (marked range in Figure 16a). Regarding the ultrasonic measurements, a continuous decline of the intensity during the heating process is visible for the in-plane, as well as the through-plane measurements. At about 90 °C, the intensity begins to drop at a high rate for both wave pathways. At peak temperature of about 112 °C at $t = 10,000$ s, shortly before the cool-down begins, there is a distinct peak visible in the in-plane component, before the intensity drops again when crossing the 100 °C. the onset of the peak lies very close to the temperature range of the boiling point of EMC of 107 °C [18]. the change in in-plane amplitude could therefore be caused by gas generation due to electrolyte vaporization. During the cool-down phase, the in-plane wave intensity begins to rise again until saturation of the amplifier, while the through-plane measurement stays at a very low residual intensity, related to measurement noise. the center of gravity of both signals stays roughly constant until about 90 °C, followed by a sharp rise. For the in-plane measurement, a similar, but inverted behavior compared to the intensity is visible.

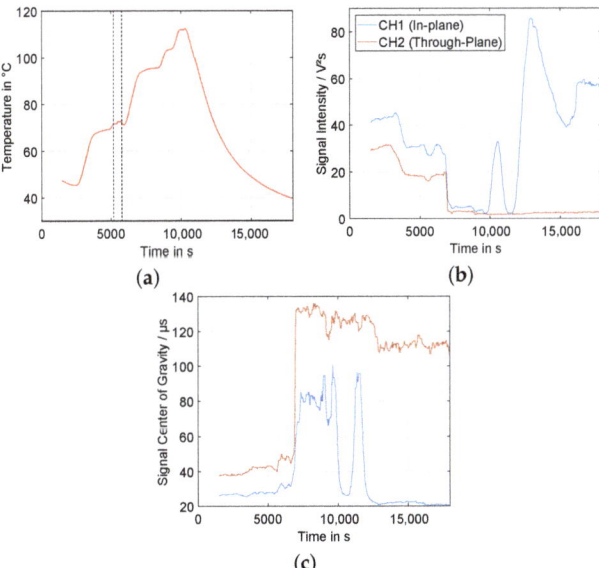

Figure 16. Left (**a**): Temperature over time during experiment. Marked range shows timeframe of temperature regulation malfunction. Center (**b**): Signal intensity over time. Right (**c**): Signal center of gravity over Time (legend same as (**b**)).

To observe the signal behavior more closely, decoupled from the non-constant slope of the heating process, Figure 17 shows the intensity and center of gravity data as a scatter plot over the actual temperature for the heat-up phase of the experiment. In this type of plot, the intensity data shows subtle differences—between 45 °C and 60 °C, the intensity of the through-plane measurement is lowering continuously, while the in-plane component stays relatively constant. At higher temperatures than 70 °C, the behavior is similar for both signal components. the center of gravity shows a very stable behavior up to 65 °C, when it begins to shift in time in positive direction. At about 85 °C, a sharp rise begins, followed by a period with relatively steady average value, although with high fluctuations. This is most likely related to the vanishingly small amplitude of the signal, and the therefore dominating noise effects.

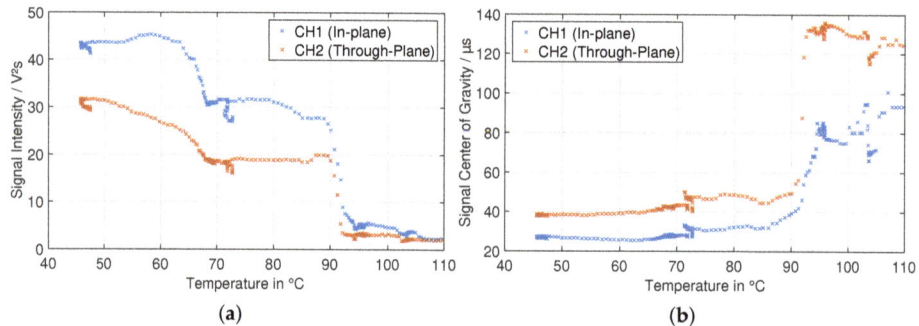

Figure 17. Signal intensity (left, (**a**)) and signal center of gravity (right, (**b**)) over temperature during the heat-up period of the experiment.

This described behavior of the ultrasonic signal is in very good compliance with the impedance spectroscopy results presented earlier. This makes it highly likely, that at least the center of gravity is a parameter with high sensitivity to degradation and gassing processes, while rejecting the changes in transducer behavior over temperature.

4. Conclusions and Outlook

In this paper, a series of high-temperature abuse experiments on a lithium-ion cell battery were performed and analyzed with different in-operando measurement methods, which can be applied without internal modifications to the battery cell design. the results show that impedance spectroscopy and ultrasonic measurements are very sensible to safety-critical degradation effects which happen during exposure to high temperatures. Impedance and ultrasonic signals show distinct changes in temperature ranges, at which SEI dissolution, evaporation of solvents and other effects are described in literature. Even at moderate temperatures of about 65 °C, beginning degradation effects can already be observed with both methods with high sensitivity. Strain gauge measurements can also be helpful to detect a possible swelling or outgassing during operation. Changes in the impedance spectrum can be cross-correlated with the thermal behavior during separator failure. These results make clear that the used methods can be very valuable during battery cell development and validation, cell characterization, but also for online diagnostics in a battery management system in safety-critical applications. In the future, further measurements with the described methods, using various other cell chemistries and housing types should be performed to assess the similarity in behaviour for different cell types. Currently, the authors are working on further improving the noise performance and precision of both methods, which will lead to an even higher sensitivity to degradation phenomena inside the cell.

Author Contributions: Conceptualization, H.Z., G.F. and D.U.S.; Formal analysis, H.Z., A.G.; Methodology, H.Z., A.G.; Software, H.Z.; Validation, G.F.; Writing—original draft, H.Z.; Writing—review & editing, G.F., A.G. and D.U.S. All authors have read and agreed to the published version of the manuscript.

Funding: Parts of this research work were funded through the ZIM ("Zentrales Innovationsprogramm Mittelstand") project *Ausfallsichere Batterie* (FKZ: KF3468301PR4) by the German *Federal Ministry of Economic Affairs and Energy*. the responsibility for this publication lies solely with the authors. the authors also acknowledge the continued funding of the *Federal Ministry of Economic Affairs and Energy* in the scope of the EXIST research transfer project *sBat - safer Battery* (FKZ: 03EFKNW181).

Conflicts of Interest: the authors declare no conflict of interest.

Appendix A. Equivalent Circuit Modelling of Impedance Measurements of Experiment HT2

For experiment HT2, a parameter fitting of the measured impedance spectra to an equivalent circuit model (ECM) was also conducted. the fitting, using a nonlinear least square (NRLS) approach, was performed with the open source Python library *PyEIS* [43]. During the the experiment, an impedance spectrum with a frequency range between 1 Hz and 6 kHz was acquired every second. Correspondingly, for each second new ECM parameters can be calculated. For each new fitting procedure, the starting parameters for the NRLS algorithm were chosen to the calculated parameters from the previous timestep.

the used ECM is shown in Figure A1 (bottom right). It consists of a series resistance element R_s and two RQ (also commonly called "ZARC") elements. the series resistance commonly models all frequency-independent loss processes of the cell, like ohmic losses in current collectors and due to contact resistance, but also the transport of Li through the electrolyte. the RQ elements are parallel connections of a resistor R and a constant phase element Q and typically model charge transfer processes. An exponential factor γ, which is typically limited to values between 0.6 and 1, controls the "depression" of the resulting semicircle, when visualizing a RQ element in a nyquist plot. A value of $\gamma = 1$ corresponds to a regular RC element. For a frequency f, the following equation describes the complex impedance of the RQ element:

$$\underline{Z}_{RQ} = \frac{R}{1 + (j2\pi f)^\gamma RQ} \tag{A1}$$

Two RQ elements are chosen in the hope to be able to model the charge transfer process of both electrodes, if these processes are visible in the impedance spectrum. However, due to the underdetermined nature of the fitting procedure, a direct physical interpretation of the fitting results is often not directly possible, as a large number of parameter sets exist which can be fitted to the impedance spectrum.

The results of the parameter fitting are shown in Figure A1. the included markings show the temperature values which correspond to a change in cell behavior, as already discussed in Section 3.2 and Figure 12. When comparing the trends for the series resistance R_s and the two resistance values of the RQ elements, R_1 and R_2, it can be seen that the series resistance only shows a weak temperature behavior at the beginning, and starts to rise at about 64 °C. R_1 behaves somewhat similar as R_s, while R_2 shows a sharp decrease at the beginning, and begins to stabilize at a nearly constant value at 64 °C. At about 95 °C, another change in trend for R_1 and R_2, and an increased rate of increased for R_s. the other ECM parameters also show distinctive changes at these temperature thresholds. Overall, the results show that the used model is also sensitive to the temperature induced changes. Such an equivalent circuit model could therefore also be used in a battery management system to detect abnormal changes in cell behavior at elevated temperature. A physical interpretation of the model parameters, for example trying to attribute parameters to either cathode or anode cannot be done with sufficient confidence with the available data.

The overall impedance of the ECM is therefore:

$$\underline{Z}_{ECM} = R_s + \underline{Z}_{RQ1} + \underline{Z}_{RQ2} \tag{A2}$$

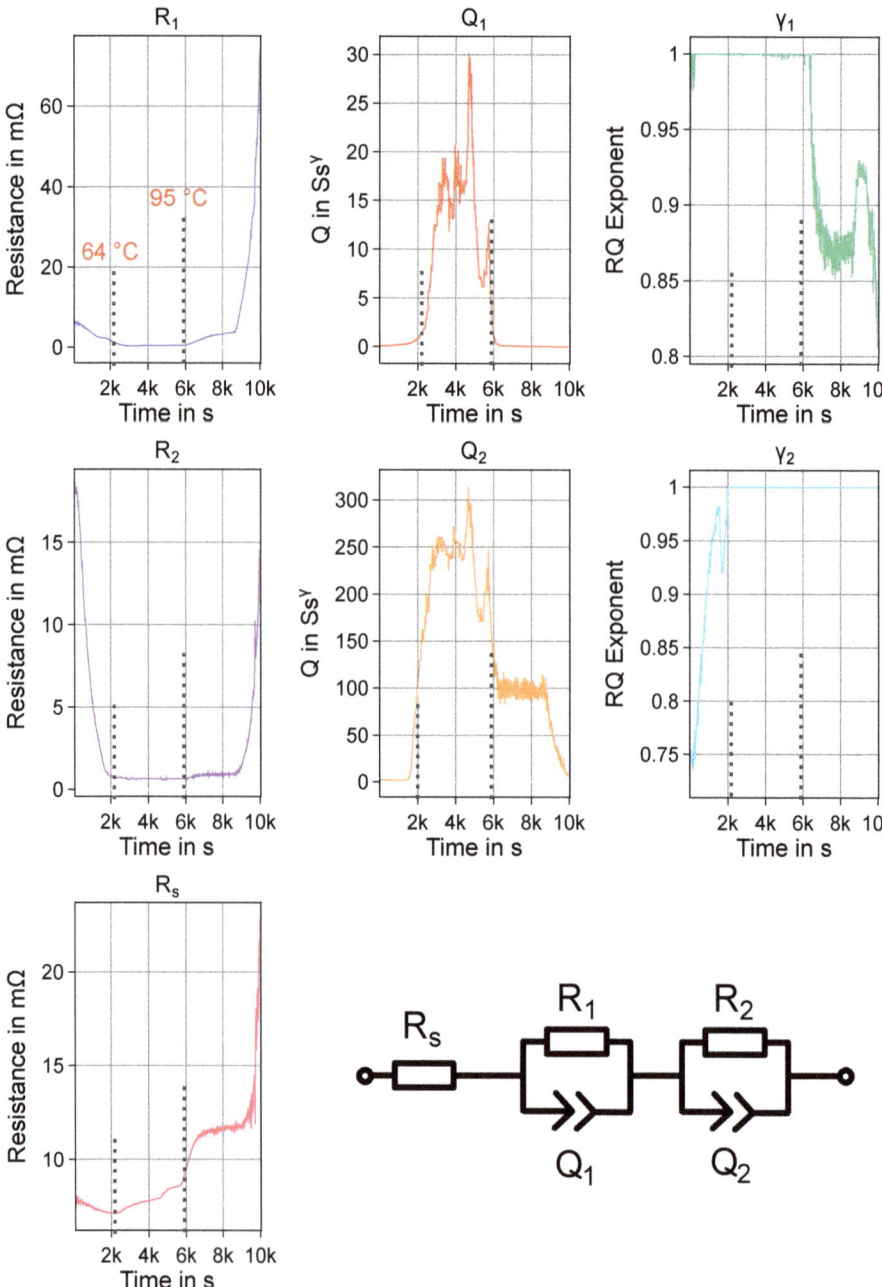

Figure A1. Used ECM (bottom right) and fitting results for impedance spectra of experiment HT2. The marked points in time and the respective temperature values of the cell correspond to the highlighted parts in Figure 12.

References

1. Ko, M.; Oh, P.; Chae, S.; Cho, W.; Cho, J. Considering Critical Factors of Li-rich Cathode and Si Anode Materials for Practical Li-ion Cell Applications. *Small* **2015**, *11*, 4058–4073. [CrossRef] [PubMed]
2. Hou, P.; Yin, J.; Ding, M.; Huang, J.; Xu, X. Surface/Interfacial Structure and Chemistry of High-Energy Nickel-Rich Layered Oxide Cathodes: Advances and Perspectives. *Small* **2017**, *13*, 1701802. [CrossRef] [PubMed]
3. Bak, S.M.; Hu, E.; Zhou, Y.; Yu, X.; Senanayake, S.D.; Cho, S.J.; Kim, K.B.; Chung, K.Y.; Yang, X.Q.; Nam, K.W. Structural Changes and Thermal Stability of Charged $LiNi_xMn_yCo_zO_2$ Cathode Materials Studied by Combined In Situ Time-Resolved XRD and Mass Spectroscopy. *ACS Appl. Mater. Interfaces* **2014**, *6*, 22594–22601. [CrossRef] [PubMed]
4. Waldmann, T.; Hogg, B.I.; Wohlfahrt-Mehrens, M. Li plating as unwanted side reaction in commercial Li-ion cells—A review. *J. Power Sources* **2018**, *384*, 107–124. [CrossRef]
5. Korthauer, R. (Ed.) *Handbuch Lithium-Ionen-Batterien*; Springer: Berlin/Heidelberg, Germany, 2013. [CrossRef]
6. Käbitz, S.; Gerschler, J.B.; Ecker, M.; Yurdagel, Y.; Emmermacher, B.; André, D.; Mitsch, T.; Sauer, D.U. Cycle and calendar life study of a graphite | $LiNi_{1/3}Mn_{1/3}Co_{1/3}O_2$ Li-ion high energy system. Part A: Full cell characterization. *J. Power Sources* **2013**, *239*, 572–583. [CrossRef]
7. Holleman, A.F.; Wiberg, E.; Wiberg, N. *Lehrbuch der Anorganischen Chemie*; de Gruyter: Berlin, Germany, 2007.
8. Waldmann, T.; Wilka, M.; Kasper, M.; Fleischhammer, M.; Wohlfahrt-Mehrens, M. Temperature dependent ageing mechanisms in Lithium-ion batteries—A Post-Mortem study. *J. Power Sources* **2014**, *262*, 129–135. [CrossRef]
9. Warnecke, A.J. Degradation Mechanisms in NMC-Based Lithium-Ion Batteries. Ph.D. Thesis, RWTH Aachen University, Aachen, Germany, 2017. [CrossRef]
10. Ecker, M.; Gerschler, J.B.; Vogel, J.; Käbitz, S.; Hust, F.; Dechent, P.; Sauer, D.U. Development of a lifetime prediction model for lithium-ion batteries based on extended accelerated aging test data. *J. Power Sources* **2012**, *215*, 248–257. [CrossRef]
11. Kim, Y. Mechanism of gas evolution from the cathode of lithium-ion batteries at the initial stage of high-temperature storage. *J. Mater. Sci.* **2013**, *48*, 8547–8551. [CrossRef]
12. Wuersig, A.; Scheifele, W.; Novák, P. CO_2 Gas Evolution on Cathode Materials for Lithium-Ion Batteries. *J. Electrochem. Soc.* **2007**, *154*, A449. [CrossRef]
13. Xiong, D.J.; Ellis, L.D.; Nelson, K.J.; Hynes, T.; Petibon, R.; Dahn, J.R. Rapid Impedance Growth and Gas Production at the Li-Ion Cell Positive Electrode in the Absence of a Negative Electrode. *J. Electrochem. Soc.* **2016**, *163*, A3069–A3077. [CrossRef]
14. Maleki, H. Thermal Stability Studies of Li-Ion Cells and Components. *J. Electrochem. Soc.* **1999**, *146*, 3224. [CrossRef]
15. Bandhauer, T.M.; Garimella, S.; Fuller, T.F. A Critical Review of Thermal Issues in Lithium-Ion Batteries. *J. Electrochem. Soc.* **2011**, *158*, R1. [CrossRef]
16. Feng, X.; Sun, J.; Ouyang, M.; He, X.; Lu, L.; Han, X.; Fang, M.; Peng, H. Characterization of large format lithium ion battery exposed to extremely high temperature. *J. Power Sources* **2014**, *272*, 457–467. [CrossRef]
17. Hess, S.; Wohlfahrt-Mehrens, M.; Wachtler, M. Flammability of Li-Ion Battery Electrolytes: Flash Point and Self-Extinguishing Time Measurements. *J. Electrochem. Soc.* **2015**, *162*, A3084–A3097. [CrossRef]
18. Lamb, J.; Orendorff, C.J.; Roth, E.P.; Langendorf, J. Studies on the Thermal Breakdown of Common Li-Ion Battery Electrolyte Components. *J. Electrochem. Soc.* **2015**, *162*, A2131–A2135. [CrossRef]
19. Yang, H.; Zhuang, G.V.; Ross, P.N. Thermal stability of $LiPF_6$ salt and Li-ion battery electrolytes containing $LiPF_6$. *J. Power Sources* **2006**, *161*, 573–579. [CrossRef]
20. Peled, E.; Menkin, S. Review—SEI: Past, Present and Future. *J. Electrochem. Soc.* **2017**, *164*, A1703–A1719. [CrossRef]
21. Roth, E.P.; Doughty, D.H.; Pile, D.L. Effects of separator breakdown on abuse response of 18650 Li-ion cells. *J. Power Sources* **2007**, *174*, 579–583. [CrossRef]
22. Barsoukov, E.; Macdonald, J.R. (Eds.) *Impedance Spectroscopy: Theory, experiment, and Applications*, 2nd ed.; Wiley-Interscience a John Wiley & Sons Inc. Publication: Hoboken, NJ, USA, 2005. [CrossRef]

23. Schiller, C.A.; Richter, F.; Gülzow, E.; Wagner, N. Validation and evaluation of electrochemical impedance spectra of systems with states that change with time. *Phys. Chem. Chem. Phys.* **2001**, *3*, 374–378. [CrossRef]
24. Zappen, H.; Ringbeck, F.; Sauer, D. Application of Time-Resolved Multi-Sine Impedance Spectroscopy for Lithium-Ion Battery Characterization. *Batteries* **2018**, *4*, 64. [CrossRef]
25. HBM Serie Y Datasheet. Available online: https://www.me-systeme.de/produkte/dehnungsmessstreifen/catalog/Serie-Y-b4710.pdf (accessed on 21 April 2020).
26. Loctite 454 Technical Datasheet. Available online: https://tdsna.henkel.com/americas/na/adhesives/hnauttds.nsf/web/A72821E60E3C0959882571870000D797/$File/454-212NEW-EN.pdf (accessed on 21 April 2020).
27. Gitis, A.; Wessel, S.; Wazifehdust, M.; Heimes, H.; Sauer, D.U.; Figgemeier, E.; Kampker, A. Vom Staub zur Elektrode: Herstellungsprozess von Elektroden für Lithium-Ionen Batteriezellen.—Teil 1. *Galvanotech. älteste Fachz. Für Die Prax. Der Oberflächentechnik* **2017**, *2017*, 7–10.
28. Hsieh, A.G.; Bhadra, S.; Hertzberg, B.J.; Gjeltema, P.J.; Goy, A.; Fleischer, J.W.; Steingart, D.A. Electrochemical-acoustic time of flight: In operando correlation of physical dynamics with battery charge and health. *Energy Environ. Sci.* **2015**, *8*, 1569–1577. [CrossRef]
29. Davies, G.; Knehr, K.W.; van Tassell, B.; Hodson, T.; Biswas, S.; Hsieh, A.G.; Steingart, D.A. State of Charge and State of Health Estimation Using Electrochemical Acoustic Time of Flight Analysis. *J. Electrochem. Soc.* **2017**, *164*, A2746–A2755. [CrossRef]
30. Gold, L.; Bach, T.; Virsik, W.; Schmitt, A.; Müller, J.; Staab, T.E.; Sextl, G. Probing lithium-ion batteries' state-of-charge using ultrasonic transmission—Concept and laboratory testing. *J. Power Sources* **2017**, *343*, 536–544. [CrossRef]
31. Arora, P.; Zhang, Z. Battery Separators. *Chem. Rev.* **2004**, *104*, 4419–4462. [CrossRef] [PubMed]
32. Bard, A.J.; Faulkner, L.R. *Electrochemical Methods: Fundamentals and Applications*, 2nd ed.; Wiley: New York, NY, USA, 2001.
33. Viswanathan, V.V.; Choi, D.; Wang, D.; Xu, W.; Towne, S.; Williford, R.E.; Zhang, J.G.; Liu, J.; Yang, Z. Effect of entropy change of lithium intercalation in cathodes and anodes on Li-ion battery thermal management. *J. Power Sources* **2010**, *195*, 3720–3729. [CrossRef]
34. Feng, X.; Fang, M.; He, X.; Ouyang, M.; Lu, L.; Wang, H.; Zhang, M. Thermal runaway features of large format prismatic lithium ion battery using extended volume accelerating rate calorimetry. *J. Power Sources* **2014**, *255*, 294–301. [CrossRef]
35. Ishikawa, H.; Mendoza, O.; Sone, Y.; Umeda, M. Study of thermal deterioration of lithium-ion secondary cell using an accelerated rate calorimeter (ARC) and AC impedance method. *J. Power Sources* **2012**, *198*, 236–242. [CrossRef]
36. Karlsruhe Institute of Technology. Lin-KK Software. Available online: https://www.iam.kit.edu/wet/Lin-KK.php (accessed on 21 April 2020).
37. Schönleber, M.; Ivers-Tiffée, E. Approximability of impedance spectra by RC elements and implications for impedance analysis. *Electrochem. Commun.* **2015**, *58*, 15–19. [CrossRef]
38. Schönleber, M.; Klotz, D.; Ivers-Tiffée, E. A Method for Improving the Robustness of linear Kramers-Kronig Validity Tests. *Electrochim. Acta* **2014**, *131*, 20–27. [CrossRef]
39. Witzenhausen, H. *Electrical Battery Models: Modelling, Parameter Identification and Model Reduction*; RWTH Aachen University: Aachen, Germany, 2017. [CrossRef]
40. Illig, J. Physically Based Impedance Modelling of Lithium-Ion Cells. In *Schriften des Instituts für Werkstoffe der Elektrotechnik, Karlsruher Institut für Technologie*; KIT Scientific Publishing: Karlsruhe, Baden, 2014; Volume 27.
41. Tanaka, N.; Bessler, W.G. Numerical investigation of kinetic mechanism for runaway thermo-electrochemistry in lithium-ion cells. *Solid State Ion.* **2014**, *262*, 70–73. [CrossRef]
42. Richard, M.N.; Dahn, J.R. Accelerating Rate Calorimetry Study on the Thermal Stability of Lithium Intercalated Graphite in Electrolyte. I. Experimental. *J. Electrochem. Soc.* **1999**, *146*, 2068. [CrossRef]
43. Knudsen, K. kbknudsen/PyEIS: PyEIS: A Python-based Electrochemical Impedance Spectroscopy Simulator and Analyzer. Available online: https://github.com/kbknudsen/PyEIS/ (accessed on 21 April 2020).

© 2020 by the authors. Licensee MDPI, Basel, Switzerland. This article is an open access article distributed under the terms and conditions of the Creative Commons Attribution (CC BY) license (http://creativecommons.org/licenses/by/4.0/).

Article

Comprehensive Hazard Analysis of Failing Automotive Lithium-Ion Batteries in Overtemperature Experiments

Christiane Essl [1,*], Andrey W. Golubkov [1], Eva Gasser [1], Manfred Nachtnebel [2], Armin Zankel [2,3], Eduard Ewert [4] and Anton Fuchs [1]

1. VIRTUAL VEHICLE Research GmbH, Inffeldgasse 21a, 8010 Graz, Austria; andrej.golubkov@v2c2.at (A.W.G.); eva.gasser@v2c2.at (E.G.); anton.fuchs@v2c2.at (A.F.)
2. Graz Centre for Electron Microscopy, Steyrergasse 17, 8010 Graz, Austria; manfred.nachtnebel@felmi-zfe.at (M.N.); armin.zankel@felmi-zfe.at (A.Z.)
3. Institute of Electron Microscopy and Nanoanalysis, NAWI Graz, Graz University of Technology, Steyrergasse 17, 8010 Graz, Austria
4. Dr. Ing. h.c. F. Porsche Aktiengesellschaft, Porschestr. 911, 71287 Weissach, Germany; eduard.ewert@porsche.de
* Correspondence: christiane.essl@v2c2.at; Tel.: +43-316-873-4017

Received: 6 April 2020; Accepted: 4 May 2020; Published: 18 May 2020

Abstract: Lithium-ion batteries (LIBs) are gaining importance in the automotive sector because of the potential of electric vehicles (EVs) to reduce greenhouse gas emissions and air pollution. However, there are serious hazards resulting from failing battery cells leading to exothermic chemical reactions inside the cell, called thermal runaway (TR). Literature of quantifying the failing behavior of modern automotive high capacity cells is rare and focusing on single hazard categories such as heat generation. Thus, the aim of this study is to quantify several hazard relevant parameters of a failing currently used battery cell extracted from a modern mass-produced EV: the temperature response of the cell, the maximum reached cell surface temperature, the amount of produced vent gas, the gas venting rate, the composition of the produced gases including electrolyte vapor and the size and composition of the produced particles at TR. For this purpose, overtemperature experiments with fresh 41 Ah automotive lithium NMC/LMO—graphite pouch cells at different state-of-charge (SOC) 100%, 30% and 0% are performed. The results are valuable for firefighters, battery pack designers, cell recyclers, cell transportation and all who deal with batteries.

Keywords: battery safety; hazard analysis; gas analysis; lithium-ion; thermal runaway; vent particle analysis; vent gas emission

1. Introduction

The market of battery electric vehicles (BEV) and hybrid electric vehicles (HEV) increases, especially in China, the U.S. and the EU [1,2]. LIBs are significantly used in the automotive sector. However, there are still challenging requirements for LIBs in the automotive sector such as costs, fast charging, lifetime, increasing energy density and safety.

It is known that battery failures can lead to critical situations inside the vehicle. The worst case is the uncontrollable exothermic chemical reaction—the TR. TR caused most of EV fires according to Sun and Huang et al., who published a review about EV fire incidents in [3]. TR is a self-accelerating exothermic reaction inside the cell which can be started by a hot spot produced inside the cell (hot spot, particle short circuit) or by a heat source outside the cell (electrical failure) [4–7]. Current methods to characterize possible battery failures are battery abuse tests like overcharge, overtemperature, over-discharge, nail penetration and fire tests. These abuse tests show the influence of cell chemistry on the failing behavior and the thermal stability of the cell [4].

Thus, the cell chemistry is an important parameter for battery safety. State-of-the-art battery chemistries used in BEVs and HEVs are based on Li-ion technology. Currently used materials are: $LiNiMnCoO_2$ (NMC), $LiNiCoAlO_2$ (NCA), $LiMn_2O_4$ (LMO), $LiFePO_4$ (LFP) and $LiCoO_2$ (LCO) as cathode; graphite and carbonaceous materials as anode; regular electrolyte mixtures of ethylene carbonate (EC), diethylene carbonate (DEC), dimethyl carbonate (DMC), ethyl methyl carbonate (EMC); a Li-salt such as $LiPF_6$ and a separator between the electrodes [8]. The cells are encased with sealed laminated foils (pouch cells) or metallic casings. During the first charge of the LIB an organic passivation layer—the so-called solid electrolyte interface (SEI)—develops on the anode.

Several decomposition stages of those cell materials in overheated LIBs have been published [9–12]. Main reactions according to literature include for the listed cell chemistries in general:

>70 °C   The conducting salt starts to decompose and reacts with solvents and the SEI [13–16].
>120 °C Reaction between intercalated lithium in the anode and electrolyte occur initiated by the SEI breakdown (90–130 °C [17]). Heat is generated [7,17]. Li and electrolyte reaction can occur between 90–230 °C [17] and produces gases like C_2H_4, C_2H_6 and C_3H_6 [5].
>130 °C Further gas develops and electrolyte vaporizes. The cell internal pressure increases until the cell casing opens at the weakest point. Accumulated gas vents from inside the cell into the battery pack (first venting). It can occur at about 120–220 °C cell surface temperature [18,19]. Separator melts between 130 °C–190 °C [6,20].
~160 °C Starting at about 160 °C the exothermic process inside the cell accelerates the self-heating and results in a TR. The TR is accompanied by violent gas and particle release (second venting). Electrolyte decomposes exothermally [5,21] between 200–300 °C [17]. At the TR, the cell temperature increases enormously due to chemical reactions inside the cell. Metal oxide cathodes decompose and produce oxygen (O_2) [22,23]. O_2 further reacts with electrolyte and produces CO_2 and H_2O [21,23].

During battery failures, like the TR, violent reactions inside the cell produce significant amounts of hot, toxic and flammable gas and the cell ejects hot particles. The released gas and particles may cause serious safety and health risks, like fire, explosion and toxic atmosphere.

These critical situations need to be analyzed in order to minimize the risks from failing LIBs and to increase safety. To reach an acceptable level of safety in EVs and to enable early failure detection, the Electrical Vehicle Safety—Global Technical Regulation (EVS-GTR) aims to harmonize vehicle regulations worldwide. These regulations discuss suitable tests to characterize safety risks [24].

It is essential to identify comparable hazards and safety parameters to evaluate the failing behavior of different cell types reliably and in order to set necessary safety measures. But which hazards need to be addressed, which safety relevant parameters need to be quantified and which methods are suitable for a comprehensive hazard analysis of a cell?

1.1. Categorized Hazards from LIBs

In literature several important hazards from failing state-of-the-art batteries are reported resulting in main five hazards, which may lead to safety and health risks (Figure 1): electrolyte vaporization, heat generation, gas emission, gas concentration and particle emission. Hazards based on high voltage and current are not considered in this study. The first venting and the TR of the cell can cause the following hazards:

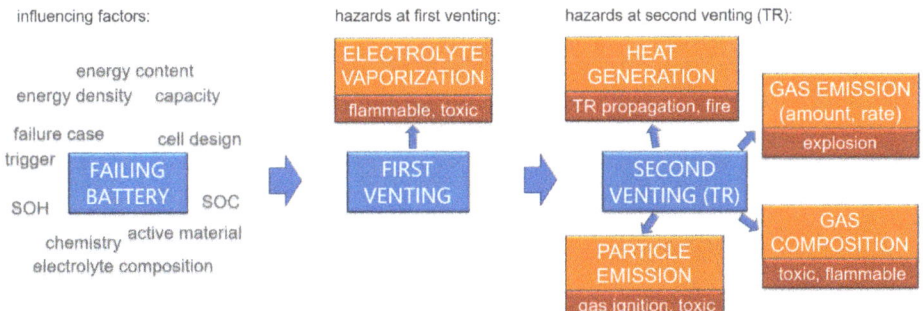

Figure 1. A failing battery can lead to hazards at the first venting and at the TR. Five categorized hazards (orange) and their consequences on safety and health (red) are presented. The battery failures are influenced by several factors.

1.1.1. Electrolyte Vaporization

Electrolyte vaporizes starting at the first venting of the cell. Contemporary electrolytes for LIBs are known to be flammable, irritant, toxic, and/or corrosive depending on the exact composition of the electrolyte mixture [4,25,26] and need to be considered as a safety and health risk. Electrolytes are assumed to be a major source of poor safety with high volume gas decomposition, large combustion enthalpy and flammability of solvent vapor [27].

1.1.2. Heat Generation

Heat generation [4,19,28–30] and significant temperature increase is one safety hazard of the TR, which may lead to TR propagation to neighboring cells or battery fire [31]. Safety relevant parameters are the cell temperature at the first venting of the cell, the TR onset temperature, the maximum reached cell surface temperature and the vent gas temperature. The temperature of the produced vent gas and the ejected particles out of the cell can reach critical high temperatures up to 1000 °C [19] and may damage the cell surrounding materials irreversibly.

1.1.3. Gas Emission

Gas emission [4,23,32,33] is another hazard with the possible consequence of explosion and rapid destruction of the pack. At the TR significant amount of gas [34,35] is produced within seconds. Safety relevant parameters are the amount of produced gas (in mol or liter) and the venting rate (in mol/s or L/s). The gas emission at TR for current state-of-the-art batteries with regular electrolytes is expected in the range of 1.3 L/Ah up to 2.5 L/Ah (at STP: 298.15 K, 100 kPa) [34]. Characteristic venting rates are (0.8 ± 0.3) mol/s at heat ramp TR experiments of 50 Ah prismatic LMO cells [19].

1.1.4. Gas Composition

Main gas compounds at TR are carbon dioxide (CO_2), carbon monoxide (CO), hydrogen (H_2) and hydrocarbons [31,32]. The produced gas is toxic and flammable [25,36]. Except for CO_2 and H_2O all produced gases are flammable, explosive and deflagration of the produced vent gas in contact with O_2 is possible. In addition, small amounts of toxic gases like hydrogen fluoride (HF) can be produced by decomposition of fluorine compounds as $LiPF_6$ [31,37].

1.1.5. Particle Emission

At TR solid hot particles of active materials and aerosols can be released by the failing cell, which are critical to ignite the combustible vent gas [4,38]. Particles should be considered as additional toxic hazard [4] and health risk. The ejected material is a mixture of solid particles, aerosols of active material, parts of current collector foil and electrolyte from the cell.

Figure 1 presents these five hazards assigned to the first venting and the second venting, the TR. The battery failing behavior on cell level and the resulting hazards are influenced by: the energy content of the cell (capacity and the energy density) [34,35,39], the chemistry/active material and separator [4,40], the electrolyte composition and additives [27,41], the failure case/trigger [4,42], the design of the cell housing (pouch versus metal can) [28], the SOC [17,23,43,44] and the state-of-health (SOH)/aging history [18,45]. Additionally, the presence of surrounding gases like O_2 changes the resulting hazards [42] due to additional chemical reactions.

Many researchers have studied single hazard categories from failing LIBs for different cell types and different chemistries [19,28,32,33], but mainly for small capacity cells with <5 Ah [32,33,42,43]. Since NMC/graphite composites are currently one of the preferred LIB chemistries in EVs and higher cell capacities and higher energy densities lead to more severe TR reaction [34,46], this study focuses on the failing behavior of modern high capacity NMC and NMC/LMO cells.

Single hazard categories from NMC and NMC/LMO cells with >20 Ah are published in [13,25,34,38,46–48]: Fang and Gao et al. concentrate on the heat generation during heat triggered TR for 25 Ah NMC [13], 1–50 Ah NMC and NMC/LMO [46] and TR propagation of 42 Ah prismatic BEV [48] cells. Ren et al. evaluate heat generation at different SOH [18]. Koch et al. focus on gas emission (amount), gas composition and mass loss at overtemperature experiments in an atmosphere of air (present O_2) [34]. Nedjalkov et al. analyze the gas composition in air (present O_2) with a nail trigger to force TR [25]. Zhang et al. focus on particle emission [38] and gas composition [47] after heating the cell.

Beside valuable information on single hazard categories, to the best of the authors knowledge, only little information is available in literature on the following hazards and safety relevant parameters of high capacity NMC or NMC/LMO cells. Nevertheless, this information is of relevance for various R&D activities towards significant safety improvements of batteries:

- The vent gas amount: Koch et al. measured an average 1.96 L/Ah at 20-81 Ah NMC cells in air (present O_2) and refer to a gas emission in the range between 1.3 L/Ah up to 2.5 L/Ah for current batteries with regular electrolytes [34] for mainly small capacity cells. Zhao et al. [35] measured for 2 Ah NMC cells at extended volume accelerating rate calorimeter (EV-ARC) abuse 1.4 L/Ah. A detailed analysis of the gas amount produced at failing high capacity NMC/LMO cells in N_2 is missing in current literature.
- The venting rate: Golubkov et al. published a characteristic venting rate of (0.8 ± 0.3) mol/s at heat ramp TR experiments of 50 Ah prismatic LMO cells [19]. A relevant analysis of the venting rate of NMC and NMC/LMO pouch cells is not available in accessible literature.
- A comprehensive gas composition analysis at heat triggered TR: Koch et al. conduct the experiments in air (present O_2) [34] and does not quantify electrolyte components and H_2O. Zhang et al. set huge effort to quantify higher hydrocarbons (1.63% of total gas amount) [47] and does not analyze electrolyte components, HF and O_2. A comprehensive gas analysis at heat triggered TR in N_2 atmosphere including electrolyte quantification is missing.
- Vent particles emission at TR: published by Zhang et al. for a prismatic 50 Ah NMC cells in N_2 atmosphere [38,47]. Since the investigated cells by Zhang et al. have a different cell design (metal can), electrolyte composition and energy density there is a need to further investigate the size and content of particles produced at TR with a nondestructive analysis method.

Additionally, a contribution of the following parameters at failing high capacity NMC or NMC/LMO cells in N_2 atmosphere would be relevant for the scientific community in this field:

- A study of the five mentioned hazards including quantification of the safety relevant parameters for the same specific cell.

- Comprehensive gas composition analysis at the first venting or at abuse experiments of cells with low SOC, where no self-heating into TR can be triggered.

Therefore, for a comprehensive hazard analysis a study on relevant parameters and measurement principles need to be addressed for all five mentioned hazards. In this study, these five hazards are characterized, safety relevant parameters are quantified, and measurement principles are provided from a large capacity NMC/LMO cell currently used in modern EV. Overtemperature experiments are conducted on three cells with different SOCs (100%, 30% and 0%). The investigated hazards (and quantified safety relevant parameters) are:

- Gas composition at first venting (gas concentrations including electrolyte vapor)
- Heat generation at TR (cell surface temperatures including maximum reached temperature)
- Gas emission at TR (amount of produced gas and venting rate)
- Gas composition at TR (gas concentrations)
- Particle emission at TR (particle size distribution and composition)

Hazards from this automotive NMC/LMO pouch cells have, to the authors' knowledge, not been the subject of any scientific publication, but, as will be shown, are important to investigate.

1.2. Structure of the Study

This study describes a comprehensive hazard analysis, safety parameter quantification and TR measurement principles of a fresh 41 Ah automotive Li-ion pouch cell. It starts with a brief investigation of initial cell material in Section 2, an introduction of the TR test bed and the applied methods in Section 3, presenting the failing behavior and hazards from the heat triggered cell in Section 4 and ending with comparing the results with existing literature in Section 5. The TR experiments of the same cell at different SOC (100%, 30% and 0%), but same TR trigger are compared to evaluate the influence of SOC to the failing behavior.

2. Investigated Cell

The investigated cell is a fresh high energy density 41 Ah Li-ion pouch cell designed for EV applications and used in a currently available EV. We extracted the cells from an EV.

The total mass of the fresh pouch cell is 865 g (Table 1). The cell consists of an electrode stack which is sealed in laminated foil. This electrode stack has 22 anode layers, 21 cathode layers and 42 separator layers. The anode layers consist of Copper (Cu) foils (current collector of the anode), which are coated with graphite on both sides. Likewise, the cathode layers consist of aluminum (Al) foils, which were coated on both sides with a mixture of NMC and LMO (spinel). The graphite particles have an average size of 25 µm and the NMC/LMO particles 12–15 µm [49]. The separator has an Al_2O_3 coating facing the cathode side. Fluorine (F) was detectable in the anode and cathode material [19].

Table 1. Specification of the automotive Li-ion pouch cell.

Parameter	Value
Nominal capacity	41 Ah
Cathode material	NMC/LMO
Anode material	graphite
Electrolyte	48% EC, 48% DEC, 4% DMC, 1 mol/L $LiPF_6$
Nominal voltage	~3.8 V
Initial mass	865 g
Volume	0.459 L
Aging state	fresh, unused
SOC	0%, 30% and 100%

The electrolyte consists of a mixture of EC, DEC and DMC solvents with 1 mol LiPF$_6$ per liter. The following molar ratios, namely 48% EC, 48% DEC and 4% DMC were determined by 1H NMR and 13C NMR analysis. No FEC and VC electrolyte additives were found by this investigation (Table 1).

The mass split of the discharged cell presented in Figure 2 is estimated based on the investigations of the cell material and considers the cell design and data from literature for NMC cells [32,43]. The mass of SEI, binder and carbon black are omitted. It is assumed that 14% of the initial mass of the cell is electrolyte and conducting salt. This corresponds to 121.5 g of electrolyte, consisting of 44 g of EC, 59 g of DEC, 3.7 g of DMC and 14.8 g of LiPF$_6$.

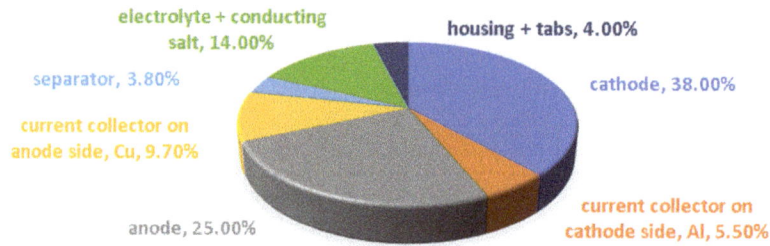

Figure 2. Estimated mass split of the investigated fresh automotive pouch cell in discharged condition.

3. Experimental Setup and Test Methods

Three experiments with fresh automotive pouch cells are conducted. In the first experiment the cell is charged to 100%. In the second experiment the cell is charged to 30% and in the third to 0%. Each single cell is triggered into the failing behavior separately by heat. During the heating phase, temperatures at several positions on the cell surface and inside the test reactor, the voltage of the cell and the pressure inside the reactor are measured.

3.1. Reactor Setup

TR experiments are carried out inside a gastight 40 bar pressure resistant stainless-steel reactor. The test-rig is published in [19,37,50] and is shown in Figure 3. The stainless-steel reactor with the implemented sample holder has a free volume of 121.5 L. The experiments can be done in N$_2$ atmosphere or in air. For safety reasons most experiments are done in N$_2$ atmosphere, as are the presented ones.

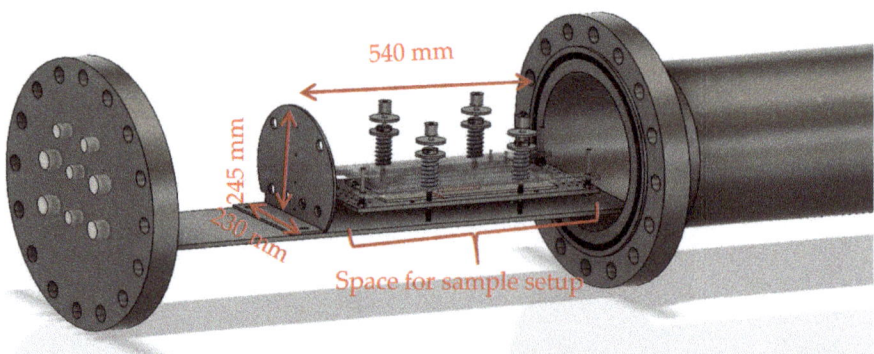

Figure 3. Test rig for thermal runaway experiments on automotive cells [19,50] designed for different cell geometries and different sample holders.

3.2. Experimental Method:

In the experiments the response of each cell (mounted inside a sample holder) to heat is measured and safety relevant parameters are quantified. The sample holder presented in Figure 4 is heated by two heater stripes (max. 500 W each) on the top stainless-steel plate and two heater stripes (max. 500 W each) on the bottom stainless-steel plate. To minimize the thermal coupling between the stainless-steel plates and the cell, insulating mica sheets (thermal conductivity of 0.23 W/mK) with 2 mm are placed between the cell and the stainless-steel plates. The mica sheets also provide channels for the thermocouple wires. Each mica sheet has positions for thermocouples. The tips of the thermocouples protrude through the mica sheets and are squeezed between the mica sheet and the cell surface. Because the mica sheets are thermal insulators, the thermocouple tips measure the cell surface temperature.

The heater increases the temperature of the cell (also compare Figure 6 heater output, black line). Though with the presented setup it is not possible to define the exact heating rate before the experiment, the average heating rate is calculated after the experiment. The heating rate is defined as the increase of the average cell surface temperature per minute between 30 °C and 200 °C.

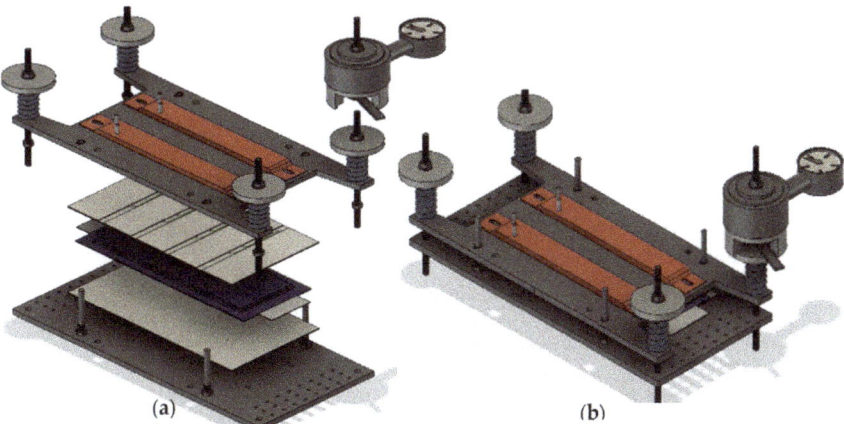

Figure 4. Cell sample holder (**a**) open and (**b**) closed; two heater stripes (red) on the top and two on the bottom side of the stainless-steel plates (dark gray), thermal insolating mica sheets (beige) between the cell (symbolic geometry and design of a pouch cell (blue)) and the stainless-steel plates, thermocouples attached on the mica sheets facing the cell surface.

The experiment method consists of several subsequent steps:

Sample and experiment preparation:

1. Insulating mica plates with thermocouples for temperature measurement are placed on the top and the bottom side of the cell (beige plates in Figure 4a).
2. The sample is fixed in the sample holder with a defined force of 3000 N (54 kPa).
3. Reactor is closed and evacuated.
4. N_2 is added until ambient pressure. Step 3 and 4 are repeated at least 2 times.
5. All gas valves are closed (the reactor is hermetically sealed).
6. Sample is charged to the desired SOC (0%, 30% and 100%).

Experimental steps:

7. The data acquisition system is started: measurement of cell surface temperature, cell voltage, temperature and pressure inside the reactor. The cell is pulsed with a battery cycler (±1 A pulses) in order to get information on the cell resistance.

8. The desired TR trigger is chosen. Here: the cell is heated by the sample holder with a constant rate of temperature increase from both sides with a specified heat ramp (0.39 °C/min at 100% SOC; 0.36 °C/min at 30% SOC until 38,000 s, then increased rate; 0.33 °C/min at 0% SOC).
9. The sample exhibits the first venting and, after being heated to the critical temperatures, the TR.
10. After reaching the maximum temperature during the exothermic reaction, the TR, the heating is switched off. The cell starts to cool down. Wait 5 min to start the experiment after-treatment.

Experiment after-treatment:

11. The valves to the gas analysis section are opened. The gas composition analysis is started.
12. After finishing the gas measurement series, the data acquisition is stopped.
13. Reactor is heated, evacuated and flushed with N_2 several times before opening. Ejected particles are sampled, and the test cell is removed.

3.3. Heat Generation Analysis/Temperature Measurement

Up to 30 thermocouples type k inside the reactor are used in each TR experiment. The temperature of the pouch cell surface is measured with twelve thermocouples on the cell top and twelve on the cell bottom positioned in defined regular distances (50 mm, arrangement 4 × 3, see Figure 5). T_{cell}^{V1} describes the average measured cell surface temperature of all thermocouples at the first venting. T_{cell}^{V2} describes the average measured cell surface temperature of all thermocouples at the second venting. The onset temperature T_{cell}^{onset} is the temperature when the temperature of the cell heating rate is faster than the heating rate of the heat ramp. The critical temperature T_{cell}^{crit} describes when the temperature rate of the selected sensor exceeds 10 °C/min (detailed description in [19]). The maximal cell surface temperature T_{cell}^{max} is the maximum recorded temperature of one of the thermocouples (depends on the position of the origin of the TR). The gas temperature is measured inside the reactor at four different positions. The average reactor temperature is used to calculate the vent gas amount produced at the battery failure.

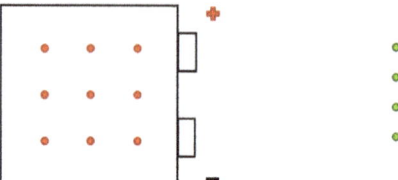

Figure 5. Scheme of the thermocouples positions on the surface of the pouch cell (red) and at different positions inside the TR reactor (green).

3.4. Gas Emission Analysis

The pressure inside the reactor is measured with a GEMS 3300B06B0A05E000 sensor. The pressure and the average gas temperature measured at equilibrium state, 5 min after the TR, are used to calculate the amount of released vent gas. The amount of released gas n_v (mol) is calculated with the ideal gas equation and is presented in liter at standard temperature and pressure (STP: 298.15 K, 100 kPa, V_{mol} = 24.465 L/mol). The amount of gas produced starting at T_{cell}^{V1} and ending at the T_{cell}^{V2} is defined as n_{v1}. n_{v2} is the gas produced after T_{cell}^{V2} and during the TR. The characteristic venting rate $\dot{n}_{ch}$ (mol/s) is calculated with the minimal duration $\Delta t_{50\%}$ (s) to produce 50% of the venting gas $n_{ch50\%}$ (mol). For the calculation of the safety relevant parameters (amount of released gas and characteristic venting rate) the same calculation is used as described in [19].

3.5. Gas Composition Analysis

The gas composition is quantified with two complementary methods in parallel: A Fourier transform infrared spectrometer (FTIR) and a gas chromatograph (GC). In contrast to [40,43] the gas analysis is enhanced with FTIR spectroscopy. The results of the two methods are combined for each measurement and—depending on expected gas components and their concentration range—the measured results of a method, either FTIR or GC, can be chosen.

The downstream connection from the reactor to the gas analysis is heated to ~130 °C. Thus, all gases with a condensation temperature below 130 °C will stay gaseous and will be detected. One converse example is the commonly used electrolyte component EC with a boiling point of 246 °C. Hence, it is very unlikely to measure EC absorbance peaks in the used test setup. The reactor gas consists of N_2 and the vent gas, which is added by the cell. Since the produced vent gas does not contain N_2, the amount of N_2 can be subtracted to calculate the concentration of each component of the vent gas only. The concentration of any gas component (c_v/%) in the vent gas is calculated with the measured concentration of this gas component in the reactor gas (c_m) and the measured N_2 concentration (c_{N2}) in the reactor gas:

$$c_v = ((c_m \times 100)/(100 - c_{N2})) \quad (1)$$

3.5.1. FTIR Spectrometer (FTIR)

A Bruker MATRIX-MG01 FTIR is used with 0.5 cm^{-1} wavenumber resolution. The MCT detector is N_2 (l) cooled. The FTIR measurement chamber itself is heated to 190 °C. The interior space of the FTIR spectrometer is purged with N_2 (g) for at least 2 h to reduce the influence of surrounding gases to the measurement. For the background measurement 100 scans are averaged. A number of 40 scans are used for each data point. To avoid contamination a cold trap and a particle filter are added in front of the FTIR gas measurement chamber. The quantification of the gas compounds is done with the software OPUS GA by Bruker. For each gas analyzed with FTIR a certain absorbance wavenumber region is chosen and compared with a reference spectrum. The setting of the software OPUS GA is optimized for the expected gases and concentrations and validated with the test gas. The FTIR spectrometer is currently optimized for: CO, CO_2, CH_4, C_2H_6, C_2H_4, C_2H_2, DEC, DMC, EC, EMC, H_2O, C_6H_{14}, HF, C_4H_{10} and C_3H_8.

3.5.2. Gas Chromatograph (GC)

For gas analysis with GC the 3000 Micro GC (G2802A) is used with three columns and TCD detectors. The three-channel system includes Molsieve (10 m × 320 μm × 12 μm), Plot U (8 m × 320 μm × 30 μm) and OV1 (8 m × 150 μm × 2.0 μm). The injector temperature and the sample inlet temperature are set to 100 °C for all three channels. The column temperature of the Molsieve channel is 80 °C (at 30 psi) and 60 °C for the Plot U and OV1 channel (40 psi each). Injection time for Molsieve and Plot U is 15 ms and 10 ms for the OV1 channel.

Since the GC uses corrosion sensitive columns, the gas is washed in water washing bottles at room temperature before entering the GC. These washing bottles are directly applied after passing the FTIR gas measurement chamber. Gases that do not dissolve or condensate in the water can be measured. The GC is calibrated for: H_2, O_2, N_2, CH_4, CO, CO_2, C_2H_6, C_2H_4, C_2H_2.

3.5.3. Accuracy of the Gas Quantification

The accuracy of the gas analysis for the presented experiments is validated with test gas of different concentrations and the systematic and statistic uncertainties for FTIR and GC analyzed gas components are added up (Table 2). The FTIR measures spectra continuously over time with a low standard deviation of the measured value (dependent on gas compound <0.2% of the measured value). The GC is calibrated with test gas at a specific uncertainty of each component Δtest gas = ±1%.

The gas quantification method of the FTIR measured spectra is optimized for the expected gas concentrations produced at first venting and during TR. FTIR measurements have advantages at low gas concentrations like for gaseous and toxic HF, but disadvantages in symmetric molecules without change of dipole moment like H_2 and if the absorption peaks of gases are at similar wavelengths.

The GC has its benefits at high concentrations of permanent gases, especially H_2, N_2 and O_2 which cannot be measured with FTIR spectrometer.

Table 2. Accuracy of the FTIR and GC gas quantification optimized for expected gas concentrations.

Gas	FTIR			GC		
	Optimized Concentration/%	Accuracy/% rel.	LOD/ppm	Calibrated Concentration/%	Accuracy/% rel.	LOD/ppm
O_2	-	-	-	0–20	±5	14
N_2	-	-	-	22–100	±3	220,000
H_2	-	-	-	0.1–35	±6	22
C_2H_2	0–10	±4	81	0.1–5	±4	200
C_2H_4	2–10	±5	14	0.1–5	±4	195
C_2H_6	0–10	±6	33	0.1–2	±5	184
CH_4	0–10	±4	114	0.1–5	±5	272
CO	0–30	±4	65	0.1–55	±6	534
CO_2	0–35	±4	121	0.1–28	±4	189
DEC	-	±4	20	-	-	-
DMC	-	±4	28	-	-	-
EC	-	±4	2	-	-	-
EMC	-	±4	25	-	-	-
H_2O	0–3	±4	120	-	-	-
C_6H_{14}	-	±4	16	-	-	-
HF	0–30	±4(min 5 ppm)	4	-	-	-
C_4H_{10}	-	±4	15	-	-	-
C_3H_8	-	-	30	-	-	-

LOD: limit of detection at the specific setting in parts per million (ppm). -: not calibrated for quantitative analysis or not possible to measure.

From the gas compounds quantified with both methods the result of one method, either FTIR or GC, is chosen depending on expected gas components and their concentration range. For small concentrations of CO, CO_2, CH_4, C_2H_6, C_2H_4, C_2H_2 the measured FTIR concentration values are chosen because of the lower LOD. If the measured concentration of C_2H_4 is significantly higher than the LOD, the GC measured value is chosen because of the higher accuracy compared to the FTIR.

3.6. Particle Collection and Particle Analysis

The ejected particles are sampled after the TR and investigated using scanning electron microscopy (SEM) at the Institute of Electron Microscopy and Nanoanalysis (FELMI) at Graz University of Technology. The analysis is focused on particle size distribution (PSD) and particle composition. A ZEISS Sigma 300 VP (Variable Pressure) and a FEI Quanta 200 ESEM (Environmental SEM) are used for the investigation of the released particles after TR. The following SEM detection modes are used:

- For material contrast: imaging with backscattered electron (BSE);
- For topographic contrast: imaging with secondary electrons (SE);
- For elemental analysis: energy dispersive X-ray spectroscopy (EDX).

For the SEM investigations the particles have to be fixed on a sample holder. The fixation must enable a homogeneous distribution without agglomeration of the particles. Gasser showed that the most reliable sampling method is to collect particles from inside the reactor with a spatula and spraying them by a jet of air on a double-sided adhesive carbon tape [51]. This method is used for the sample preparation and subsequently the particles are analyzed with SEM/EDX to measure particle size and particle elemental composition.

Prior to the investigation, EDX simulations are performed with the public access program NIST DTSA-II [52]. Therewith the electron beam interaction was simulated, to be able to assess the best beam energy for SEM-EDX measurements of particles with the measured particle sizes [51].

3.7. Mass Reduction Analysis

The weight of the test sample is measured before and after the experiment using a scale (KERN K8) with a measurement uncertainty of ±0.01 g. After the experiment after-treatment including the heating of the reactor, the vacuum and the N_2 flushing the weight of the remaining cell and large parts (>30 mm length) of the cell outside the cell housing are measured.

4. Results

Three experiments with fresh automotive pouch cells were conducted. In the first experiment the cell is charged to 100%. In the second experiment the cell is charged to 30% and in the third to 0%. The first venting of the cell could be observed at all three test samples. The TR could only be triggered at the fully charged cell.

4.1. Heat Generation/Temperature Response

One critical hazard of a failing cell is heat generation, which can be detected by measuring the temperature response of the cell to the trigger (Figure 6). The experiment of the fresh automotive pouch cell at 100% SOC is compared to the 30% SOC cell in Figure 6a,c during the whole heat ramp experiment and Figure 6b,d at the main exothermic event.

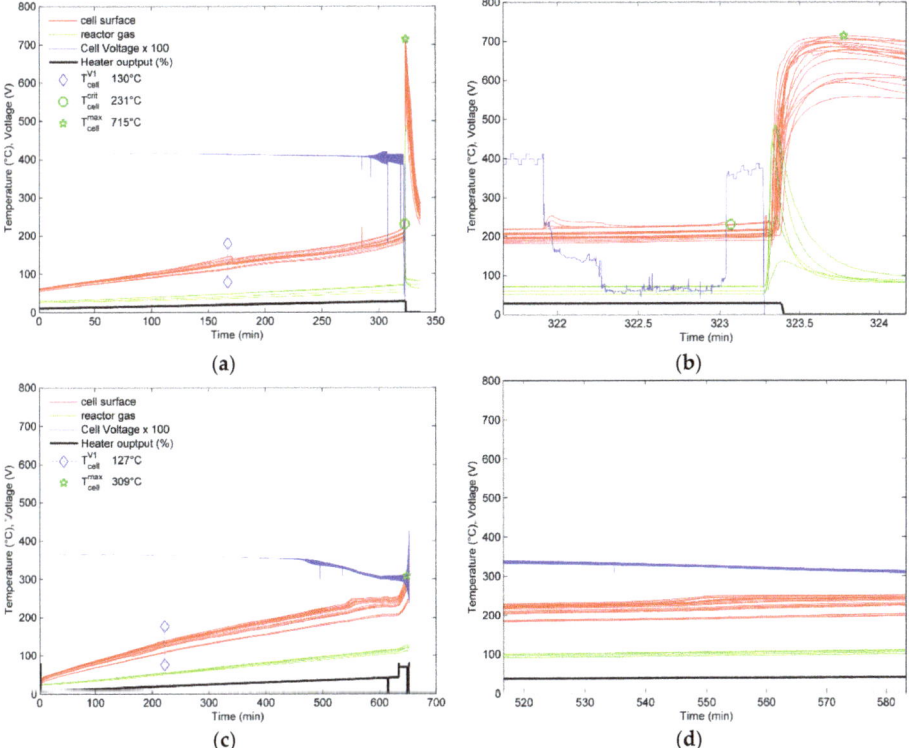

Figure 6. Overtemperature experiments of a fresh automotive pouch cell at (**a,b**) 100% SOC and (**c,d**) 30% SOC: (**a,c**) show the temperatures at up to 30 different positions during the heat ramp experiment measured on the cell surface (red) and inside the reactor (green). The heater output of the sample holder in% is plotted (black line). The cell voltage times 100 is plotted in blue. (**b,d**) show the temperature measured at the main exothermic event. In (**b**) ±1 A pulses are visible (blue).

4.1.1. Experiment with the 100% SOC Cell

As the fully charged cell is heated it shows a minor temperature excursion in the range of T_{cell}^{V1} = 130 °C—the first venting of the cell—10,300 s after activating the heat ramp (Figure 6a). The pouch cell opens. If the cell gets heated up further, the cell reaches the onset temperature. The onset of the main exothermic reaction is detected at T_{cell}^{onset} = 170 °C. The voltage of the cell started decreasing during the heating phase at 70 °C and dropped completely to 0 V at 203 °C cell surface temperature. The second venting starts at T_{cell}^{V2} = 212 °C. The main exothermic reaction developed to a rapid TR at T_{cell}^{crit} = 231 °C (self-heating beyond 10 °C/min). At 100% SOC the cell exhibited an exothermic reaction after 19,397 s and reached a maximum temperature of T_{cell}^{max} = 715 °C on the cell surface. The main exothermic reaction begun at a location between the center of the cell and the positive tab of the cell. Within 4.28 s the exothermic reaction propagated through the cell (time between the rapid increase of the first thermocouple and the increase of the last thermocouple in Figure 6b).

4.1.2. Experiment with the 30% and 0% SOC Cell

Compared to the fully charged fresh cell, the cell with 30% SOC behaves differently using the same overtemperature setup (Figure 6c,d). After reaching the first venting at about T_{cell}^{V1} = 127 °C, no exothermic reaction can be detected even by heating beyond 231 °C. The 30% SOC cell is heated with a constant rate of 0.36 °C/min until 38,000 s and afterwards with an increased rate up to 309 °C (Figure 6c). After reaching the 309 °C maximum cell surface temperature, the heat ramp is stopped.

The 0% SOC cell also could not be triggered into TR by heat. At T_{cell}^{V1} = 120 °C cell surface temperature, the first venting is detected. The experiment is stopped heating up to 240 °C.

4.2. Gas Emission

4.2.1. Experiment with the 100% SOC Cell

The pressure inside the reactor increases slowly at the first venting of the pouch cell and abruptly at the TR (Figure 7a). Figure 7b shows that the gas emission of the cell at the TR takes in total about 4 s. About 50% of the gas is produced in $\Delta t_{50\%}$ = 1.44 s and 90% in $\Delta t_{90\%}$ = 3.22 s.

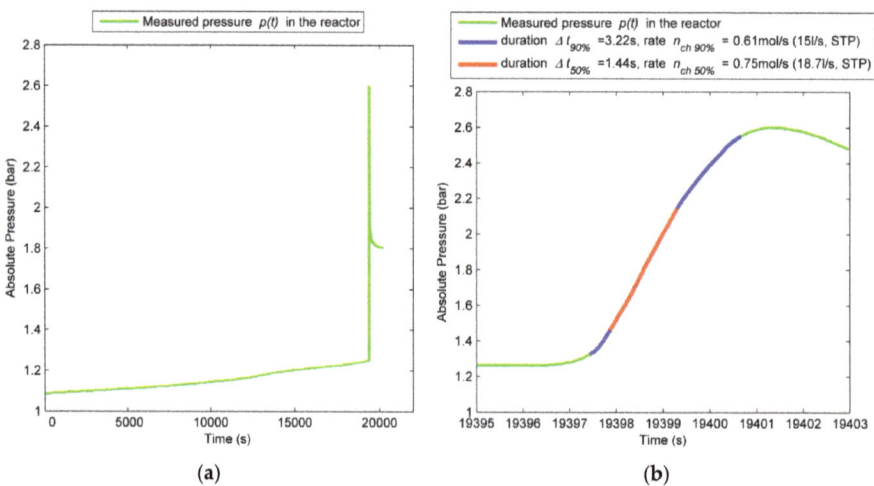

Figure 7. Absolute pressure (green) versus time of the fully charged cell (**a**) during the whole experiment and (**b**) at the TR only. The maximum pressure is reached 4 s after the TR starts. 50% of the gas is produced in 1.44 s (red line). 90% of the gas is produced in 3.22 s (blue).

The fully charged cell released during the first venting $n_{v1} = 0.14$ mol of gas (Figure 8a). During the main TR reaction, the cell released additional $n_{v2} = 2.17$ mol of gas with a characteristic venting rate of $\dot{n}_{ch} = 0.8$ mol/s (18.7 L/s). The calculated produced vent gas amount is shown in Figure 8a. At 100% SOC in total $n_v = 2.31$ mol gas, which is equivalent to 52 norm liters (at 0 °C, 1013.25 hPa) and 57 L at STP, are produced. The fully charged cell produced 0.06 mol/Ah (equivalent to 15 mol/kWh, 1.3 L/Ah) during the overtemperature TR experiment.

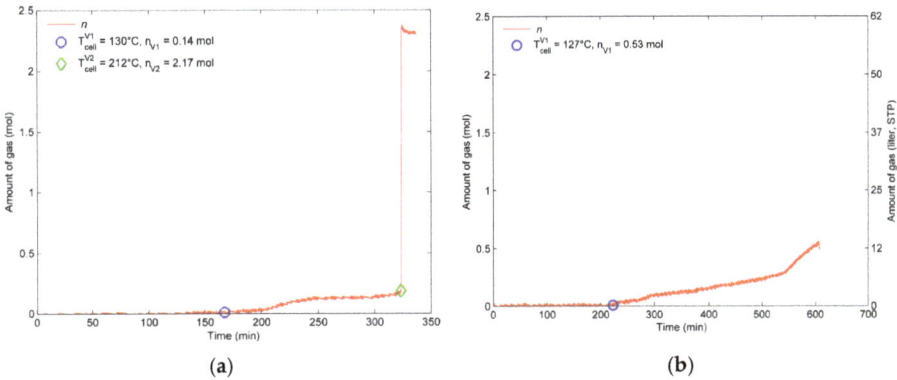

Figure 8. Produced vent gas amount n_v in mol and liter at STP during the experiments of the (**a**) 100% and (**b**) 30% SOC cell. At the 100% SOC cell two venting stages are measured: A first venting starting at T_{cell}^{V1} and a second venting starting at T_{cell}^{V2}. The 30% SOC cell released gas starting at the first venting at T_{cell}^{V1} until the heating was stopped.

4.2.2. Experiment with the 30% and 0% SOC Cell

The 30% SOC cell released $n_v = 0.53$ mol (13 L) gas during the first venting and constant evaporation of electrolyte until the heating is stopped at 309 °C (Figure 8b). Compared with n_{v1} of the fully charged cell, the 30% cell released $n_v = 0.11$ mol until $T_{cell} = 212$ °C. The discharged cell shows a similar behavior and produces $n_v = 0.41$ mol (10 L) gas until the heating is stopped at 240 °C. In these cases, after the first venting, additional gas is produced during the heating phase.

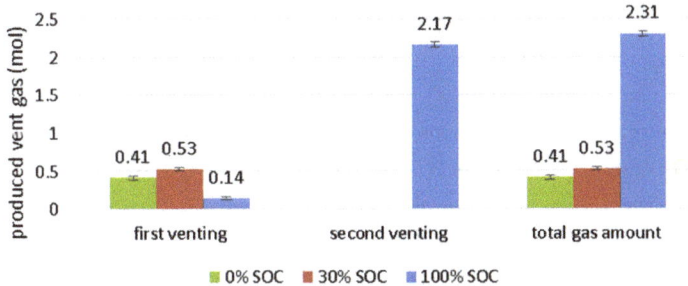

Figure 9. Produced vent gas amount in mol for 0% (green), 30% (red) and 100% SOC (blue) pouch cell at overtemperature experiments at first venting and second venting in comparison.

Figure 9 shows the produced gas amount in mol of the 0%, 30% and 100% charged cell for the first venting, the second venting and the total gas emission. In case of the 0% and 30% cell no second venting could be triggered, therefore, the gases produced until the heating is stopped are added up to the first venting. Hence, the amount of produced gas at the first venting is higher at the 0% and the 30% SOC cells than at the 100% SOC cell.

4.3. Vent Gas Composition

The main gas components at the heat triggered cell at 0% and 30% SOC are CO_2, DEC, H_2O with minor components like CO, H_2, C_2H_4, CH_4, C_3H_8, C_2H_6, C_2H_2 (Figure 10). The main gas components of the fully charged cell are in descending order at the first venting DEC, H_2O, CO_2, CO, C_2H_6, H_2, C_2H_4 and at the TR CO_2, H_2, CO, H_2O, C_2H_4, CH_4, DEC, C_4H_{10}, C_2H_6, C_2H_2 (Table 3, Figure 10). In Table 3 the measured gas concentration values of the experiment at 100% and 30% SOC are listed as well as the vent gas composition in% and mol according to Equation (1).

Table 3. Measured gas concentration values at heat triggered fresh automotive pouch cell at 100% SOC versus 30% SOC in N2.

Gas	100% SOC			30% SOC		
	Measured Gas	Vent Gas (without N_2)	Vent Gas (without N_2)	Measured Gas	Vent Gas (without N_2)	Vent Gas (without N_2)
	c_m/%vol	c_v/% vol	c_v/mol	c_m/%vol	c_v/% vol	c_v/mol
O_2	0.01	0.04	0.00	0.00	0.00	0.00
N_2	69.21			89.01		
H_2	7.06	22.93	0.53	0.41	4.47	0.02
C_2H_2	0.02	0.05	0.00	0.01	0.12	0.00
C_2H_4	1.81	5.88	0.14	0.27	2.93	0.02
C_2H_6	0.30	0.99	0.02	0.03	0.36	0.00
CH_4	1.06	3.46	0.08	0.05	0.52	0.00
CO	5.11	16.59	0.38	0.47	5.15	0.03
CO_2	11.80	38.33	0.89	4.39	47.73	0.25
DEC	0.83	2.69	0.06	1.91	20.72	0.11
DMC	0.00	0.00	0.00	0.00	0.00	0.00
EC	0.00	0.00	0.00	0.00	0.00	0.00
EMC	0.00	0.00	0.00	0.00	0.00	0.00
H_2O	2.32	7.55	0.17	1.61	17.50	0.09
C_6H_{14}	0.00	0.00	0.00	0.00	0.00	0.00
HF	0.00	0.00	0.00	0.00	0.00	0.00
C_4H_{10}	0.39	1.26	0.03	0.00	0.00	0.00
C_3H_8	0.00	0.00	0.00	0.05	0.50	0.00
Gas amount			2.31 mol			0.53 mol

c_m: measured gas concentration including N_2 atmosphere; c_v/% Vol: vent gas in volume%, according to Equation (1); c_v/mol: vent gas in mol.

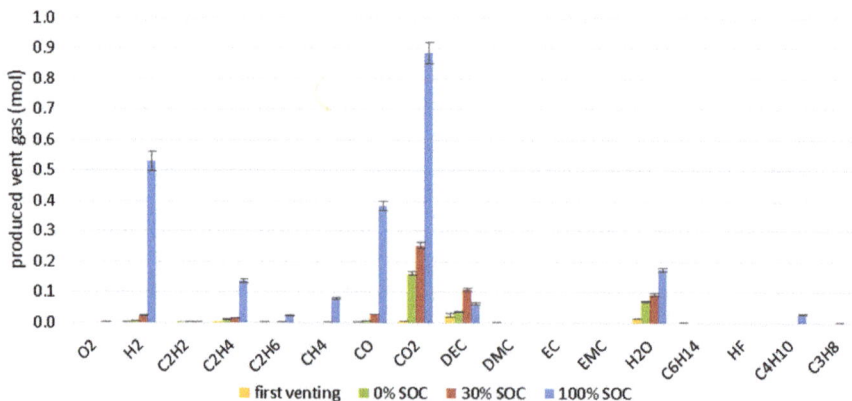

Figure 10. Measured gas composition in mol: immediately after the first venting at T_{cell}^{V1} of the 100% SOC cell (yellow); after the heat ramp was stopped at the 0% (green) and 30% SOC cell (red); and after the TR of the 100% SOC cell (blue); experimental setup in N_2.

The measured gas components at the 30% SOC and 0% SOC cell match with the gas compounds measured at the beginning of the first venting of the 100% SOC cell at about 120-130 °C cell surface temperature. Additionally, it is assumed that the quantified gases at the 30% and 0% SOC cell are dominated by SEI decomposition, electrolyte vapor and decomposition reaction of the electrolyte above 200 °C [5]. At the experiments of the 100%, 30% and 0% SOC cell no HF is detected.

The FTIR spectra of vent gases produced at the 100% (blue) and the 30% (red) charged cell are compared directly in Figure 11. The absorbance spectrum shows for the 30% SOC cell significant higher absorption peaks of the used electrolyte DEC between 1000–1850 cm^{-1} than at the venting of the fully charged cell. In the spectrum of the gas produced at the 100% SOC cell the electrolyte absorption peaks decreased (decomposition of the electrolyte, TR reaction and less long heating time at the 100% SOC cell) and CO, CO_2, CH_4 and C_2H_4 increased.

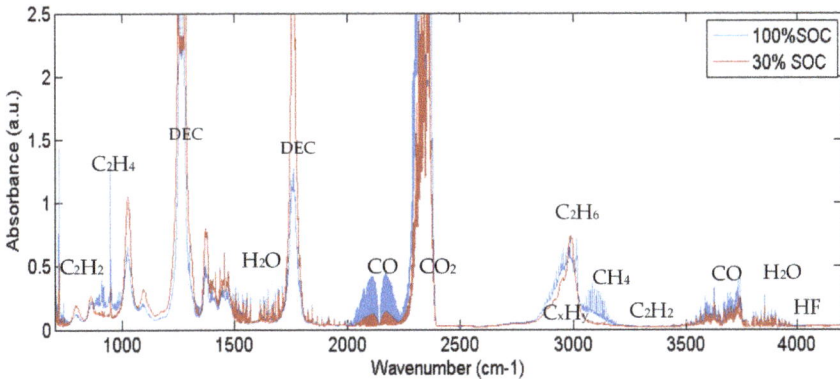

Figure 11. FTIR spectrum of the gas composition measured after the TR of the 100% SOC cell (blue) in comparison to the spectrum measured after stopping the heat ramp at the 30% SOC cell (red).

4.4. Particle Emission

Imaging of particles collected after the TR is performed using SEM. SE images deliver topographic contrast (Figure 12a). Although BSE imaging enables material contrast (Figure 12b), where particles with higher mean atomic number appear comparatively brighter and particles of different composition could be discerned by different gray levels, SE imaging is used to enhance the visibility of carbonated particles on the carbon substrate. To determine the PSD, SE images are binarized by gray value thresholding. Results of the measured average particle areas are presented in Table 4. Due to different reasons, like image noise or image resolution, particles segmented with the threshold method which are beneath 2 µm^2 in area have a big relative uncertainty. The investigation of the particle size shows that most of the particles have an area smaller than 10 µm^2 and about half of the particles are smaller than 5 µm^2.

Table 4. Average measured area (a) of particles and average number of particles produced from an automotive pouch cell (at 100% SOC) at overtemperature.

Area of Particles/µm^2	Average Number of Particles/%
1 < a ≤ 2	21.8 ± 7.6
2 < a ≤ 3	11.6 ± 2.2
3 < a ≤ 5	12.2 ± 2.7
5 < a ≤ 10	15.8 ± 0.6
10 < a ≤ 50	26.2 ± 5.5
50 < a ≤ 100	6.6 ± 3.4
100 < a	5.9 ± 5.5

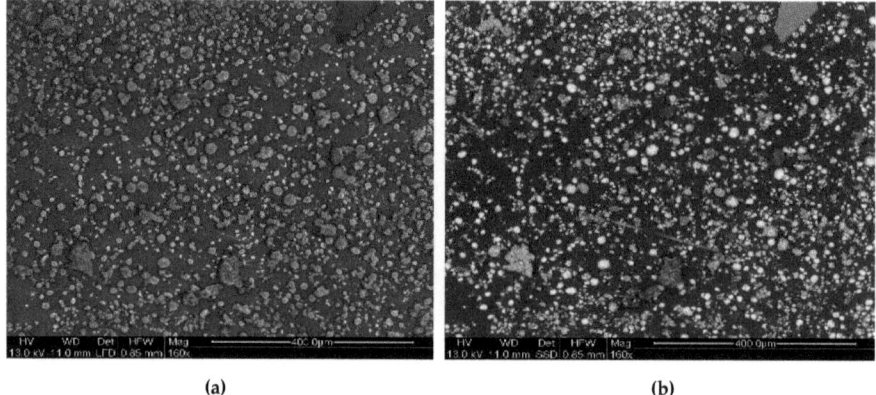

Figure 12. SEM images of particles assembled after the TR. (**a**) SE image shows the topographic contrast; (**b**) BSE measurement shows the material contrast of the same area of the sample. Particles were positioned on a carbon adhesive tape.

To obtain a precise particle composition EDX analysis is used. Therefore, the combination of the SEM with an Oxford XMax 80 EDX detector is applied using the software AZtec for EDX control an evaluation. Therewith it is possible to simultaneously obtain the PSD and the elemental composition of every individual particle. With this setup five different categories of particles are identified and assigned the following classes:

1. Particles mainly consisting of Al and O. Their assumed chemical formula is Al_2O_3 (Figure 13).
2. Particles with huge amounts of nickel (Ni), manganese (Mn), O and smaller amounts of cobalt (Co). The assumed chemical formula is $(Li + NMC)_3O_4$.
3. Particles mainly consisting of Mn and O. The average elemental composition has the estimated chemical formula Mn_2O_3 or its decomposition products.
4. Particles with a high content of C. Very small EDX peaks of O, fluorine (F) and phosphorus (P) were measured.
5. The fifth particle class describes agglomerates with several different material composites which do not fit into one of the listed classes.

The identified particles were parts of the cell active material and were ejected by the cell due to the exothermic reaction. The Mn rich particles (class 2 and 3) result from the cathode. The C rich particles originate from the anode. F and P may result from the salt $LiPF_6$. A small amount of C measured at almost every particle can result from the used carbon tape, the conducting carbon in the cathode or the carbon coating which was performed prior to the investigation in order to get an electrically conductive surface of the specimen.

In the Supplementary Materials SEM images of particles of the listed classes and the correlated spectra are explained. Exemplarily Figure 13 shows (a) the SE image, (b) the BSE image and (c) the EDX spectrum of a particle of class 1. The main elements in this particle are O and Al, as shown in the EDX spectrum. For the most particles of this class the chemical formula Al_2O_3 can be assumed.

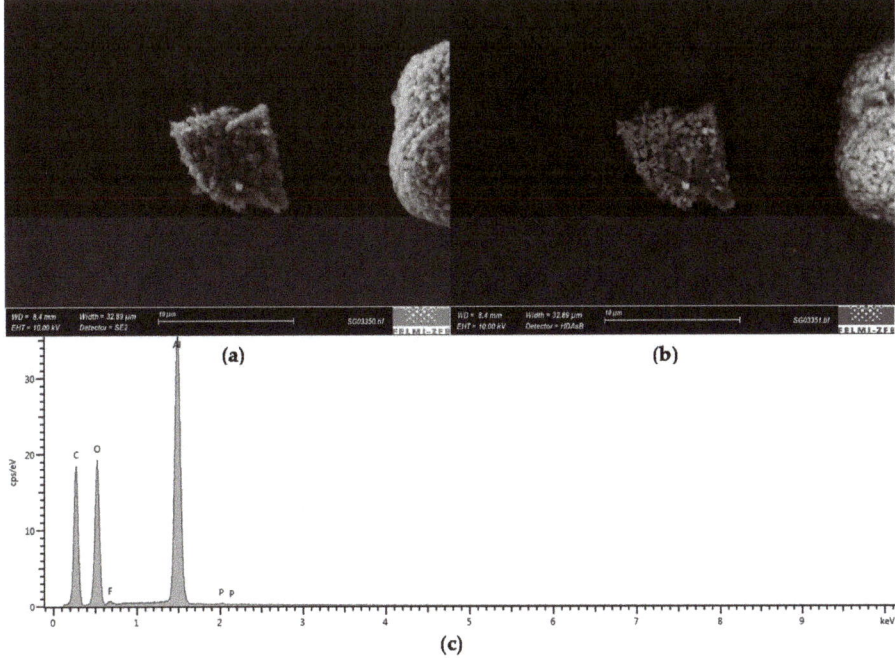

Figure 13. Analysis of a particle of class 1: (**a**) SE image, (**b**) BSE image, (**c**) EDX spectrum. The presented scale in (**a**) and (**b**) is 10 μm.

4.5. Mass Reduction

Since no TR could be triggered at the 0% and 30% SOC cell, the initial cell mass of 865 g is reduced by 15% during the whole experimental test including the aftertreatment. Considering the amount of vent gas and the molar mass of the measured main gas components produced until the heat ramp was stopped, the 30% SOC cell released in total 27 g uncondensed gas during the heat ramp experiment. We assume that the mass reduction of 15% is due to the measured gas, condensed gas and additional gases produced at the experiment after-treatment.

At the 100% SOC overtemperature experiment the initial cell mass of 868 g reduced to 491 g after the TR. This means a cell mass reduction by 43%. This mass reduction can be explained as the sum of released gas, liquids and ejected particles at the TR. Considering the amount of vent gas and the molar mass of the measured main gas components H_2, CO and CO_2 and the side products CH_4, C_2H_4, DEC, H_2O, C_2H_6, C_4H_{10} in total 74 g not condensed gas is released during the TR experiment. The measured gas components are about 20% of the lost cell mass during TR and about 9% of the initial cell mass. The result of the total mass of produced gas is used to assume the mass of the produced particles at the TR. The total mass loss (377 g) minus the gas amount (74 g) results in ~300 g particles. We assume that EC, one of the main electrolyte components, condensed after the TR. Gas with high boiling temperature will condensate on the colder reactor walls, but the amount of condensed gas is not the focus of this study.

4.6. Optical Observation of the Cell after TR

The pouch foil of the fully charged cell is heavily damaged on the top and bottom side after the TR and the Cu foil is visible on the top. The foil opened on all three welded sides except for the side with the terminals. In Figure 14 the cell stack including metallically glossy droplets are visible. We assume that these are Al droplets from the Al current collector. At the 30% and the 0% SOC cell no visible openings of the pouch foil surface are observed. The pouch is still closed on the sides of the terminals. An opening is observed opposite the terminals.

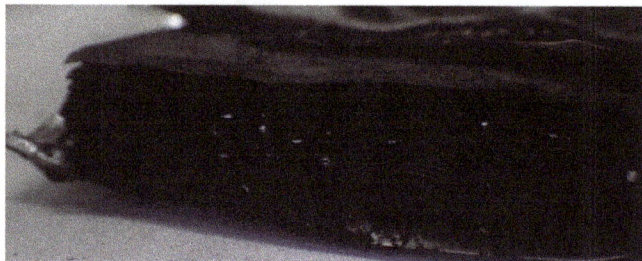

Figure 14. The pouch cell after TR was opened on the welded sides. Droplets were visible between the stacked cell layers.

5. Discussion

The heat triggered TR experiments of a currently used high capacity cell—extracted from a modern mass-produced EV—enables studying hazards and quantify safety relevant parameters from this automotive cell. Since there are few papers available for failing high capacity NMC/LMO cells, the study of those hazards is even more important. Respective papers concentrate on single hazard categories. We concentrate on all five categorized hazards and the safety relevant parameters at different SOC. Table 5 sums up all safety relevant findings of the heat triggered battery failures of the fresh automotive pouch cell at 100%, 30% and 0% SOC.

Table 5. Summary of safety relevant parameters of overtemperature experiment of the fresh automotive pouch cell at 100%, 30% and 0% SOC.

Safety Relevant Parameter	100% SOC	30% SOC	0% SOC
First venting (electrolyte vapor)	yes	yes	yes
Thermal runaway	yes	no	no
Start voltage (V)	4.18	3.67	3.11
Heat ramp (°C/min)	0.39	0.36	0.33
T_{cell}^{V1} (°C)	130	127	120
$T_{voltage=0}$ (°C)	203	190	190
T_{cell}^{crit} (°C)	231	-	-
T_{cell}^{max} (°C)	715 (self-heating)	309 (external heating)	242 (external heating)
Duration of TR (s)	4	-	-
Amount of vent gas n_v (mol)	2.31 (57 l)	0.53 (13 l)	0.41 (10 l)
Characteristic venting rate $\dot{n}_{ch}$ (mol/s)	0.8 (18.7 L/s)	-	-
Main gas compounds	CO_2, CO, H_2	DEC, CO_2, H_2O	DEC, CO_2, H_2O
C_{H2} (vol%)	23	4	1
C_{CO} (vol%)	17	5	1
Particle release (g)	~300	-	-

The fully charged cell can be triggered thermally into TR. At 30% SOC and lower, it is not possible to trigger the cell into TR with the same heat setup (Table 5). If the cell is fully charged during thermal abuse the electrolyte reacts with the lithiated anode after the SEI breakdown [7,17]. Additionally, the stability of the delithiated cathode material is decreased [44]. If the cell is at 0% or 30% SOC the reaction of the lithiated anode with the electrolyte is reduced due to the lack of Li in the anode. No exothermal decomposition of those cells is observed. Increased safety with decreasing SOC is consistent with [12,17,43,44], although referenced literature describes different chemistries and cell components: NCA and LFP [43]; NCA [44]; NMC/LTO [12]. The thermal interactions between several binder materials and anode carbon at 50% and 100% SOC is reported in [17].

Still one question is remaining: Which SOC is the minimum to trigger TR thermally? SOC_{crit} is defined as the lowest SOC to trigger TR. For this investigated cell it seems to be >30%, but there is no general answer for other cells, especially not for higher energy density cells. The SOC influences hazards, consequently safety and health risks from failing LIB. At failing cells with SOC < SOC_{crit} the vaporizing electrolyte and the electrolyte decomposition has the risk of flammable, toxic and corrosive gases. At cells with SOC > SOC_{crit} additional serious risks from heat generation, hot gas and particle emission due to the uncontrollable exothermal reaction need to be considered.

5.1. Hazard Analysis of Failing Automotive Pouch Cells

5.1.1. Heat Generation/Temperature Increase

Temperature sensors on the cell surface show the TR propagation through the cell in 4.28 s. This rapid exothermal reaction and maximal cell temperatures above 700 °C can challenge prevention of TR propagation to neighboring cells and increase resulting risks and damage.

The comparison of the experiments at 100%, 30% and 0% SOC illustrates that the first venting of the investigated cell begins between T_{cell}^{V1} = 120 °C–130 °C cell surface temperature. The deviations between the measured T_{cell}^{V1} values may not be connected to the SOC and is explained as a measurement uncertainty. T_{cell}^{V1} is comparable with the measured temperature rate change (first venting) of overheated NMC pouch cells at about 120 °C plotted by Ren et al. [18]. Ren et al. shows in [17] (Figure 11) that the first venting appeared almost at the same temperature ~120 °C independent of the four different degradation paths and SOH. This would mean that aging effects, like SEI growth and electrolyte consumption, does not influence the first venting. For 50 Ah LMO prismatic metal can cells at our test stand the first venting was observed between T_{cell}^{V1} = 194 °C–220 °C [19]—far apart from our measured values for the pouch cell. This may indicate the influence of different cell design (metal can), vent design and chemistry (LMO) to T_{cell}^{V1}.

The next important temperature is the critical temperature T_{cell}^{crit}, where the temperature rate of the hottest sensor exceeds 10 °C/min, immediately before the full TR. At the fully charged cell T_{cell}^{crit} = 231 °C is comparable with the defined temperature T_2 by Feng et al. [46]. Feng et al. correlated the influence of gravimetric energy density to the maximum reached temperature in [45] (Figure 6). Our result of T_{cell}^{max} = 715 °C fits the presented maximum temperature of NMC/LMO and NMC cells with similar energy density measured in [18] and [46]. At the TR, the cell temperature increases enormously due to chemical reactions inside the cell mainly produced by NMC degradation and reaction of the cathode and the solvent according to [12,17]. The maximum reached temperature can be significantly higher than 715 °C on the surface of the cell and even more inside the cell itself as demonstrated by [13]. The exothermic decomposition of the delithiated cathode material and the reaction between the released O_2 with the solvent is speculated to be the reason for reaching the maximum cell surface temperature [17,22] at the fully charged cell.

Energy density, cathode material and cell design seem to be a main influencing factor for safety relevant and critical temperatures like the first venting as well as the maximum reached cell surface temperature.

5.1.2. Gas Emission

Pressure increase at the first venting does not present any hazards. But the abrupt gas production at the TR and the venting rate of 18.7 L/s can lead to explosion of a battery pack.

The soft pouch packaging ruptured at T_{cell}^{V1} and the cell started to release gas continuously until the TR happens or the heating is stopped. The 100% SOC cell released 0.14 mol gas before the TR. During the TR, the cell released abrupt additional 2.17 mol of gas within 4 s. The 4 s reaction time is observed in the measured temperature and pressure data at the TR. The characteristic venting rate is 0.8 mol/s (18.7 L/s) is comparable with the published results of Golubkov et al. for heated 50 Ah prismatic LMO cells (0.8 ± 0.3) mol/s [19]. This parameter is a relevant parameter for battery pack design and vent design. For higher energy densities and higher capacities increased maximum gas rates are expected. In addition, the reaction time of 4 s observed by the pouch cell may be different for prismatic metal can cells.

The measured 1.3 L/Ah vent gas for this cell is barely within the literature review of Koch et al. of 1.3 L/Ah–2.5 L/Ah for current state-of-the-art batteries [34] and shows that the presented cell produced less gas compared with cells of similar capacity, energy density and chemistry, but the vent gas emission still needs to be considered as a serious safety risk. Compared to other state-of-the-art automotive pouch and metal can cells analyzed in our test setup, this investigated cell produces less gas per Ah at 100% SOC heat trigger, although no gas reducing electrolyte additives could be found. Roth et al. investigated the vent gas amount at different cathode materials (LCO, NCA, NMC, LFP, LMO) and found that all cells produce about 1.2 L/Ah and that a main factor of predicting gas generation is the volume of the used electrolyte [27]. It needs to be mentioned that more vent gas is expected at the presence of O_2 (as measured by Koch et al. as 1.96 L/Ah [34]) and at increasing SOC, like published at overcharge experiments of NCA and LFP cells in [43]. Additional published gas emission values are for NMC 1.2 L/Ah (0.9 Ah NMC) [27], 1.4 L/Ah (2 Ah NMC) [35] and 0.9 L/Ah (2.6 Ah NMC in air) [42]. Deviations from [42] may be explained due to different vent gas amount calculation. The literature source reporting of 2.5 L/Ah is not experimentally determined.

Therefore, we assume that NMC/LMO cells produce between 1.2 L/Ah-2 L/Ah gas at thermal abuse. If the cell goes into TR (SOC ≥ SOC_{crit}) main influencing factors seem to be the capacity of the cell, the electrolyte amount, the SOC and present O_2. According to Roth et al. cathode material has a minor influence on the gas amount.

5.1.3. Gas Composition below SOC_{crit}—30% and 0% SOC

Vent gases measured at the 30% and 0% SOC cell and the first venting are dominated by CO_2, H_2O and electrolyte vapor. At this cell EC (irritant, PAC-1: 30 mg/m^3) and DEC (flammable, PAC-1: 2 mg/m^3) are the main electrolyte components. Lebedeva et al. state clear that most of the currently used LIB electrolytes are toxic, irritant or harmful in addition to being flammable and may even be carcinogenic [26]. Therefore, the opening of the cell and first venting below SOC_{crit} need to be handled as a serious risk due to irritant, toxic and flammable composites, especially at the early opening soft pouch packing and the vaporization of electrolyte inside a closed system (pack, garage, tunnel).

Beside significant electrolyte vapors the following gas components were measured at the heated 30% and 0% SOC cell in descending order: CO_2, H_2O, DEC, CO, H_2, C_2H_4, CH_4, C_3H_8, C_2H_6, C_2H_2. There are many studies reporting gas generation from electrolyte at cycling, formation and heating. The main gas components are similar to the measured gas components in this experiment (CO_2, CO, C_2H_4, CH_4, C_3H_8, H_2, C_2H_6 [53–55]), although the exact gas concentration depends highly on the used electrolyte composition and the additives.

Gas generated at overheating of cells below SOC_{crit} are rarely published. Literature on high capacity NMC or NMC/LMO cells concerning the first venting or gassing at cells with SOC < SOC_{crit} is missing. Literature from small capacity cells: For a 3.35 Ah NCA cell Golubkov et al. presented on 25% SOC 18,650 cells at heating similar main gas compounds: CO_2, H_2, CH_4, C_2H_4, CO [43] (electrolyte and higher hydrocarbons were not quantified). For a 1 Ah LCO cell with 50% PC, 20% EMC, 15% DEC and 10% DMC Kumai et al. measured before and after cycling tests significant different gas compositions, but also the same main gas components: CH_4, CO_2, CO, C_2H_6, C_3H_8 and C_3H_6 [23] (H_2 and electrolyte compounds were not quantified). The produced gases can also be compared with gases produced at the formation process and cycling of NMC cells: At a NMC(422)/graphite cell with 3:7 EC:EMC and $LiPF_6$ at 100% SOC CO_2, C_2H_4, C_2H_6, C_2H_5F, C_3H_8 and CH_4 are measured in decreasing order [53]. Wu et al. investigated at LTO/NMC cells the gas generation at different electrolyte compositions with and without cell formation (SEI) and found significant reduction in CO_2 compared to cells with SEI [55].

Possible sources of the identified gases are therefore: for CO_2: electrolyte [54] and SEI decomposition [5,55], for CO: EC [54], for C_2H_4: EC [54], SEI decomposition [5], for C_2H_6: DEC [54] and DMC [5], for H_2: linear carbonates [55], C_3H_8 and CH_4: DMC [55].

It seems that the cathode material plays a minor role for the gas composition at the first venting and at thermal abuse of cells below SOC_{crit}. The major influence appears to be the electrolyte composition.

5.1.4. Gas Composition—100% SOC

Main components after TR are: 38% CO_2, 23% H_2, 17% CO, 8% H_2O, 6% C_2H_4, 4% CH_4 and electrolyte vapor 3% DEC. TR vent gas consists—apart from CO_2 and H_2O—of mainly toxic (CO) and flammable (H_2, CH_4, DEC) gases. Beside the risk of toxic and flammable atmosphere, fire and explosion are serious consequences.

CO_2 is the most abundant gas component in the vent gas at the heat triggered TR at 100%, 30% and 0% SOC. At the 100% charged cell a 3.9 times higher CO_2 amount was measured than at the 30% SOC cell. The ratio of CO_2:CO = 9.3:1 for the 0% SOC and 30% SOC cell and CO_2:CO = 2.3:1 for the 100% SOC cell. This observation can change at TR of LIBs with higher energy density, where CO_2:CO ratios less than one are possible at TR [34] and more CO than CO_2 is produced due to incomplete combustion reaction. Similar CO_2:CO ratios of measured gases at heat triggered TR of NMC cells are observed in [40], although the investigated cell is a 1.5 Ah 18,650 cells with DMC:EMC:EC:PC (7:1:1:1) and an energy density of 133 Wh/kg (only CO_2, H_2, CO, CH_4 and C_2H_4 were analyzed). In addition, perfect comparable main gas concentrations were measured for NMC cells with different electrolyte compositions by Koch et al. The mean substance concentration values over 51 NMC LIBs fit perfectly for the presented results in this study: 37% CO_2, 22% H_2, 6% C_2H_4 and 5% CH_4 [34] with the difference in CO amount (28% CO by Koch et al.). The different CO amount can be explained by the lower energy density at our NMC/LMO cell. Koch et al. did not quantify gaseous H_2O and electrolyte [34]. For different cathode materials similar gases, but different gas concentrations, were observed [40]. If the same cell chemistry is analyzed, but different triggers are used (like overcharge or nail penetration instead of overtemperature), different preferred chemical reactions take place ending up in different gas compositions [32].

As stated by Zhang et al. in literature no more than 10 gas species in the vent gas are quantified except for their own study [47]. Thus, in this study, 18 possible gas compounds during battery failures are presented. Additional gases identified by other authors, but not listed in this study, for instance C_3H_6 [34] and other higher hydrocarbons (less than 1.7% of the total gas emission according to [47]), were not identified. The deviations may be explained by different cell chemistry, different reaction probability, the test setup and the gas analysis methods. Commonly used electrolytes as EC, DEC, DMC and EMC absorb at similar wavenumber regions and can only be identified clearly at certain wavenumber regions with the FTIR.

Although for the presented experiments no hydrogen fluoride (HF) could be detected, HF is expected to be released by the cell in small amounts [32,36] and to undergo further reactions with the materials inside the reactor, the analysis region and the released particles. Beside the HF production, F may also remain in the cell itself and LiF can be formed. For another aged 18 Ah cell with NMC/LTO chemistry in our test setup, 66 ppm (0.396 mmol) HF were measured [37].

Adding up all quantified gas components at the presented results does not sum up to 100% in total. Possible reasons of the deviation are the sum of uncertainties of each gas component and gases which could not be identified/measured in this experiment.

In addition to the listed gases produced at the venting of cells with SOC < SOC_{crit}, at TR an increase of especially H_2, CO_2 and CO were observed. Though the total amount of measured electrolyte at the fully charged cell is reduced in comparison to the cell at 30% SOC (Figure 11), parts of the vent gas result from decomposing parts of 44 g EC, 59 g DEC, 3.7 g DMC according to [7,56,57] and result in mainly CO_2 and H_2O. Further sources for the gases are for H_2: the reaction of binder material and Li in the anode [42]; for CO_2: oxidation of the electrolyte on the negative electrode surface and $LiPF_6$ and further reaction with the released O_2 of the decomposing cathode [5,21,27,54].

Concluding, the vent gas composition of a failing LIB may be highly sensitive to the SOC, the failure mode/trigger, the used electrolyte composition (especially for cells with SOC < SOC_{crit}), the chemistry and the energy density. This NMC/LMO cell produces similar gases and concentrations as published NMC cells.

5.1.5. Particle Emission

The ejected particles contain elements that are potentially toxic and could act as an ignition source of the emitted burnable gasses, due to their high temperature [4,38]. Furthermore, most of the particles are smaller than 10 μm^2 and can therefore be inhaled deeply into the lungs [58].

Challenges to the particle analysis were the sampling method and the evaluation of the exact particle size and composition. Sampling is the bottleneck of any analytic method and may compromise the results, even when using a measurement method with high precision. During sampling, the material of interest should not be altered, and the sample should be representative. Several methods were tested and are described in [51]. However, the jet of air sampling method used in the end provides a uniform distribution of the particles on the carbon tape used in the SEM measurements, allowing the individual analysis of the particles regarding their size and composition. It has to be mentioned that the air sampling method is selective concerning the dimensions of the particles, but we assume that it is representative for these particles, which are relevant concerning hazards during inhalation.

The particles contain elements that are potentially toxic for humans including Al, Ni, F. Those elements were also reported in [38]. Thus, safety equipment for people handling cells after TR is important such as particle masks and protective clothing. However, the measured major particle size (<10 μm^2) and the reported mass loss does not match with the observations of [38,47]. Zhang et al. show in [38] for a fully charged metal can cell particle matter account for 11.20% of the cell mass. Measured particle sizes were less than 0.85 mm at nearly 45% of particles. In [47] Zhang et al. report a mass loss of 28.53% at a 50 Ah cell due to gas and particle emission with a near 90% of the particles with a size of 0.5 mm in diameter. Zhang et al. measures lower maximum cell surface temperature (438 °C) [47]. The deviation in particle size may be explained due to differences in the cell design (metal can versus pouch), the chemical composition, the sample preparation techniques and the analysis methods.

In [38,47] four different methods were used for the characterization of settleable particulate matter in the chamber, where the thermal runaway was investigated. In fact, very precise methods were applied, which have the drawback, that not one and the same sample can be used for each method. This is a great advantage of SEM combined with EDX, because after getting a specimen holder with disjunct fixated (carbon tape) particles the number, morphology, size and elemental composition (from the element boron (B) to uranium (U)) can be measured using only one methodical approach on the same sample. Hence a good statistic can be achieved, and even individual information of each particle is enabled. Additionally, it has to be highlighted that the only alteration of the sample is the application of a thin carbon layer on the particles, which is fundamental for imaging without charging, but is not compromising the elemental assessment. Thus, using SEM/EDX no heating of the material or dilution in a supporting liquid is needed as is prerequisite at several chemical or elemental analytical methods.

Beside elemental analysis using EDX even chemical analysis via Raman spectroscopy would help to identify particles. Especially organic materials (e.g., carbon rich particles) could be assessed. A new system called RISE (Raman Imaging and SEM) combines high resolution imaging using an SEM with chemical analysis by an integrated Raman microscope [59]. Thus, correlative microscopy combining morphologic, elemental and chemical investigation could be realized. In this special case the application of a carbon layer would be obstructive since it would mask the signal for Raman measurements.

However, the used SEM enables a special vacuum mode (Variable Pressure), where imaging without charging and subsequent EDX and Raman analysis can be realized.

5.1.6. Mass Reduction

At the TR, the investigated cell reduces the initial mass by 43% due to gas and particle emission. This result is comparable with pouch and hard case cells at 100% SOC overtemperature experiments by [34] reporting mass loss of 15–60% for NMC cells with 20–81 Ah. Zhang et al. measured significant lower mass loss (29%) for overheated prismatic NMC cell [47]. The mass loss of the 0% and 30% charged cells after the experiment after-treatment (15%) is comparable with the assumed amount of electrolyte (14%). Therefore, it is assumed that the mass loss of the 0% and 30% charged cell is mainly due electrolyte vaporization and decomposition of SEI, electrolyte and synthetic material.

The quantified mass reduction seems to depend on the SOC, the energy content of the cell and the cell design (metal can prismatic or cylindrical versus pouch cell).

5.2. Forecast for Failing Behavior of Future Cells

Cells with higher energy density than the investigated cell, which are currently planned for the next generation of EVs, may behave differently and it is possible that a TR even below 30% SOC can be triggered by heat. In TR experiments with different cell generations and increasing gravimetric energy density at our test bench, the failing event results in more heat, higher mass loss, more gas and the gas composition changes towards increased toxic components (CO) compared to the presented results as indicated in [34]. New cell technology with increased Ni-content in NMCs are also supposed to have a reduced thermal stability and therefore failing behavior is supposed to change [47].

For comparability of experiments it is important to highlight influencing factors like the cell capacity, the SOC, the SOH, the energy density and the chosen TR trigger for each experiment. It is expected, that for instance the impact of overcharge triggered cells is higher than in heat triggered cells: gas amount and toxicity (CO) increase with SOC [33].

5.3. Forecast for Failing Behavior of Aged Cells

Aged cells (without Li plating) with increased SEI thickness and decreased electrolyte content are supposed to have a decreased heat generation and gas emission as observed by [18] and [35]. For pouch cells the first vent was observed at the same mean surface temperature for aged cells as for fresh cells [18], in contrast to a different cell design in [35], where the first venting started at a lower temperature at the investigated cylindrical cell. Further investigations on the first venting at different cell designs need to be done for early failure detection.

5.4. Recommended Failure Detection

As a result of the presented hazards and risks, special safety equipment and failure detection methods are recommended. For instance, temperature, pressure and gas monitoring is recommended at battery applications, especially inside the EV battery pack. This may enable failure detection at an early stage, as aimed by EVS–GTR. An unwanted opening of the cell could be detected with the proposed monitoring. Early failure detection is gaining more importance due to increasing cell energy density.

6. Conclusions

A comprehensive hazard analysis of modern automotive high capacity NMC/LMO—graphite pouch cells was performed at three overtemperature TR experiments. The investigated cells are currently used in commercially available mass-produced EVs.

In the first experiment the cell is charged to 100%, in the second to 30% and in the third to 0% SOC. The results confirm the influence of the SOC on the failing behavior of the LIB. The fully charged cell could be triggered into TR, but the cells with SOC ≤ 30% could not. The experiments show that there are serious risks (safety and health) at failing state-of-the-art Li-ion cells resulting from electrolyte vapor, generated heat, gas and particles at TR as toxic and flammable gas, explosion and fire. Safety relevant hazards are electrolyte vaporization, heat generation, gas emission including gas rate, gas composition including electrolyte and particle emission including size and content of the particles.

Main findings of the investigated automotive cells are:

- The first venting is measured at 120–130 °C cell surface temperature independent of the SOC.
- At the 30% and 0% SOC cell:

    o The main gas components are after the first venting and constant gas production until the heating is stopped in descending order CO_2, DEC, H_2O, CO, H_2, C_2H_4, CH_4, C_3H_8.
    o One presented hazard is electrolyte vaporization. Commonly used electrolyte components such as EC, DEC, DMC, EMC in an unsealed cell are critical due to the consequential irritant, toxic, cancerogenic and flammable atmosphere. At this cell EC (irritant, PAC-1: 30 mg/m^3) and DEC (flammable, PAC-1: 2 mg/m^3) are the main electrolyte components. It is important to address this hazard especially in large traction battery EV applications, where significant amounts of electrolyte may vaporize inside a closed system (pack, garage, tunnel).

- At the fully charged (100% SOC) pouch cell two venting stages were observed: A first venting and a second venting (TR). The second venting starts above average cell temperature of 212 °C. The TR has the following hazards and consequences, which end up as safety and health risks:

    o Enormous heat is generated by the cell, the cell surface temperatures increased above 700 °C. The main exothermic reaction developed to a rapid TR when the hottest measured part of the cell reached 231 °C. Within 4.28 s the TR propagated through the cell. This high surface temperature can lead to TR propagation to neighboring cells and irreversible damage of the battery pack.
    o Overall, 2.31 mol (57 L, 1.3 L/Ah) of *gas* is produced. The cell released 0.14 mol before the TR. During the TR, the cell released in 4 s additional 2.17 mol with a characteristic rate of 0.8 mol/s (18.7 L/s). 50% of the gas is produced in 1.4 s. The abrupt pressure increase at the TR is a serious risk inside a closed volume.
    o The cell mass reduces by 43% of the initial mass. This mass reduction can be explained as the sum of released gas and ejected particles at TR.
    o The main gas components are: 38% CO_2, 23% H_2, 17% CO, 8% H_2O, 6% C_2H_4, 4% CH_4 and 3% electrolyte vapor (DEC). The measured gas components are about 20% of the lost cell mass during TR and 9% of the initial cell mass. Toxic (CO) and flammable (H_2, CH_4, DEC, etc.) gas components are dangerous when entering the passenger compartment.
    o A large number of ejected particles are smaller than 10 µm^2. Novel nondestructive sampling and analysis methods were used to evaluate the particle parameters: The smallest analyzed particles have an area of 0.1 µm^2, thus a circle equivalent diameter of roughly 6 nm. A total of twelve elements were detected in the particles, including elements like Al, Ni or F. These ejected hot particles (~35% of the initial cell mass) may ignite the vent gas, are carcinogenic and respirable for humans.

- The NMC/LMO cell is comparable to results of failing NMC cells concerning heat generation (max. reached temperature), gas emission and main gas components. Although, the exact gas composition is highly sensitive to the electrolyte mixture.

To reach an acceptable level of safety in EVs a comprehensive analysis of hazards is very important. In order to define testing standards, the battery hazard influencing factors (such as energy content of the cell, chemistry, the failure case/trigger, cell design, SOC and SOH) must be characterized clearly. The five presented hazards addressed in this study should also be considered in future work for different cell types. We recommend to include in the quantification of safety relevant parameters such as the maximum reached cell surface temperature, the amount if produced vent gas, the venting rate, the composition of the produced gases at the first venting and the TR including electrolyte vapor and the size and composition of the produced particles to cover the most significant hazards at battery failures.

Our future work is aimed to evaluate the influence of different triggers, cell design (pouch versus prismatic metal can) and aging on the failing behavior of large automotive Li-ion cells with higher capacity than the presented sample. To guarantee safety at LIB applications it is important to be aware of potential safety and health risks originated from failing cells.

Supplementary Materials: The following are available online at http://www.mdpi.com/2313-0105/6/2/30/s1, Document S1: SEM images of particle classes.

Author Contributions: Conceptualization, C.E.; methodology for gas analysis, C.E.; methodology for particle analysis, E.G., A.Z., M.N., C.E.; software, C.E.; validation, C.E., A.W.G.; formal analysis, C.E., A.W.G., E.G.; investigation, C.E.; data curation, C.E., E.G.; writing—original draft preparation, C.E.; writing—review and editing, A.W.G., A.F., M.N., A.Z., E.G., E.E.; visualization, C.E., A.W.G., E.G.; supervision, A.F.; All authors have read and agreed to the published version of the manuscript.

Funding: The publication was written at VIRTUAL VEHICLE Research GmbH in Graz and is partially funded by the COMET K2—Competence Centers for Excellent Technologies Program of the Federal Ministry for Transport, Innovation and Technology (BMVIT), the Federal Ministry for Digital and Economic Affairs (BMDW), the Austrian Research Promotion Agency (FFG), the Province of Styria and the Styrian Business Promotion Agency (SFG). The study shows the results of the FFG project SafeBattery. The K-project SafeBattery is funded by the BMVIT, BMDW, Austria and Land Steiermark within the framework of the COMET—Competence Centers for Excellent Technologies program. The COMET program is administered by the FFG.

Acknowledgments: Special thanks to the institution ICTM for the cooperation in the project SafeBattery, particularly Ilie Hanzu and Petra Kaschnitz for the analysis of the electrolyte of the investigated cell. Special thanks to the Institute of Electron Microscopy and Nanoanalysis (FELMI) of Graz University of Technology for the cooperation at the particle analysis in the course of Eva Gassers Master's thesis, especially to Werner Grogger. The setup of the electron microscope Sigma 300 VP was enabled by the project "HRSM-Projekt ELMINet Graz — Korrelative Elektronenmikroskopie in den Biowissenschaften" (i.e., a cooperation within "BioTechMed-Graz", a research alliance of the University of Graz, the Medical University of Graz and Graz University of Technology), which was financed by the Austrian Federal Ministry of Education, Science and Research. The combination with the EDX detector could be realized by the project "Innovative Materialcharakterisierung" (SP2016-002-006), which is part of "ACR Strategisches Projektprogramm 2016" of the Austrian Cooperative Research (ACR), where a support by the Austrian Federal Ministry for Digital and Economic Affairs is to be mentioned. The thermal runaway test stand was developed with technical and financial support by AVL List GmbH.

Conflicts of Interest: The authors declare no conflict of interest. The funders had no role in the design of the study; in the collection, analyses or interpretation of data; in the writing of the manuscript or in the decision to publish the results.

Abbreviations

LIB	Lithium-ion battery	EV	Electric vehicle
TR	Thermal runaway	SOC	State-of-charge
BEV	Battery electric vehicle	HEV	Hybrid electric vehicles
EC	Ethylene carbonate	DEC	Diethylene carbonate
DMC	Dimethyl carbonate	EMC	Ethyl methyl carbonate
SEM	Scanning electron microscope	EDX	Energy dispersive X-ray spectroscopy
FTIR	Fourier-transform infrared (spectroscopy)	BSE	Backscattered electron
GC	Gas chromatograph	SE	Secondary electrons
ppm	Parts per million	PSD	Particle size distribution
SOH	State-of-health	IC	Ion chromatography
SEI	Solid electrolyte interface	LOD	Limit of detection
ICTM	Institute of Chemistry and Technology of Materials, Graz University of Technology	FELMI	Institute of Electron Microscopy and Nanoanalysis, Graz University of Technology

References

1. Wagner, I. 2017—Statista-Statistic_id264754_Worldwide-Vehicle-Sales-by-Propulsion-Technology-2017-2030 (1).pdf. Statista. 2017. Available online: https://www.statista.com/statistics/264754/worldwide-vehicle-sales-by-propulsion-technology-2025/ (accessed on 2 March 2020).
2. Ahlswede, A. 2020—Statista-Statistic_id681259_Absatz-von-Elektroautos-in-Ausgewaehlten-Maerkten-Weltweit-Bis-2019.pdf. Statista. 2020. Available online: https://de.statista.com/statistik/daten/studie/681259/umfrage/absatz-von-elektroautos-in-ausgewaehlten-maerkten-weltweit/ (accessed on 2 March 2020).
3. Sun, P.; Bisschop, R.; Niu, H.; Huang, X. *A Review of Battery Fires in Electric Vehicles*; Springer: New York, NY, USA, 2020. [CrossRef]
4. Pfrang, A.; Kriston, A.; Ruiz, V.; Lebedeva, N.; di Persio, F. *Safety of Rechargeable Energy Storage Systems with a Focus on Li-Ion Technology*; Elsevier Inc.: Amsterdam, The Netherlands, 2017. [CrossRef]
5. Spotnitz, R.; Franklin, J. Abuse behavior of high-power, lithium-ion cells. *J. Power Sources* **2003**, *113*, 81–100. [CrossRef]
6. Bandhauer, T.M.; Garimella, S.; Fuller, T.F. A Critical Review of Thermal Issues in Lithium-Ion Batteries. *J. Electrochem. Soc.* **2011**, *158*, R1–R25. [CrossRef]
7. Wang, Q.; Sun, J.; Yao, X.; Chen, C. Thermal Behavior of Lithiated Graphite with Electrolyte in Lithium-Ion Batteries. *J. Electrochem. Soc.* **2006**, *153*, A329. [CrossRef]
8. Blomgren, G.E. The development and future of lithium ion batteries. *J. Electrochem. Soc.* **2017**, *164*, A5019–A5025. [CrossRef]
9. Yang, H.; Zhuang, G.V.; Ross, P.N. Thermal stability of LiPF6 salt and Li-ion battery electrolytes containing LiPF6. *J. Power Sources* **2006**, *161*, 573–579. [CrossRef]
10. Jhu, C.Y.; Wang, Y.W.; Wen, C.Y.; Shu, C.M. Thermal runaway potential of LiCoO2 and Li(Ni1/3Co1/3Mn1/3)O2 batteries determined with adiabatic calorimetry methodology. *Appl. Energy* **2012**, *100*, 127–131. [CrossRef]
11. Orendorff, C.J.; Lamb, J.; Steele, L.A.M.; Spangler, S.W.; Langendorf, J. *Quantification of Lithium-Ion Cell Thermal Runaway Energetics*; Sandia Report; Sandia National Laboratories (SNL-NM): Albuquerque, NM, USA, 2016; p. 0486.
12. Huang, P.; Wang, Q.; Li, K.; Ping, P.; Sun, J. The combustion behavior of large scale lithium titanate battery. *Sci. Rep.* **2015**, *5*, 7788. [CrossRef]
13. Feng, X.; Fang, M.; He, X.; Ouyang, M.; Lu, L.; Wang, H.; Zhang, M. Thermal runaway features of large format prismatic lithium ion battery using extended volume accelerating rate calorimetry. *J. Power Sources* **2014**, *255*, 294–301. [CrossRef]
14. Finegan, D.P.; Scheel, M.; Robinson, J.B.; Tjaden, B.; Hunt, I.; Mason, T.J.; Millichamp, J.; Di Michiel, M.; Offer, G.J.; Hinds, G.; et al. In-operando high-speed tomography of lithium-ion batteries during thermal runaway. *Nat. Commun.* **2015**, *6*, 6924. [CrossRef]
15. Harris, S.J.; Timmons, A.; Pitz, W.J. A combustion chemistry analysis of carbonate solvents used in Li-ion batteries. *J. Power Sources* **2009**, *193*, 855–858. [CrossRef]
16. Mikolajczak, C.; Michael Kahn, P.; White, K.; Thomas Long, R. *Lithium-Ion Batteries Hazard and Use Assessment Final Report*; Fire Protection Research Foundation: Quincy, MA, USA, 2011.
17. Roth, E.P.; Doughty, D.H.; Franklin, J. DSC investigation of exothermic reactions occurring at elevated temperatures in lithium-ion anodes containing PVDF-based binders. *J. Power Sources* **2004**. [CrossRef]
18. Ren, D.; Hsu, H.; Li, R.; Feng, X.; Guo, D.; Han, X.; Lu, L.; He, X.; Gao, S.; Hou, J.; et al. A comparative investigation of aging effects on thermal runaway behavior of lithium-ion batteries. *eTransportation* **2019**, *2*, 100034. [CrossRef]
19. Golubkov, A.W.; Planteu, R.; Krohn, P.; Rasch, B.; Brunnsteiner, B.; Thaler, A.; Hacker, V. Thermal runaway of large automotive Li-ion batteries. *RSC Adv.* **2018**, *8*, 40172–40186. [CrossRef]
20. Wang, Q.; Ping, P.; Zhao, X.; Chu, G.; Sun, J.; Chen, C. Thermal runaway caused fire and explosion of lithium ion battery. *J. Power Sources* **2012**, *208*, 210–224. [CrossRef]
21. Gachot, G.; Ribière, P.; Mathiron, D.; Grugeon, S.; Armand, M.; Leriche, J.B.; Pilard, S.; Laruelle, S. Gas chromatography/mass spectrometry as a suitable tool for the li-ion battery electrolyte degradation mechanisms study. *Anal. Chem.* **2011**, *83*, 478–485. [CrossRef]
22. Dahn, J.R.; Fuller, E.W.; Obrovac, M.; von Sacken, U. Thermal stability of LixCoO2, LixNiO2 and λ-MnO2 and consequences for the safety of Li-ion cells. *Solid State Ion.* **1994**, *69*, 265–270. [CrossRef]

23. Kumai, K.; Miyashiro, H.; Kobayashi, Y.; Takei, K.; Ishikawa, R. Gas generation mechanism due to electrolyte decomposition in commercial lithium-ion cell. *J. Power Sources* **1999**, *81–82*, 715–719. [CrossRef]
24. Kenichiroh, K. Journey to a New Regulatory Option. OICA Submission to IWG for GTR 20, Phase 2 2019(IWG#19). pp. 1–31. Available online: https://wiki.unece.org/download/attachments/86311372/EVS19-E1TP-0300%5BOICA%5DJourneytoaNewRegulatoryOption.pdf?api=v2 (accessed on 23 February 2020).
25. Nedjalkov, A.; Meyer, J.; Köhring, M.; Doering, A.; Angelmahr, M.; Dahle, S.; Sander, A.; Fischer, A.; Schade, W. Toxic Gas Emissions from Damaged Lithium Ion Batteries—Analysis and Safety Enhancement Solution. *Batteries* **2016**, *2*, 5. [CrossRef]
26. Lebedeva, N.P.; Boon-Brett, L. Considerations on the Chemical Toxicity of Contemporary Li-Ion Battery Electrolytes and Their Components. *J. Electrochem. Soc.* **2016**, *163*, A821–A830. [CrossRef]
27. Roth, E.P.; Orendorff, C.J. How electrolytes influence battery safety. *Electrochem. Soc. Interface* **2012**, *21*, 45–49. [CrossRef]
28. Larsson, F. *Assessment of Safety Characteristics for Li-Ion Battery Cells by Abuse Testing*; Chalmers University of Technology: Göteborg, Sweden, 2014.
29. Hollmotz, L.; Hackmann, M. *Lithium Ion Batteries for Hybrid and Electric Vehicles—Risks, Requirements and Solutions out of the Crash Safety Point of View*; 11–0269; International Technical Conference on the Enhanced Safety of Vehicles (ESV): Washington, DC, USA, 2011.
30. Feng, X.; Ouyang, M.; Liu, X.; Lu, L.; Xia, Y.; He, X. Thermal runaway mechanism of lithium ion battery for electric vehicles: A review. *Energy Storage Mater.* **2018**, *10*, 246–267. [CrossRef]
31. Larsson, F.; Andersson, P.; Mellander, B.E. Lithium-Ion Battery Aspects on Fires in Electrified Vehicles on the Basis of Experimental Abuse Tests. *Batteries* **2016**, *2*, 9. [CrossRef]
32. Fernandes, Y.; Bry, A.; de Persis, S. Identification and quantification of gases emitted during abuse tests by overcharge of a commercial Li-ion battery. *J. Power Sources* **2018**, *389*, 106–119. [CrossRef]
33. Somandepalli, V.; Marr, K.; Horn, Q. Quantification of Combustion Hazards of Thermal Runaway Failures in Lithium-Ion Batteries. *SAE Int. J. Altern. Powertrains* **2014**, *3*, 98–104. [CrossRef]
34. Koch, S.; Fill, A.; Birke, K.P. Comprehensive gas analysis on large scale automotive lithium-ion cells in thermal runaway. *J. Power Sources* **2018**, *398*, 106–112. [CrossRef]
35. Zhao, C.; Sun, J.; Wang, Q. Thermal runaway hazards investigation on 18650 lithium-ion battery using extended volume accelerating rate calorimeter. *J. Energy Storage* **2020**, *28*, 101232. [CrossRef]
36. Larsson, F.; Andersson, P.; Blomqvist, P.; Mellander, B.E. Toxic fluoride gas emissions from lithium-ion battery fires. *Sci. Rep.* **2017**, *7*, 10018. [CrossRef]
37. Essl, C.; Golubkov, A.W.; Planteu, R.; Rasch, B.; Fuchs, A. Transport of Li-Ion Batteries: Early Failure Detection by Gas Composition Measurements. In Proceedings of the 7th Transport Research Arena (TRA), Vienna, Austria, 16–19 April 2018. [CrossRef]
38. Zhang, Y.; Wang, H.; Li, W.; Li, C.; Ouyang, M. Size distribution and elemental composition of vent particles from abused prismatic Ni-rich automotive lithium-ion batteries. *J. Energy Storage* **2019**, *26*, 100991. [CrossRef]
39. Feng, X.; Zheng, S.; Ren, D.; He, X.; Wang, L.; Liu, X.; Li, M.; Ouyang, M. Key characteristics for thermal runaway of Li-ion batteries. *Energy Procedia* **2019**, *158*. [CrossRef]
40. Golubkov, A.W.; Fuchs, D.; Wagner, J.; Wiltsche, H.; Stangl, C.; Fauler, G.; Voitic, G.; Thaler, A.; Hacker, V. Thermal-runaway experiments on consumer Li-ion batteries with metal-oxide and olivin-type cathodes. *RSC Adv.* **2014**, *4*, 3633–3642. [CrossRef]
41. Nagasubramanian, G.; Fenton, K. Reducing Li-ion safety hazards through use of non-flammable solvents and recent work at Sandia National Laboratories. *Electrochim. Acta* **2013**, *101*, 3–10. [CrossRef]
42. Diaz, F.; Wang, Y.; Weyhe, R.; Friedrich, B. Gas generation measurement and evaluation during mechanical processing and thermal treatment of spent Li ion batteries. *Waste Manag.* **2019**, *84*, 102–111. [CrossRef] [PubMed]
43. Golubkov, A.W.; Scheikl, S.; Planteu, R.; Voitic, G.; Wiltsche, H.; Stangl, C.; Fauler, G.; Thaler, A.; Hacker, V. Thermal runaway of commercial 18650 Li-ion batteries with LFP and NCA cathodes—Impact of state of charge and overcharge. *RSC Adv.* **2015**, *5*, 57171–57186. [CrossRef]
44. Perea, A.; Paolella, A.; Dubé, J.; Champagne, D.; Mauger, A.; Zaghib, K. State of charge influence on thermal reactions and abuse tests in commercial lithium-ion cells. *J. Power Sources* **2018**, *399*, 392–397. [CrossRef]
45. Waldmann, T.; Wohlfahrt-Mehrens, M. Effects of rest time after Li plating on safety behavior—ARC tests with commercial high-energy 18650 Li-ion cells. *Electrochim. Acta* **2017**. [CrossRef]

46. Feng, X.; Zheng, S.; Ren, D.; He, X.; Wang, L.; Cui, H.; Liu, X.; Jin, C.; Zhang, F.; Xu, C.; et al. Investigating the thermal runaway mechanisms of lithium-ion batteries based on thermal analysis database. *Appl. Energy* **2019**, *246*, 53–64. [CrossRef]
47. Zhang, Y.; Wang, H.; Li, W.; Li, C. Quantitative identification of emissions from abused prismatic Ni-rich lithium-ion batteries. *eTransportation* **2019**, *2*, 100031. [CrossRef]
48. Gao, S.; Feng, X.; Lu, L.; Ouyang, M.; Kamyab, N.; White, R.E.; Coman, P. Thermal Runaway Propagation Assessment of Different Battery Pack Designs Using the TF5 Draft as Framework. *J. Electrochem. Soc.* **2019**, *166*, A1653–A1659. [CrossRef]
49. Kovachev, G.; Schröttner, H.; Gstrein, G.; Aiello, L.; Hanzu, I. Analytical Dissection of an Automotive Li-Ion Pouch Cell. *Batteries* **2019**, *5*, 67. [CrossRef]
50. Golubkov, A.W.; Planteu, R.; Rasch, B.; Essl, C.; Thaler, A.; Hacker, V. Thermal runaway and battery fire: Comparison of Li-ion, Ni-MH and sealed lead-acid batteries. In Proceedings of the 7th Transport Research Arena (TRA), Vienna, Austria, 16–19 April 2018; Volume 43. [CrossRef]
51. Gasser, E. Characterization of Gas and Particle Released during Thermal Runaway of Li-Ion Batteries. Master's Thesis, Graz University of Technology, Graz, Austria, 2019; pp. 2–92.
52. Goldstein, J.I.; Newbury, D.E.; Michael, J.R.; Ritchie, N.W.M.; Scott, J.H.J.; Joy, D.C. *Scanning Electron Microscopy and X-Ray Microanalysis*, 4th ed.; Springer: New York, NY, USA, 2017. [CrossRef]
53. Self, J.; Aiken, C.P.; Petibon, R.; Dahn, J.R. Survey of Gas Expansion in Li-Ion NMC Pouch Cells. *J. Electrochem. Soc.* **2015**, *162*, 796–802. [CrossRef]
54. Onuki, M.; Kinoshita, S.; Sakata, Y.; Yanagidate, M.; Otake, Y.; Ue, M.; Deguchi, M. Identification of the Source of Evolved Gas in Li-Ion Batteries Using [sup 13]C-labeled Solvents. *J. Electrochem. Soc.* **2008**, *155*, A794. [CrossRef]
55. Wu, K.; Yang, J.; Liu, Y.; Zhang, Y.; Wang, C.; Xu, J.; Ning, F.; Wang, D. Investigation on gas generation of Li4Ti5O 12/LiNi1/3Co1/3Mn1/3O2 cells at elevated temperature. *J. Power Sources* **2013**, *237*, 285–290. [CrossRef]
56. Gachot, G.; Grugeon, S.; Jimenez-Gordon, I.; Eshetu, G.G.; Boyanov, S.; Lecocq, A.; Marlair, G.; Pilard, S.; Laruelle, S. Gas chromatography/Fourier transform infrared/mass spectrometry coupling: A tool for Li-ion battery safety field investigation. *Anal. Methods* **2014**, 6120–6124. [CrossRef]
57. Sun, W.; Yang, B.; Hansen, N.; Westbrook, C.K.; Zhang, F.; Wang, G.; Moshammer, K.; Law, C.K. An experimental and kinetic modeling study on dimethyl carbonate (DMC) pyrolysis and combustion. *Combust. Flame* **2016**, *164*, 224–238. [CrossRef]
58. Spangl, W.; Kaiser, A.; Schneider, J. *Herkunftsanalyse der PM10-Belastung in Österreich—Ferntransport und Regionale Beiträge*; Umweltbundesamt: Wien, Austria, 2006.
59. Schmidt, R.; Fitzek, H.; Nachtnebel, M.; Mayrhofer, C.; Schroettner, H.; Zankel, A. The Combination of Electron Microscopy, Raman Microscopy and Energy Dispersive X-Ray Spectroscopy for the Investigation of Polymeric Materials. *Macromol. Symp.* **2019**, *384*, 1–10. [CrossRef]

© 2020 by the authors. Licensee MDPI, Basel, Switzerland. This article is an open access article distributed under the terms and conditions of the Creative Commons Attribution (CC BY) license (http://creativecommons.org/licenses/by/4.0/).

Article

Theoretical Impact of Manufacturing Tolerance on Lithium-Ion Electrode and Cell Physical Properties

William Yourey

College of Engineering, Penn State University—Hazleton Campus, Hazleton, PA 18202, USA; wxy40@psu.edu

Received: 18 March 2020; Accepted: 8 April 2020; Published: 15 April 2020

Abstract: The range of electrode porosity, electrode internal void volume, cell capacity, and capacity ratio that result from electrode coating and calendering tolerance can play a considerable role in cell-to-cell and lot-to-lot performance variation. Based on a coating loading tolerance of ±0.4 mg/cm^2 and calender tolerance of ±3.0 μm, the resulting theoretical range of physical properties was investigated. For a target positive electrode porosity of 30%, the resulting porosity can range from 19.6% to 38.6%. To account for this variation during the manufacturing process, as much as 41% excess or as little as 59% of the target electrolyte quantity should be added to cells to match the positive electrode void volume. Similar results are reported for a negative electrode of 40% target porosity, where a range from 30.8% to 48.0% porosity is possible. For the negative electrode as little as 72% up to 28% excess electrolyte should be added to fill the internal void space. Although the results are specific to each electrode composition, density, chemistry, and loading the presented process highlight the possible variability of the produced parts. These results are further magnified as cell design moves toward higher power applications with thinner electrode coatings.

Keywords: porosity; manufacturing; tolerance; Lithium-Ion; capacity ratio; electrolyte volume

1. Introduction

Lithium ion cells have been the pinnacle method of providing energy for portable electronics, with numerous manufacturers around the world providing batteries of different chemistries [1], dimensions, capacity [2], and power [3]. With numerous positive electrode active materials available to cell manufactures, lithium cobalt dioxide (LiCoO$_2$) has historically been the material of choice due to its proven performance and reliability [4,5]. In manufacturing LiCoO$_2$ cells at both the commercial and laboratory scale, variability is introduced. These tolerances on produced parts, present during any manufacturing process, can have a large impact on the final product's reliability, repeatability, and functionality. Lithium ion cells are no exception to this and thus, manufacturing variation must be considered during cell performance evaluations [6]. Whether it be the formed aluminum laminate package dimensions for prismatic cells, electrode dimensions, electrode coating mass loading, or electrode calender thickness, as well as other production steps, these variations inevitably affect the final product [7]. Numerous authors have indirectly investigated the impact of these manufacturing tolerances on cell performance. Investigating the effect of both anode and cathode porosity on thick lithium ion electrodes, Singh et al. [8] demonstrated that variation in cathode and anode porosity for a constant heavy loading play a considerable role in cell performance, effecting both electrode integrity and cell rate capability, while also concluding that peak performance occurs at an electrode specific target porosity, where small deviations effect performance. The effect of anode porosity and thickness on capacity fade was investigated by Suthar et al. [9], where low porosity yields high electrode tortuosity, a significant reduction in rate capability, and increased capacity fade. The influence of positive [10], and negative [11] electrode density was demonstrated to show that as electrode density is increased, internal electrode electrolyte volume is decreased, leading to increased polarization and poor high rate performance, as well as influencing irreversible capacity loss during formation [12]. The effect of a negative to positive electrode matching ratio on various performance characteristics have also been investigated, showing the effect of area ratio [13], mass ratio [14], and areal capacity ratio [15], and concluding that maintaining an optimal ratio (mass or capacity) is critical to performance, and small deviations result in possible lithium metal plating or increased irreversible capacity loss. Authors have also shown the effects of electrolyte volume on cell performance as a function of electrolyte to electrode void volume [16,17], highlighting the critical importance of sufficient electrolyte volume to cell performance. It is the author's goal to look at the resulting electrode and cell physical properties, namely electrode porosity, change in electrode internal void volume, and capacity ratio that result as a function of electrode coating and calendering tolerance. These tolerances and resulting physical property variations have a direct effect on the resulting cell capacity, rate capability, and cycle life and should be considered during the cell design and evaluation process.

As electrode coating and calendering are performed through various techniques, from laboratory scale doctor-blading to large commercial scale web coating, a relatively large tolerance was used for the evaluated electrode coating (± 0.4 mg/cm^2) and calendering process (± 3.0 µm). Regardless of technique and scale, variation is present in each process [18].

For lab scale electrochemical analysis, coin cells are primarily used as these cells are a quick and cost-effective method to acquire electrochemical results, compared to more elaborate, typically more reproducible, lithium ion pouch or hardware cells. Coin cells have the additional advantage of containing a relatively large void volume inside the crimped cell and outside the typical single pair electrode stack [16]. This additional space allows for an excess of electrolyte to be added, ensuring adequate volume and proper electrode wetting. For analysis purpose, the electrode porosity change would be more critical compared to the change in electrolyte volume which should be added for full electrode saturation, as this porosity change affects cell performance [9,10]. During the manufacturing of lithium ion pouch cells for commercial applications, the void volume outside the electrode stack is minimized, with the goal of producing a cell or battery with the greatest volumetric energy density possible. The idea of maximizing cell volumetric energy density, while still containing enough void space for electrolyte, is no trivial matter, especially when taking into account the large variations in stack void volume that occur during the manufacturing process based on coating and calendering tolerance.

Relating to the variations of electrode and cell physical parameters resulting from manufacturing tolerances, the following theoretical results are provided for single positive and negative electrode coatings absent of foil. The methods and process presented, through small alterations, can be applied to both single and double side-coated foils of any thickness, as well as supercapacitor electrode manufacturing [19]. Various coating methods, drying procedures, binder types, active materials, and calender methods all affect electrode "spring back" or relaxation following electrode processing. Due to the numerous calendering, coating, and composition options available, as well as alternative non-commercial electrode materials under development [20], this has been omitted. For this case study, one pair of lithium ion electrode compositions at three loadings have been selected as representative cells. With these representing a high-energy, standard, and high-power mass loading, this selection is an attempt to replicate commercial lithium ion cells in production today, where the focus of the manuscript is to provide a possible explanation into lot-to-lot variation which occurs in cells where all manufactured parts meet design specifications and tolerances. The resulting large range of porosity and cell matching ratio can account for this variation. The presented results are scalable for any coating formulation, thickness, electrode size, and capacity ratio, with the goal being to highlight the considerations which should be investigated during the design and manufacturing process and the large impact of process variation. Through an extensive literature review, the author has found no similar work published, including one where both electrode coating and calendering variations are considered. Two processes are present in all lithium ion electrodes manufactured commercially today.

2. Materials and Methods

Generic or standard $LiCoO_2$ positive electrode and graphite negative electrode formulation were selected for the investigation of manufacturing tolerance on electrode and cell physical characteristics, namely porosity, cell capacity, matching ratio, and void volume. Table 1 highlights the formulations used for both positive and negative electrodes along with the resulting mixture density used in determining the target electrode calender thickness. The selected formulations, required for analysis, can be altered to any desired formulation and evaluation.

Table 1. Generic positive and negative electrode formulations.

Positive Electrode		
Material	Weight Percent	Density (g/cm^3)
$LiCoO_2$	93	5.00
Conductive Additive	4	2.00
PVDF Electrode Binder	3	1.80
Positive Mixture	-	4.49
Negative Electrode		
Material	Weight Percent	Density (g/cm^3)
Active Carbon	92	2.20
Conductive Additive	1	2.00
PVDF Electrode Binder	7	1.80
Negative Mixture	-	2.16

Three different electrode coating weights or loadings (mg/cm^2) were used to highlight the effect manufacturing tolerances have on the resulting electrode and cell physical properties. Using a positive electrode specific capacity of 150 mAh/g for $LiCoO_2$ (4.2 V vs. Graphite), the three selected loadings correspond to a high power/low energy, mid-range, and high energy/low power loading of 1.40, 2.79 and 4.19 mAh/cm^2, respectively. This range of electrode loadings was selected as it represents the range of the majority of coatings used in commercial lithium ion cells today and are displayed in Table 2 [21].

Table 2. Target positive electrode loadings and corresponding areal capacity.

Design	Electrode Loading	$LiCoO_2$ Loading	mAh/cm^2 @ 150 mAh/g
High Power: Low Energy	10 mg/cm^2	9.3 mg/cm^2	1.40
Mid-Range	20 mg/cm^2	18.6 mg/cm^2	2.79
High Energy: Low Power	30 mg/cm^2	27.9 mg/cm^2	4.19

The amount of negative electrode active material present in a lithium ion cell must correctly match lithium content in the positive electrode for a specific charge voltage. This results in a favorable negative to positive equal area active material capacity ratio. Table 3 illustrates the corresponding negative electrode loadings used to match the previously outlined positive electrodes from Table 2. The weight ratios of positive to negative active components are shown along with the areal capacity loading of the active graphite. The areal loading represents a 300 mAh/g reversible capacity negative electrode material. Also shown is the reversible capacity ratio of the negative and positive electrode, assuming 300 and 150 mAh/g, respectively.

Table 3. Negative electrode loadings and corresponding areal capacity, Positive to negative active material weight ratio, and reversible capacity ratio (graphite = 300 mAh/g and $LiCoO_2$ = 150 mAh/g).

Design	Electrode Loading	Graphite Loading	mAh/cm^2 @ 300 mAh/g	P:N Active Weight Ratio	N:P Capacity Ratio
High Power	5.56 mg/cm^2	5.12 mg/cm^2	1.53	1.82:1.00	1.1:1.0
Mid-Range	11.12 mg/cm^2	10.23 mg/cm^2	3.07	1.82:1.00	1.1:1.0
High Energy	16.68 mg/cm^2	15.35 mg/cm^2	4.60	1.82:1.00	1.1:1.0

Given the electrode target loadings, material density and target porosity, the corresponding calender density and target calender thickness can be calculated from Equations (1) and (2), where rho, ρ, represents density.

$$\text{Calender } \rho = (\text{Material } \rho) \times (1 - \text{porosity}) \quad (1)$$

$$\text{Calender Thickness (cm)} = (\text{Electrode Loading (g cm}^{-2})) \times (\text{Calender } \rho \text{ (g cm}^{-3}))^{-1} \quad (2)$$

The following results are compared to a target calender thickness for both the positive and negative electrodes for each cell design. Variation in electrode loading, during the coating process and calendering to a target thickness, will result in a range of cell capacity, electrode capacity ratio, and electrode porosity. The resulting variation in electrode porosity from the nominal design results in a range of electrode void volume. For optimal cell performance, this electrode internal void volume should be saturated with electrolyte during the electrolyte addition step of the manufacturing process. Saturation ensures proper cell function and safety during subsequent cycling and evaluation.

Using an electrode coating tolerance of ±0.4 mg/cm^2 and a calender tolerance of ±3.0 µm, analysis was performed to determine the possible range in electrode porosity and the resulting variation in positive and negative electrode void volume. The corresponding electrolyte volume changes required to fully saturate the lithium ion electrodes, cell capacity change, and negative to positive capacity ratio change are also presented. Combining the tolerances of the loading and calender process, the range of electrode porosity can be calculated from Equation (3), and from this the electrode void volume can be calculated for any electrode size.

$$\text{Porosity} = 1 - x/y \quad (3)$$

where x = Target Electrode Loading ± Loading Tolerance (g cm^{-2}), y = Electrode ρ (g cm^{-3}) × Target Calender Thickness ± Calender Tolerance (cm).

3. Results and Discussion

3.1. Effect of Electrode Calendering Tolerance

With the assumption that during the coating process the target positive and negative electrodes loadings are correct, the resulting cell should possess the correct negative to positive capacity ratio per the design specifications. Focusing solely on the impact calender tolerance will have on electrode porosity for the three previously highlighted positive and negative electrode loadings, as expected, it can be seen in Figure 1 that the largest impact occurs for the lightest electrode loading. The target electrode calender thickness for these high power/low energy electrodes is the thinnest of the three designs. With these lighter electrode loadings, if the electrode is calendered to a thinner value than the target, a larger percentage of the porosity is removed, or vice versa, for a higher value than target calender thickness.

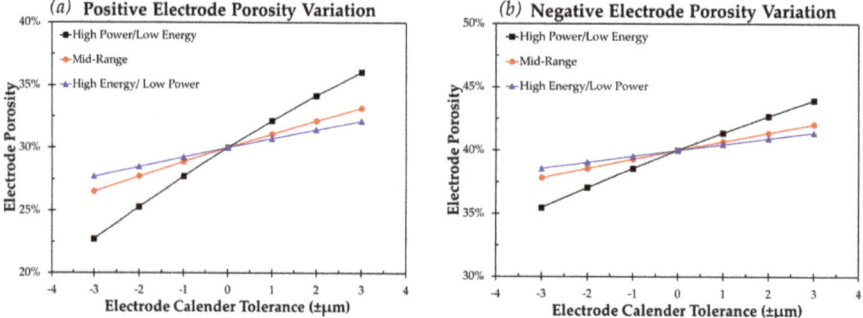

Figure 1. Variation in (**a**) positive electrode porosity and (**b**) negative electrode porosity for a calender tolerance range of ±3.0 µm and a target porosity of 30% and 40% respectively.

The porosity range highlighted in Figure 1 for a thin, high power, low energy electrode translates to 22.7–36.0% for a positive electrode and 35.5–43.9% for a negative electrode with target values of 30% and 40%, respectively. Although the internal electrode void volume change per area from the target porosity is equal across all cell and electrode types for the same calender tolerance, when adding electrolyte to a cell the percent less or excess electrolyte required to account for this variation is a critical number to consider. Table 4 demonstrates the percent change in electrode internal void volume resulting from variation in electrode calender thickness.

Table 4. Percent change in electrode void volume resulting from electrode calendering tolerance.

Design	Calender Range from Target Value (μm)						
Electrode Loading	−3	−2	−1	0	1	2	3
Low Energy (Positive)	−31.4%	−21.0%	−10.5%	0.0%	10.5%	21.0%	31.4%
Low Energy (Negative)	−17.5%	−11.7%	−5.8%	0.0%	5.8%	11.7%	17.5%
Mid-Range (Positive)	−15.7%	−10.5%	−5.2%	0.0%	5.2%	10.5%	15.7%
Mid-Range (Negative)	−8.8%	−5.8%	−2.9%	0.0%	2.9%	5.8%	8.8%
High Energy (Positive)	−10.5%	−7.0%	−3.5%	0.0%	3.5%	7.0%	10.5%
High Energy (Negative)	−5.8%	−3.9%	−1.9%	0.0%	1.9%	3.9%	5.8%

For the data presented in Table 4, the importance and impact of electrode design can be noted. Electrodes with a lower composite density (negative electrodes) see a smaller change in internal void volume for the same variation in calender thickness. Also attributing to this effect is the chosen target porosity of the electrode. A higher target porosity sees a smaller change in electrode void volume percentage as electrode calender thickness is changed. As cell designs move toward higher energy density, either through reduced porosity or increased active material content, this effect is magnified, and consideration becomes increasingly vital.

3.2. Effect of Electrode Coating Tolerance

Similar to the effect of calender variation described in Section 3.1, analysis was performed with electrode coating tolerance. For the following results, a coating tolerance of ±0.4 mg/cm^2 is applied to determine the impact on not only electrode porosity and void volume, but also capacity and cell capacity ratio. As typical commercial lithium ion cells are designed with excess negative electrode material acting as a safety factor to eliminate lithium plating, variation from the target coating (mg/cm^2) directly affects the mass of the active material. This results in variation of the negative to positive capacity ratio along with cell capacity. Table 5 represents the negative to positive capacity ratio for a coating tolerance of ±0.4 mg/cm^2. These following results are shown for a high power-low energy electrode pair, as this low mass loading shows the largest range of capacity ratio of the three.

Table 5. Variation in negative to positive equal area capacity ratio resulting from an electrode coating tolerance of ±0.4mg/cm^2.

High Power/Low Energy	Negative Electrode Coating (mg/cm^2)								
Pos. Electrode (mg/cm^2)	−0.4	−0.3	−0.2	−0.1	0.0	0.1	0.2	0.3	0.4
−0.4	1.06	1.08	1.10	1.13	1.15	1.17	1.19	1.21	1.23
−0.3	1.05	1.07	1.09	1.11	1.13	1.15	1.17	1.20	1.22
−0.2	1.04	1.06	1.08	1.10	1.12	1.14	1.16	1.18	1.20
−0.1	1.03	1.05	1.07	1.09	1.11	1.13	1.15	1.17	1.19
0.0	1.02	1.04	1.06	1.08	1.10	1.12	1.14	1.16	1.18
0.1	1.01	1.03	1.05	1.07	1.09	1.11	1.13	1.15	1.17
0.2	1.00	1.02	1.04	1.06	1.08	1.10	1.12	1.14	1.16
0.3	0.99	1.01	1.03	1.05	1.07	1.09	1.11	1.13	1.14
0.4	0.98	1.00	1.02	1.04	1.06	1.08	1.10	1.11	1.13

The negative to positive capacity ratios range from 0.98 to 1.23, 1.04 to 1.16, and 1.06 to 1.14 for high power/low energy, mid-range, and low power/high energy coatings, respectively, again highlighting that as thinner, higher active material content loadings are utilized, the impact of manufacturing tolerance is magnified. For a higher power electrode, the target loading is a lower value; with the same coating tolerance applied to all coatings, the lightest loadings see the largest impact of coating variation.

Electrodes calendered to the correct thickness still demonstrate a range of porosity resulting from loading variation. Figure 2 and Table 6 represent both the range of porosity as well as the percent change in electrode void volume. These results are provided for a positive and negative electrode calendered to the target thickness.

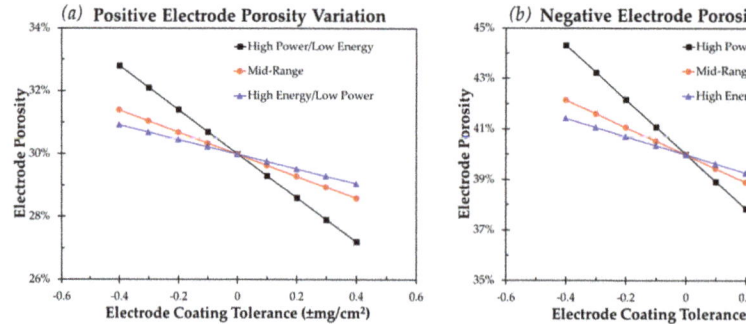

Figure 2. Variation in (**a**) positive electrode porosity and (**b**) negative electrode porosity for a coating tolerance of ±0.4 mg/cm^2 and a target porosity of 30% and 40%, respectively.

Table 6. Percent change in electrode void volume resulting from electrode loading tolerance.

Electrode Loading	Electrode Percent Internal Void Volume Change								
	Electrode Loading Range from Target Value (mg/cm^2)								
	−0.4	−0.3	−0.2	−0.1	0	0.1	0.2	0.3	0.4
Low Energy (Positive)	9.33%	7.00%	4.67%	2.33%	0.00%	−2.33%	−4.67%	−7.00%	−9.33%
Low Energy (Negative)	10.79%	8.09%	5.40%	2.70%	0.00%	−2.70%	−5.40%	−8.09%	−10.79%
Mid-Range (Positive)	4.67%	3.50%	2.33%	1.17%	0.00%	−1.17%	−2.33%	−3.50%	−4.67%
Mid-Range (Negative)	5.40%	4.05%	2.70%	1.35%	0.00%	−1.35%	−2.70%	−4.05%	−5.40%
High Energy (Positive)	3.11%	2.33%	1.56%	0.78%	0.00%	−0.78%	−1.56%	−2.33%	−3.11%
High Energy (Negative)	3.60%	2.70%	1.80%	0.90%	0.00%	−0.90%	−1.80%	−2.70%	−3.60%

Not only does the coating variation affect electrode porosity and cell matching ratio, but also the resulting cell capacity. The percentage range of capacity variation resulting from coating tolerance is highly dependent on target loading. Based solely on positive electrode active material loading, the areal capacity change corresponding to ±0.4 µm loading tolerance is ±4.0%, ±2.0% and ±1.3% for high power/low energy, mid-range, and high energy/low power loadings, respectively.

3.3. "Worst Case Scenario"—Combination of Coating and Calender Tolerance

Performing analysis on coating and calendering processes independently allows for investigation into the greatest impact on electrode physical parameters, allowing the cell manufacturer to understand this impact and focus on one process at a time to minimize cell-to-cell or lot-to-lot variation. Looking at a realistic process of combining both manufacturing operations, the possible "worst case scenario" results in a larger variation of electrode physical properties. From Equation (3), the resulting porosities of a high power/low energy electrode are shown in Table 7.

Table 7. Porosity range for high power/low energy positive and negative electrodes showing combined.

Coating Variation (mg/cm²)	Positive Electrode						
	Calender Tolerance (μm)						
	−3.0	−2.0	−1.0	0.0	1.0	2.0	3.0
−0.4	25.8%	28.3%	30.6%	32.8%	34.8%	36.8%	38.6%
−0.3	25.0%	27.5%	29.9%	32.1%	34.2%	36.1%	38.0%
−0.2	24.3%	26.8%	29.2%	31.4%	33.5%	35.5%	37.3%
−0.1	23.5%	26.1%	28.5%	30.7%	32.8%	34.8%	36.7%
0	22.7%	25.3%	27.7%	30.0%	32.1%	34.1%	36.0%
0.1	21.9%	24.6%	27.0%	29.3%	31.5%	33.5%	35.4%
0.2	21.2%	23.8%	26.3%	28.6%	30.8%	32.8%	34.8%
0.3	20.4%	23.1%	25.6%	27.9%	30.1%	32.2%	34.1%
0.4	19.6%	22.3%	24.8%	27.2%	29.4%	31.5%	33.5%

Coating Variation (mg/cm²)	Negative Electrode						
	Calender Tolerance (μm)						
	−3.0	−2.0	−1.0	0.0	1.0	2.0	3.0
−0.4	40.1%	41.6%	43.0%	44.3%	45.6%	46.8%	48.0%
−0.3	39.0%	40.5%	41.9%	43.2%	44.5%	45.8%	47.0%
−0.2	37.8%	39.3%	40.8%	42.2%	43.5%	44.7%	45.9%
−0.1	36.6%	38.2%	39.7%	41.1%	42.4%	43.7%	44.9%
0	35.5%	37.1%	38.6%	40.0%	41.4%	42.7%	43.9%
0.1	34.3%	35.9%	37.5%	38.9%	40.3%	41.6%	42.9%
0.2	33.2%	34.8%	36.4%	37.8%	39.3%	40.6%	41.9%
0.3	32.0%	33.7%	35.3%	36.8%	38.2%	39.6%	40.9%
0.4	30.8%	32.5%	34.1%	35.7%	37.2%	38.6%	39.9%

For a high power/low energy electrode pair, a large range of porosities are possible from relatively small variations in coating and calendering: 19.6% to 38.6% and 30.8% to 48.0% for a nominal 30% and 40% porosity positive and negative electrode, respectively. This porosity variation in a high power/low energy electrode results in a full saturation electrolyte volume range of 59% to 141% compared to an electrode coated and calendered to the target value. Similar results for a high power/low energy negative electrode as shown with an electrolyte volume range of 72% to 128%. These data also highlight the fact that although there are numerous outcomes which result in the correct porosity, the matching ratio, thickness, and cell capacity will be varied, affecting cell performance.

4. Conclusions

Although physical experimental data are not presented for the theoretical design and resulting variation, it is understood that all manufactured parts, whether in industry or academia, will have an associated deviation from the target value. This variation in the produced parts may be the result of the non-uniformity of laboratory scale calender rolls, or the industrial manufacturing process where parts are presented as having a target value and accepted tolerance range. In either scenario, the produced parts may be deemed acceptable. Progressing forward, the manufacturing tolerance can be reduced through the use of high precision coating and calendering equipment. The reduction in these process tolerances will have a dramatic effect on lithium ion cell electrode porosity and matching ratio consistency, which inevitably affects repeatable cell capacity, cycle life, rate capability, safety, as well as many other important characteristics of lithium ion cells.

A simple look at the electrode manufacturing process and the correlation to electrolyte volume may offer a quick explanation into cell-to-cell or lot-to-lot variation observed in different lithium ion cell manufacturing and quality control processes. While this process was applied to a generic positive and negative electrode formulation at different loadings, the analysis and resulting porosity, void volume, and matching ratio variation calculations can be applied to any electrode manufacturing process. As electrode design moves toward higher energy designs with a higher percentage of active material or thinner, higher power electrodes, the effect of these variations is increased.

For the selected 93% LCO positive electrode and 92% active carbon negative electrode, a coating tolerance of ±0.4 mg/cm^2 and a calender tolerance of ±3.0 μm was used. For the high power/low energy positive and negative electrode target loading of 10 and 5.56 mg/cm^2, respectively, a porosity range of 19.6% to 38.6% and 30.8% to 48.0% for a nominal 30% and 40% porosity positive and negative electrode is possible. Also shown is an equal area negative to positive cell matching ratio range of 0.98 to 1.23, for a target value of 1.1.

Funding: This research received no external funding.

Acknowledgments: Work has been partially supported by the Penn State Hazleton Research Development Grant.

Conflicts of Interest: The authors declare no conflict of interest.

References

1. Blomgren, G.E. The Development and Future of Lithium Ion Batteries. *J. Electrochem. Soc.* **2017**, *164*, A5019–A5025. [CrossRef]
2. Manthiram, A. An Outlook on Lithium Ion Battery Technology. *ACS Cent. Sci.* **2017**, *3*, 1063–1069. [CrossRef] [PubMed]
3. Deng, D. Li-ion batteries: Basics, progress, and challenges. *Energy Sci. Eng.* **2015**, *3*, 385–418. [CrossRef]
4. Nitta, N.; Wu, F.; Lee, J.T.; Yushin, G. Li-ion battery materials: Present and future. *Mater. Today* **2015**, *18*, 252–264. [CrossRef]
5. Berckmans, G.; Messagie, M.; Smekens, J.; Omar, N.; Vanhaverbeke, L.; Van Mierlo, J. Cost Projection of State of the Art Lithium-Ion Batteries for Electric Vehicles Up to 2030. *Energies* **2017**, *10*, 1314. [CrossRef]
6. Shin, D.; Poncino, M.; Macii, E.; Chang, N. A statistical model of cell-to-cell variation in Li-ion batteries for system-level design. In Proceedings of the International Symposium on Low Power Electronics and Design (ISLPED), Beijing, China, 4–6 September 2013; pp. 94–99.
7. Asif, A.A.; Singh, R. Further Cost Reduction of Battery Manufacturing. *Batteries* **2017**, *3*, 17. [CrossRef]
8. Singh, M.; Kaiser, J.; Hahn, H. Effect of Porosity on the Thick Electrodes for High Energy Density Lithium Ion Batteries for Stationary Applications. *Batteries* **2016**, *2*, 35. [CrossRef]
9. Suthar, B.; Northrop, P.W.C.; Rife, D.; Subramanian, V.R. Effect of Porosity, Thickness and Tortuosity on Capacity Fade of Anode. *J. Electrochem. Soc.* **2015**, *162*, A1708–A1717. [CrossRef]
10. Smekens, J.; Gopalakrishnan, R.; Steen, N.V.; Omar, N.; Hegazy, O.; Hubin, A.; Van Mierlo, J. Influence of Electrode Density on the Performance of Li-Ion Batteries: Experimental and Simulation Results. *Energies* **2016**, *9*, 104. [CrossRef]
11. Novák, P.; Scheifele, W.; Winter, M.; Haas, O. Graphite electrodes with tailored porosity for rechargeable ion-transfer batteries. *J. Power Sources* **1997**, *68*, 267–270. [CrossRef]
12. Shim, J.; Striebel, K.A. Effect of electrode density on cycle performance and irreversible capacity loss for natural graphite anode in lithium-ion batteries. *J. Power Sources* **2003**, *119–121*, 934–937. [CrossRef]
13. Son, B.; Ryou, M.-H.; Choi, J.; Kim, S.-H.; Ko, J.M.; Lee, Y.M. Effect of cathode/anode area ratio on electrochemical performance of lithium-ion batteries. *J. Power Sources* **2013**, *243*, 641–647. [CrossRef]
14. Xue, R.; Huang, H.; Li, G.; Chen, L. Effect of cathode:anode mass ratio in lithium-ion secondary cells. *J. Power Sources* **1995**, *55*, 111–114. [CrossRef]
15. Kim, C.S.; Jeong, K.M.; Kim, K.; Yi, C.W. Effects of capacity ratios between anode and cathode on electrochemical properties for lithium polymer batteries. *Electrochim. Acta* **2015**, *155*, 431–436. [CrossRef]
16. An, S.J.; Li, J.; Mohanty, D.; Daniel, C.; Polzin, B.J.; Croy, J.R.; Trask, S.E.; Wood, D.L. Correlation of Electrolyte Volume and Electrochemical Performance in Lithium-Ion Pouch Cells with Graphite Anodes and NMC532 Cathodes. *J. Electrochem. Soc.* **2017**, *164*, A1195–A1202. [CrossRef]
17. Kang, S.-J.; Yu, S.; Lee, C.; Yang, D.; Lee, H. Effects of electrolyte-volume-to-electrode-area ratio on redox behaviors of graphite anodes for lithium-ion batteries. *Electrochim. Acta* **2014**, *141*, 367–373. [CrossRef]

18. Nanjundaswamy, K.S.; Friend, H.D.; Kelly, C.O.; Standlee, D.J.; Higgins, R.L. Electrode fabrication for Li-ion: Processing, formulations and defects during coating. In Proceedings of the IECEC-97 Thirty-Second Intersociety Energy Conversion Engineering Conference (Cat. No.97CH6203), Honolulu, HI, USA, 27 July–1 August 1997; Volume 1, pp. 42–45.
19. Repp, S.; Harputlu, E.; Gurgen, S.; Castellano, M.; Kremer, N.; Pompe, N.; Wörner, J.; Hoffmann, A.; Thomann, R.; Emen, F.M.; et al. Synergetic effects of Fe^{3+} doped spinel $Li_4Ti_5O_{12}$ nanoparticles on reduced graphene oxide for high surface electrode hybrid supercapacitors. *Nanoscale* **2018**, *10*, 1877–1884. [CrossRef] [PubMed]
20. Genc, R.; Alas, M.O.; Harputlu, E.; Repp, S.; Kremer, N.; Castellano, M.; Colak, S.G.; Ocakoglu, K.; Erdem, E. High-Capacitance Hybrid Supercapacitor Based on Multi-Colored Fluorescent Carbon-Dots. *Sci. Rep.* **2017**, *7*, 1–13. [CrossRef] [PubMed]
21. Lain, M.J.; Brandon, J.; Kendrick, E. Design Strategies for High Power vs. High Energy Lithium Ion Cells. *Batteries* **2019**, *5*, 64. [CrossRef]

 © 2020 by the author. Licensee MDPI, Basel, Switzerland. This article is an open access article distributed under the terms and conditions of the Creative Commons Attribution (CC BY) license (http://creativecommons.org/licenses/by/4.0/).

Article

Influence of Laser-Generated Cutting Edges on the Electrical Performance of Large Lithium-Ion Pouch Cells

Tobias Jansen [1,*], Maja W. Kandula [1], Sven Hartwig [1], Louisa Hoffmann [2], Wolfgang Haselrieder [3] and Klaus Dilger [1]

1. Institute of Joining and Welding, Technische Universität Braunschweig, Langer Kamp 8, 38106 Braunschweig, Germany; m.kandula@tu-braunschweig.de (M.W.K.); s.hartwig@tu-braunschweig.de (S.H.); k.dilger@tu-braunschweig.de (K.D.)
2. Institute for High Voltage Technology and Electrical Power Systems, Technische Universität Braunschweig, Schleinitzstraße 23, 38106 Braunschweig, Germany; louisa.hoffmann@tu-braunschweig.de
3. Institute for Particle Technology, Technische Universität Braunschweig, Volkmaroder Straße 5, 38104 Braunschweig, Germany; w.haselrieder@tu-braunschweig.de
* Correspondence: tobias.jansen@tu-braunschweig.de; Tel.: +49-531-391-95796

Received: 5 November 2019; Accepted: 26 November 2019; Published: 3 December 2019

Abstract: Laser cutting is a promising technology for the singulation of conventional and advanced electrodes for lithium-ion batteries. Even though the continuous development of laser sources, beam guiding, and handling systems enable industrial relevant high cycle times, there are still uncertainties regarding the influence of, for this process, typical cutting edge characteristics on the electrochemical performance. To investigate this issue, conventional anodes and cathodes were cut by a pulsed fiber laser with a central emission wavelength of 1059–1065 nm and a pulse duration of 240 ns. Based on investigations considering the pulse repetition frequency, cutting speed, and line energy, a cell setup of anodes and cathodes with different cutting edge characteristics were selected. The experiments on 9 Ah pouch cells demonstrated that the cutting edge of the cathode had a greater impact on the electrochemical performance than the cutting edge of the anode. Furthermore, the results pointed out that on the cathode side, the contamination through metal spatters, generated by the laser current collector interaction, had the largest impact on the electrochemical performance.

Keywords: production strategies; laser cutting; cell manufacturing; automotive pouch cells

1. Introduction

Due to continuing human-induced CO_2 emissions, the global warming of the earth and associated negative consequences of extreme weather are steadily increasing [1]. The impact of climate change has increased awareness of the population in industrialized countries of environmentally friendly or carbon-neutral behavior. These interests are driving the development of new, greener, and more efficient technologies for the major CO_2 producing sectors, which are electrical energy production and individual mobility. To reduce CO_2 emissions in the individual mobility sector, electric mobility is considered a key technology [2,3]. In order to maintain the positive CO_2 balance of electric mobility, the use of electrical energy from renewable energy sources is required, but also the production of cells must be made more efficient and, therefore, more ecological. In particular, a more efficient large-scale cell production can be achieved by reducing rejects by optimizing existing or developing new technologies, as material costs dominate over operating and investment costs [4].

Lithium-ion pouch cells are considered to be the most effective electrochemical technology. Due to their advantages regarding the high volumetric utilization of the installation space of the battery pack, they are especially well suited for automotive batteries. Because of the possibility to customize the cell geometry, the pouch cell leads to less dead volume compared to conventional cylindrical 18650 cells [5]. Besides, the high volumetric energy densities on battery level pouch cells have high volumetric energy of 466 WhL^{-1} and a specific energy density of 241 Whkg^{-1} on cell level (Cathode: LG Chem. NCM 111, Anode: LG Chem. Graphite) [6]. Even though cylindrical 18650 cells have a volumetric energy density, which is about 20% higher than those of pouch cells [5], the homogeneous mechanical behavior during charging and discharging of pouch cells lead to longer cycle lifetime, which is to be considered a long term advantage over the higher energy density [7]. For pouch cells, in general, it is fundamental to cut the endlessly coated electrodes and make them suitable for the following stacking process. It is imperative to look more closely at the cutting process itself since each cut could lead to manufacturing errors in terms of contamination and cutting edge quality [8]. Due to a large number of processes, it is necessary to optimize every step to improve ecological productivity. Figure 1 shows the evolution of the accumulated production rejects of a conventional cell production line schematically.

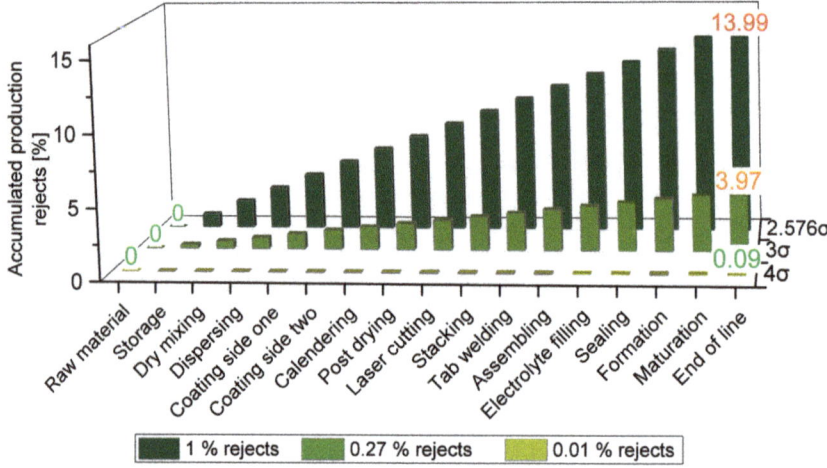

Figure 1. Influence of the reject rate of every single process on the total reject rate of a cell production line.

As shown in Figure 1, each process can contribute to the overall efficiency of the production line. Due to the numerous processes of battery production, even small reject rates can lead to a high overall reject rate and, therefore, to low utilization of raw materials. Already at a reject rate of only 1% per process, the accumulated rejects at the end of the line reach almost 14%. Therefore, each process should be at least in the range of a 4σ reject rate, respectively, with a reject rate of under 0.09% with regard to ecological production. Investigating and improving the singulation process can contribute to accomplishing these high standards.

Besides the established shearing or die-cutting (DIN 8588) [9], laser cutting is becoming increasingly common in cell production lines due to its process-immanent advantages over the contact-based singulation method [10]. Especially for very thick and fragile electrodes, or pure lithium metal anodes, laser cutting is no longer an alternative but state of the art technology due to the lack of contact and the associated lack of mechanical stress [11]. In the following section, we have given a summary of the numerous studies in this field and laser ablation generally.

State of the Art Laser Cutting of Electrodes

The interactions between the material and the laser-generated photons during the cutting process is very dynamic and very complex due to a large number of possible and partly mutually dependent influencing factors. The key factors of a laser cutting plant are wavelength (λ), average power (P_{avg}), spot size (d_{spot}), laser profile and Rayleigh length in focus, cutting speed (v_c), the number of passes, and cutting angle. In the case of pulsed laser beam sources, the additional factors are pulse peak power (P_{peak}), pulse energy (E_P), pulse repetition frequency (PRF), and pulse shape, as well as pulse duration (τ). The relevant material properties of the electrode are the composition of the coating (anode/cathode), collector material, coating thickness, collector thickness, absorption coefficient, and degree of compaction of the electrode (Figure 2).

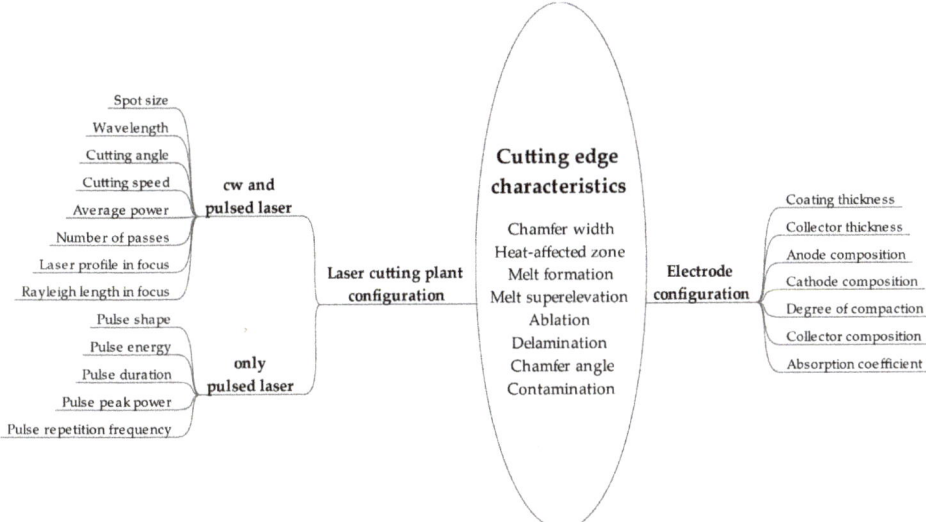

Figure 2. Relevant primary process parameters and material properties of a continuous wave (cw) and pulsed laser cutting process for electrodes [10,12].

For a simplified description of the influence of the primary process parameters in Figure 2, the secondary parameters energy density (ED) [13], intensity (I), pulse fluence (H_p), and number of laser pulses per surface increment (n_{line}) [13] for pulsed beam sources can be used to explain the ablation or cutting behavior. Considering a certain wavelength, average power, and spot size, these parameters can be adjusted by v_c, PRF, and τ. The secondary parameters are defined by the following equations:

$$ED = \frac{P_{avg}}{v_c\, d_{spot}} \left[\frac{j}{cm^2}\right] \tag{1}$$

$$I = \frac{4\, P_{avg}}{\pi\, d_{spot}^2} \text{ for cw and } I_P = \frac{4\, P_{Peak}}{\pi\, d_{spot}^2} \text{ for pulsed laser with } P_{Peak} \approx \frac{P_{avg}}{PRF\, \tau} \left[\frac{W}{cm^2}\right] \tag{2}$$

$$H_p = \frac{E_P}{\pi\, d_{spot}^2} \left[\frac{j}{cm^2}\right] \tag{3}$$

$$n_{line} = \frac{PRF\, (d_{spot} + \tau\, v_c)}{v_c} \tag{4}$$

The energy density describes the energy input per surface increment (cutting length times const. spot size) on the material to be processed, depending on the cutting speed, photonic power, and spot size. This parameter can be used to describe the scalability of the cutting speed for a defined laser/scanner system and material or to define the necessary energy input for a quality cut and the cut-through limit. Since the energy density represents the average power input independent of the peak power, the intensity is necessary to further describe the cutting process with a pulsed laser beam source. In addition, the pulse fluence and the number of pulses, which hit per surface increment, must be specified. Based on the energy density, the intensity, the pulse fluence, and the number of hits per area increment, the amount of material removal and the material removal behavior can be derived. Figure 3 schematically shows the number of hits per surface increment as a function of the v_c, PRF, τ, and the d_{spot}, as well as the effects on the electrode cutting edge characteristics. The ablation thresholds a and b, as well as the heat-affected zone c shown in Figure 3, can be derived from the mentioned secondary parameters.

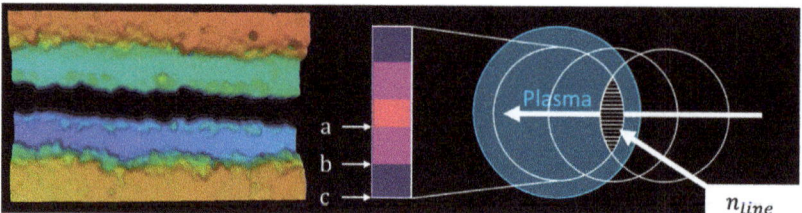

Figure 3. Laser scanning microscope image of a laser-cut electrode (**left**); Ablation thresholds depending on the number of hits, intensity, pulse fluence, energy density, as well as the laser- and corresponding plasma-intensity-profile (**right**): (a) cut-through threshold, (b) ablation threshold, and (c) thermic influencing threshold [10].

As the type of energy input can lead to different removal mechanisms, a comparison of continuous wave (cw) and pulsed systems based on the energy density is only conditionally possible. Here, a distinction can be made between thermal (cw/pulsed) and athermal (pulsed) dominant removal processes. The thermal ablation generally proceeds in three successive phases, and, in the case of a continuous cut, also in parallel phases. In the first phase, the photons are absorbed by the surface and penetrate a near-surface area. In the following second phase, the temperature increases at the surface as a result of the absorption, and in deeper zones by thermal conductive effects. The rising temperature leads to a transformation of the state of matter from solid to liquid and liquid to gas, or directly from solid to gas. In the third phase, the penetration depth increases and thus the melting or evaporation zone, wherein the material is expelled in liquid or gaseous form from the kerf. The material ejection can be further distinguished into fusion, sublimation, and photochemical cutting/ablation. The athermal removal process is characterized by the fact that the duration of the photonic energy input is too short for initiating heat conduction. This removal process can be achieved with pulse durations of less than 10 ps. Furthermore, due to the short exposure time of pulses in the ps and fs range, the spatial extent of the resulting plasma is negligibly small [14]. Considering the high intensities, athermal processes can be further distinguished between sublimation cutting and photochemical ablation.

Figure 4 shows the processes that occur after the first impact of the photons on the processed material. On average, the photons are absorbed in 10 fs, where the photonic energy is converted into thermal energy within 100 fs by electron-electron relaxation of the electrode systems of the covalent bonds. This is followed by an electron-phonon relaxation after about 1 to 10 ps, which leads to a heat transfer of the electrons into the lattice structure. Parallel to this process, the ablation begins, and after about 100 ps, the phonon-phonon relaxation leads to heat conduction [15].

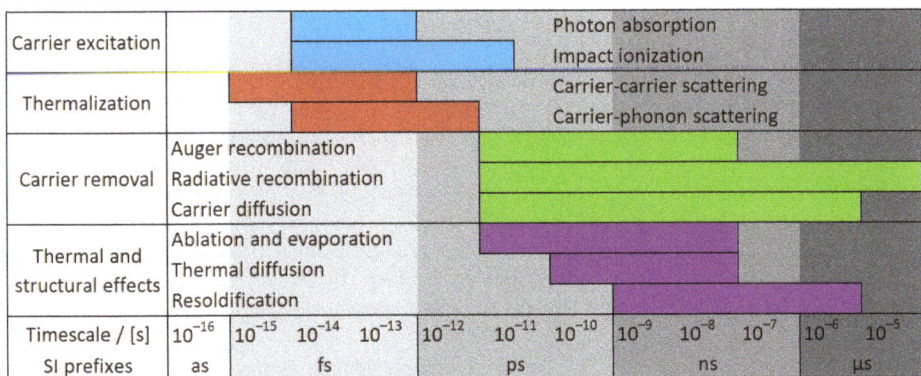

Figure 4. Time-scaled processes during a laser pulse material interaction, based on [16,17].

By means of the beam source used in this study, it is possible to generate pulses in the ns range. Due to the relatively long pulse duration of 240 ns, we can only realize a thermally affected cutting process (thermal removal process) with this system [14]. As a result of the composite structure and the associated varying material properties over the thickness of the electrode, it can be assumed that several removal mechanisms take place simultaneously and/or serially during cutting. The cutting of the porous coating will be characterized by a photochemical and sublimation proportion, whereas the cutting of the metallic collector will be characterized by sublimation and a fusion proportion. In addition to the direct interaction between laser and material, a plasma which correlates to the intensity distribution of the laser will also interact with the electrode. This can lead to material removal or thermal loading of the active material, in addition to thermal conduction effects.

The laser-induced plasma is caused by the high intensities of the individual pulses, which leads to strong oscillations of the free electrons, enabling them to knock bounded electrons out of neutrally charged atoms. The avalanching increase of free and strong oscillating electrons leads to a large number of free electrons and positively charged species. The resulting high-energy plasma absorbs the photons of the laser radiation by the inverse Bremsstrahlung (IB) and the photoionization (PI). In this case, IB is considered to be the main absorption mechanism. The photons in this mechanism are absorbed by free electrons as they collide with neutral or ionized species. This leads to an increase in the energy of the electrons and thus to an increase in the degree of ionization and the temperature of the plasma. If the temperature and density of the plasma rise above a certain level, the plasma can shield the area from being cut from the laser radiation. This effect is referred to as the plasma shielding effect [18].

Investigations on cw and pulsed laser beam cutting of electrodes have already been carried out in previous studies. The results showed that the use of single-mode cw fiber lasers made it possible to achieve very high, industry-relevant cutting speeds due to the high average power and the achievable low spot sizes. Studies showed that it was possible to cut an anode (120 µm) with 11,666 mms^{-1} and a cathode (130 µm) with 10,000 mms^{-1} with a single-mode cw fiber laser at a wavelength in the infrared range (1070 nm), an average power of 5000 W, and a spot size of 25 µm [19]. This resulted in an energy density of 2000 jcm^{-2} for the cut-through limit of the cathode and an energy density of 1714 jcm^{-2} for the anode. The lower required energy density for the anode was probably due to the low collector thickness and the lower degree of compaction. In another study, an anode (50 µm) with a collector thickness of 30 µm was cut with a cutting speed of 2000 mms^{-1} using a single-mode cw Ytterbium fiber laser (1070 nm, 250 W, and 23 µm spot size). The much lower required energy density of 543 jcm^{-2} could be explained by the lower material thickness and by the higher intensity [20]. Based on these results, further investigations using the same system showed that compacted (56 µm) and non-compacted anodes (70 µm) with a collector thickness of 20 µm could be cut at a speed of 5000 mms^{-1} [21]. The low energy density of 217 jcm^{-2} required for the singulation suggested that the collector thickness was the dominating speed-influencing parameter of the electrode. Considering the strong influence of the collector thickness, as well as the intensity of the focused laser spot, it was plausible that an energy density of 818 jcm^{-2} at 5000 mms^{-1} led to a cut-through of an anode (100 µm) with a current collector thickness of 10 µm by using a single-mode cw fiber laser (1070 nm, 450 W, 11 µm spot size) [22]. In principle, the investigations with cw laser systems showed that the achievable chamfer width (total ablated area) was less than 50 µm and increased with increasing energy densities [21].

The trend of these dependencies was also found in investigations with pulsed laser beam sources. In addition to this, the investigations showed that with pulsed beam sources in the nanosecond range, higher pulse repetition frequencies enabled higher cutting speeds and led to a smaller chamfer width [19]. By using an ns pulsed fiber laser (1070 nm, 100 W, 50 µm spot size, 500 kHz, and 30 ns), it was possible for Kronthaler et al. [23] to cut an anode (114 µm) with a collector thickness of 10 µm at a speed of 1200 mms^{-1}. Using the same system parameter, a slightly higher cut-through limit of 1250 mms^{-1} could be achieved for a 124 µm thick cathode with a collector thickness of 20 µm [23]. From the given parameters, an energy density of 160 jcm^{-2} resulted, for the cut-through limit, in an intensity per pulse of 3.4×10^8 Wcm^{-2} and a hit number of 20. Lutey et al. showed in their research that the number of hits per area increment caused an increase in the plasma shielding effect. With higher numbers of hits, a higher average power was needed to realize a cut. At 125 hits per area increment (500 kHz), an energy density of 660 jcm^{-2} was needed, whereas, with a hit number of 5 (20 kHz), only an energy density of 352 jcm^{-2} sufficed [24].

Further results on laser cutting of electrodes showed that low energy densities and intensities were necessary for singulation for a smaller wavelength. At a wavelength of 1064 nm, an energy density of 448 jcm^{-2} and an intensity of 28.5×10^8 Wcm^{-2} were necessary [24], whereas, at 355 nm, an energy density of 340 jcm^{-2} and an intensity of 1.7×10^8 Wcm^{-2} were sufficient to cut a 120 µm thick anode [19]. These results could not be confirmed by studies with a laser in the green electromagnetic spectrum (532 nm, 1 ns, 6 W). For the singulation of an anode (130 µm), with the same collector thickness, an energy density of 560 jcm^{-2} and an intensity of 50.8×10^8 Wcm^{-2} were needed. This could be attributed to the fact that the number of hits per surface increment of 165 increased the plasma shielding effect, and thus reduced the maximum cutting speed [25].

In order to realize the most dynamic and fast cutting processes possible in a cutting plant, the beam guidance on the workpiece is conventionally carried out by means of a remote scanner system. As a result of the low mass and the associated low inertia of the deflection mirrors, very high speeds and repetition accuracies can be achieved in the horizontal plane. Due to the varying distance of the electrode to the scanner system, a focus adjustment in the vertical direction is necessary. This adjustment can be produced by means of additional lenses (focus-shifter) or by a static f-theta objective. Considering the static beam refocusing, the advantage of an f-theta objective over a focus shifter is the higher repetition accuracy and wear-freedom. However, the dynamic focus by means of a focus shifter allows us to adjust the working field and to customize the focus in certain areas.

2. Experimental

2.1. Materials

To investigate the influence of the laser process parameters on the properties of the cutting edge and the influence on the electrochemical performance, double-sided coated electrodes with industrially available material components were used. For the anode, the active material SMGA4 (91 wt.%; Hitachi, Japan), with a specific capacity of 360 mAhg^{-1}, was coated on a 10 µm thick copper collector (Sumisho Metallex, Japan). Subsequently, the coating was compacted to a density of 1.5 gcm^{-3}, which results in the porosity of 32.25% and a total thickness of 123 µm. On the cathode side, the active material NMC 111 (90 wt.%; BASF, Germany), with a specific capacity of 165 mAhg^{-1}, was coated on 20 µm thick aluminum collector (Hydro Aluminum Rolled Products, Germany) and compacted to a degree of 2.8 gcm^{-3}. The described compaction led to porosity of 31.35% and a total electrode thickness of

143 µm. For the anode and cathode, a conductivity additive SFG6L (2 wt.% anode, 2 wt.% cathodes; Imerys, Switzerland), a carbon black C65 (2 wt.% anode, 4 wt.% cathodes; Imerys, Switzerland), and a PVDF binder (5 wt.% anode, 4 wt.% cathodes; Solvay, Italy) were utilized. The cell manufacturing was carried out with a 27 µm thick separator (Separion) and a conventional $LiPF_6$ electrolyte (UBE Industries Ltd., Japan). The conductive salt $LiPF_6$ was solved in a concentration of one mole in a solvent consisting of ethylene carbonate (EC), dimethyl carbonate (DMC), and ethyl methyl carbonate (EMC), with a volumetric ratio of the solvent components of 1:1:1. To suppress the evolution of gas during the first charging, the electrolyte contained 2 wt.% of Vinylene Carbonate (VC). As further additives for reducing hydrogen formation at high voltages, the electrolyte contained 3 wt.% of cyclohexylbenzene (CHB).

2.2. Analysis of the Cutting Edge Characteristics

The prescriptive characteristics of the electrode cutting edge are shown in Figure 5a by means of a microsection. These characteristics were determined by light microscopy (VHX 2000 light microscope (LM), Keyence, Osaka, Japan) and laser scanning microscope (VK-X Series 3D Laser Scanning Confocal Microscope (LSM), Keyence, Osaka, Japan). Here, the parameters chamfer width and heat-affected zone (HAZ) were considered to be the significant influencing factors on the electrochemical performance and, therefore, investigated further. The HAZ is defined as an area where the active material is thermally stressed but not removed. The chamfer width is characterized by active material removal and a melt formation zone.

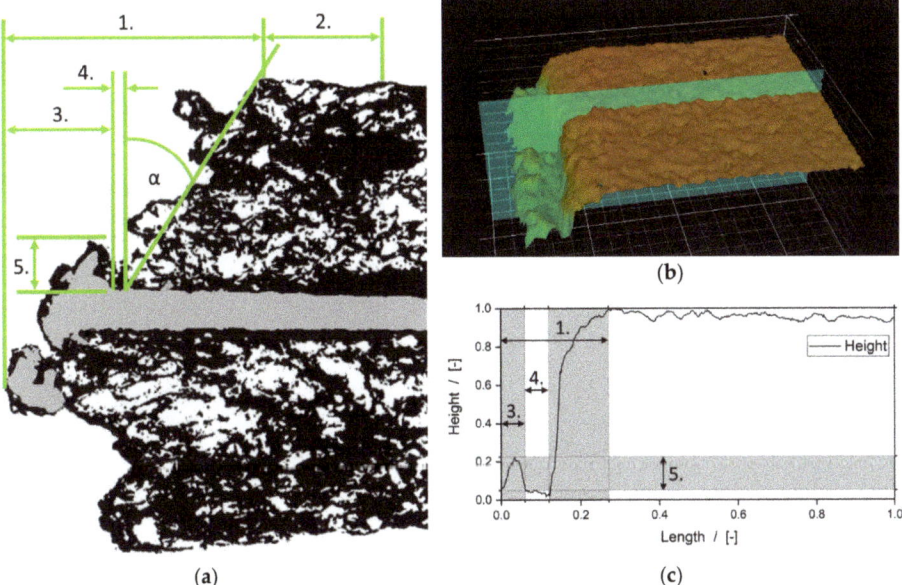

Figure 5. Analysis of the prescriptive cutting edge characteristics: (**a**) Schematic microsection of a laser-generated cutting edge, 1. Chamfer width (chw), 2. The heat-affected zone (HAZ), 3. Melt formation, 4. Ablation, 5. Melt superelevation, α Chamfer angle; (**b**) LSM image of a cutting edge; (**c**) Analysis of a cutting edge by LSM data, based on [10].

To measure these characteristics, LM and LSM images were taken of the upper side of the electrodes at the cutting edges. The areas of the heat-affected zone and the chamfer width could be clearly separated by a combined measuring method. By means of the LSM topography images (Figure 5b,c), the chamfer width was measured. By subtracting the chamfer width from the entire affected area (LM images), the heat-affected zone could be quantified. For the determination of contaminant products as a result of the laser-material interaction, SEM (FEI Quanta 650, Thermo Fisher, Waltham, MA, USA) and EDX (Oxford X-Max 80 mm², Oxford Instruments, Abingdon, England) images of the cut electrodes were taken.

2.3. Cell Format and Manufacturing

To evaluate the influence of the cutting edge on the electrochemical performance, pouch cells with 15 compartments were built (total cell capacity about 9 Ah). The relatively high number of compartments was chosen to emphasize the effect of the cutting edge properties of the electrodes, maximizing the effects of the cutting edge characteristics on the electrochemical performance. The surface of the anode coated with active material was 16,484 mm² and, based on the geometry shown in Figure 6a, gave a cutting edge to surface ratio of 0.030 mm^{-1}. The cathode is defined by an area of 15,209 mm² and a ratio of cutting edge to the surface of 0.031 mm^{-1}. For the anode, a circumferential overlap of 2.5 mm resulted from the illustrated geometries for the anode and the cathode. This overlap guaranteed the total stress of the cathode as a reference in these examinations and ensured the correct balancing of the compartment.

By means of a z-folding process (prototype plant, Jonas & Redmann, Berlin, Germany), the singularized electrodes were stacked alternating between a separator to form an electrode-separator-composite (ECS). Subsequently, the individual collectors of the electrodes were joined to a tab via ultrasonic welding (Ultraweld F20, Branson Ultraschall, Hannover, Germany). For this purpose, the 15 single anode collectors were welded to a nickel tab with an energy of 200 j, and cathodes collectors to an aluminum tab with an energy of 100 j at an oscillating sonotrode amplitude of 30 µm. Subsequently, the ECS was dried under vacuum for 120 °C for 16 h. In the following step, the ECS was inserted into the pouch bag and filled under argon atmosphere with electrolyte and sealed. Finally, the filled cells were tempered for 4 h at 60 °C to support the complete wetting of the electrodes. The finished assembled cell is shown in Figure 6b [12].

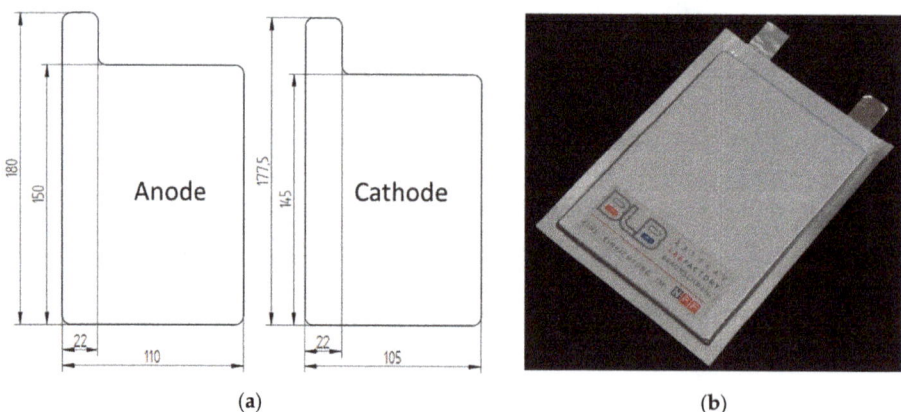

Figure 6. Cell format and pouch cell design: (**a**) Electrode format, (**b**) Complete pouch cell.

2.4. Cell Diagnostic

After assembly and wetting, the manufactured cells were placed in a climate chamber (WKM Inc., Lachendorf, Germany) at 20 °C and connected to a battery tester (Series XCTS, Basytec Inc., Asselfingen, Germany) with a fixed torque of 2.54 Nm. Due to the high capacity of the battery cells, the tests required a high safety environment. Therefore, the climate chambers were equipped with a fire extinguishing system (Wagner Group Inc., Langenhagen, Germany). In the event of an accident, the climate chamber is flooded with nitrogen gas. Furthermore, an activated carbon filter (Stöbich technology Inc., Goslar, Germany) will filter the exhaust air in the pipe duct.

In our experiments, the cells were formed in two cycles. They were first charged and discharged at 1/10 C, and in the second formation cycle with 1/2 C. Upper and lower cut-off voltages were 4.2 V and 2.9 V, respectively, for all charge-discharge cycles. To characterize the cells, a capacity test at 1/10 C and a pulse test (1 C for 1 s) to determine the internal resistance were performed. After the formation process, the cells were matured over eight days with a state of charge (SOC) of 50% at 20 C. Then, the aging of the cells began with a C-rate test with different discharge-rates from 1/5 to 2 C, which lasted 20 cycles. Long-term cycling was then started at 1 C for 100 cycles. After this, the cyclization was paused, and the internal resistance was measured in a pulse test before the C-rate test was repeated. These aging investigations were repeated periodically until at least 450 cycles were reached. In this study, 5 cells per laser variation were analyzed, and only the long-term cycling was considered.

2.5. Laser Cutting Plant and Key Parameter

The laser cutting plant used in this study was an in-house development and construction. Due to its modular structure and the process-immanent advantages of the scanner system, it is suitable for a large number of different electrode formats. The beam source used was a nanosecond pulsed fiber laser with a central emission wavelength of 1059–1065 nm (G4 Pulsed Fiber Laser, SPI Lasers UK Ltd., Southampton, UK). The average power of the fiber laser was 72 W with a peak pulse power of up to 13 kW and an M^2 of <1.6. Guidance and focusing of the laser beam were performed by a 3-axis laser beam deflection system (AXIALSCAN 30/FOCUSSHIFTER, Raylase AG, Wessling, Germany) with a working field of 400 × 400 mm^2. The first two dimensions of the scanner were needed to drive the spot over the workpiece to create the cutout. The third dimension was needed to ensure a constant spot size of ~74 µm with a focus depth of 0.6 mm on one level over the entire working field. All laser cuts were made in one pass. The fully automated handling system was carried out by simple roll to roll and pick and place operation (Figure 7a), controlled by an Arduino Mega 2560. Since the cut could only be realized in the focus level of the laser spot, a special negative form for the positioning of the electrode had to be built for each electrode format (Figure 7b). The positioning of the electrode was done by negative pressure on holes surrounding the cutting curve. Even though a cutting on the fly was possible with the used remote scanner system, a static cutting operation was used to guarantee constant cutting speed over the complete cutting length and an easy format change.

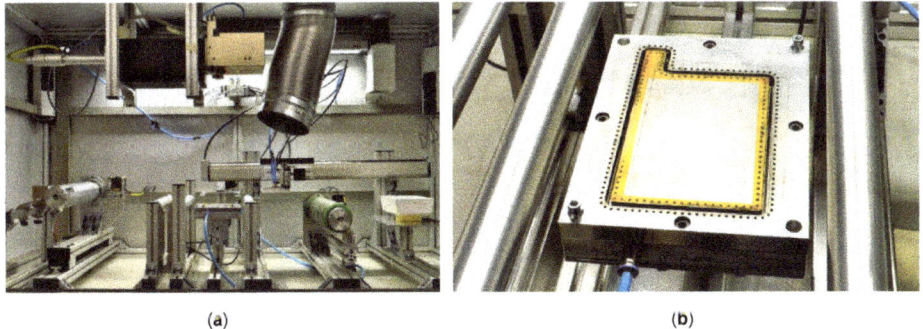

Figure 7. Laser cutting plant: (**a**) Front view of the laser cutting plant; (**b**) Negative form.

3. Results and Discussion

The presentation of the results has been divided into three sections. In the first section, the results of the influence of the laser parameters on the cutting edge characteristics, chamfer width, and heat-affected zone have been considered and discussed. The investigations focused on the pulse repetition frequency, the cutting speed, and the pulse length, as well as the influence of the number of hits per surface increment, and intensity at a constant energy density. In the second section, the influence of the cutting edge characteristics and corresponding laser process parameters on the electrochemical performance has been examined. Building on these results, the last section would present further investigations of the cutting process and the cutting edge, explaining the electrochemical behavior.

3.1. Influence of the Laser Process Parameters on the Cutting Edge Characteristics

In a first step, we investigated the influence of the laser parameters on the cutting edge properties. For this purpose, the pulse repetition frequency (PRF) and the cutting speed (V_c) were deliberately varied with constant power and pulse duration. The resulting cut edges were evaluated according to the analysis methods presented. On the basis of the obtained data, models could be developed by means of the analysis program Design Expert 11, which describes the examined parameter space. The experimental design was made with a D-optimal strategy and comprised 13 experiments with five repeated measurements each. The response surface model was adapted to the measured values by fitting it to a quadratic polynomial of the form of Response = Intercept * + A + B + AB + AB^2 + A^2B A^2 B^2. The results for the anode (Figure 8) showed that both parameters influenced the formation of the chamfer width and the heat-affected zone. In Figure 8a model, it could be seen that the formation of the chamfer width steadily decreased with increasing PRF. This tendency could also be observed with increasing cutting speed. The smallest chamfer width for this model was obtained at the maximum achievable speed of 700 mms^{-1}, with a pulse repetition frequency of 490 kHz. Considering the additional laser parameters, this resulted in a number of hits per area increment of 51 and an energy density of 141 jcm^{-2}. The decrease in the chamfer width with increasing speed or decreasing energy density could be explained by the lower energy input per area. The dependence on energy density has already been confirmed by previous publications [24]. The decrease of the chamfer width with increasing pulse repetition frequency could be attributed to different mechanisms, which result from the adjusted mode of the energy input. Due to the constant average power and the constant pulse duration, the pulse peak power had to drop with increasing pulse repetition frequencies to reduce the energy per pulse. As a consequence, the intensity and the energy decreased with increasing pulse repetition frequency, and thus the width of the threshold intensity or the spot diameter, which led to an ablation, becoming smaller due to the Gaussian intensity distribution. Besides, the increased number of hits of 360 (at 490 kHz and 100 mms^{-1}) could lead to a more intensive harmonic laser-plasma interaction, which reduced the energy impinging on the target by shielding effects and thus reduced the energy density. In addition, the reduced pulsed peak power could lead to a smaller broadening of the plasma formation. Since the plasma was also involved in the removal of material and the thermal load on the surface, the proportion of the total material removal became less at higher frequencies. Literature regarding ns laser-induced breakdown spectroscopy describes threshold pulse fluences for forming a plasma of 1.01 jcm^{-2} for aluminum and 1.46 jcm^{-2} for copper. As the smallest pulse fluence at 490 kHz for this System was 3.45 jcm^{-2}, the plasma formation, in general, would always occur for the examined parameter space [26].

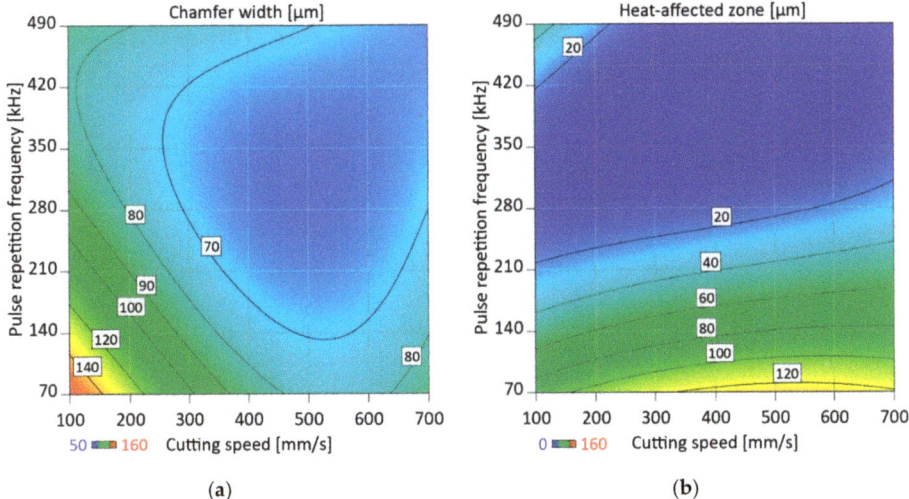

Figure 8. Model for the influence of the pulse repetition frequency and cutting speed on the anode cutting edge: 72 w, 240 ns; (**a**) Influence on the chamfer width (Cubic fitting: $R^2 = 0.77$, Adjusted $R^2 = 0.75$, Predicted $R^2 = 0.73$); (**b**) Influence on the heat-affected zone (Cubic fitting: $R^2 = 0.94$, Adjusted $R^2 = 0.93$, Predicted $R^2 = 0.93$).

From the heat-affected zone model (Figure 8b), it could be seen that increasing the PRF could reduce the heat-affected zone. The cause for the dependence of the thermal load on the PRF could be explained by the reasons given above for the dependence of the chamfer width on the PRF. The correlation of the HAZ with the cutting speed showed the opposite behavior to the chamfer width. As the cutting speed increased, the HAZ increased in the range of 70–350 kHz. This could be explained by the fact that with increasing cutting speed, the chamfer width and the kerf were becoming steadily smaller. This means that less material was removed for higher cutting speeds. Due to the Gaussian intensity distribution and the increasing speed, the energy input profile changed to the effect that the material which was no longer ablated underwent such high thermal stress that there was an optical change. This means that the final product cut with high speed contained a larger active material area that is thermally stressed, which would be completely removed at lower speeds. The investigations on the cathode were carried out in smaller parameter space (Figure 9a) with respect to the speed (100–400 mms^{-1}) because due to the higher material thickness of the collector already at 455 mms^{-1}, the cut-through limit for high PRF was reached. The results for the formation of the chamfer width as a function of the PRF and V_c showed similar tendencies as the anodic model. Both with increasing PRF and with increasing V_c, the chamfer width decreased significantly. The results showed that the smallest chamfer widths could only be achieved through the combination of low energy densities and high PRF. Possible causes for these dependencies could be transferred from the explanations to the anode. The model presented for the development of the heat-affected zone for cathodes (Figure 9b) could be adjusted with an R^2 of 0.93 and allowed a 92% reliable prediction. With increasing PRF, the HAZ decreased significantly until it approached zero at 490 kHz. Here, the influence of the cutting speed played only a minor role. Despite the low significance, an increase in the cutting speed led to a reduction in the HAZ. These results were opposite to the results for the anode. The difference in behavior was explained by two facts. Firstly, the intensity and energy difference between the ablation threshold and the thermal stress threshold were smaller for the cathode active material than for the anode active material. Secondly, the ablation threshold of the cathode active material was higher than that of the anode active material.

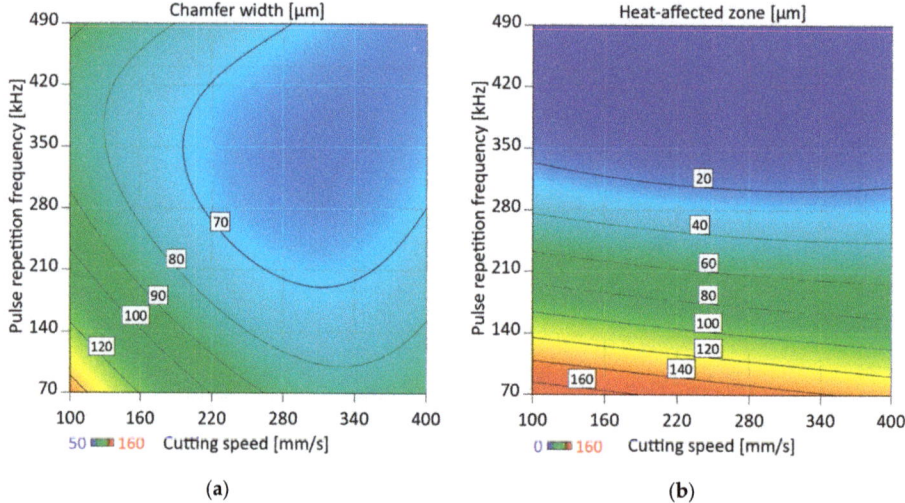

Figure 9. Model for the influence of the pulse repetition frequency and cutting speed on the cathode cutting edge: 72 w, 240 ns; (**a**) Influence on the chamfer width (Cubic fitting: $R^2 = 0.91$, Adjusted $R^2 = 0.90$, Predicted $R^2 = 0.88$); (**b**) Influence on the heat-affected zone (Cubic fitting: $R^2 = 0.94$, Adjusted $R^2 = 0.93$, Predicted $R^2 = 0.92$).

Further investigations outside of the considered parameter space in the range of very low cutting speeds (50 mms^{-1}) and very high energy densities, respectively, showed that on the cathode and anode cutting edge, either no or very small HAZ could be identified. This was caused by the slope of the intensity distribution and the very high energy input. As a result, the areas that were previously only thermally stressed at lower intensities were subjected to material removal at higher energy densities. Furthermore, the high energy density at 50 mms^{-1} led to an increase in the ablation area, since the ablation thresholds of the collector and active material differ significantly.

With regard to the strong influence of the PRF on the chamfer width and the HAZ, the influence of the pulse duration and the pulse peak power on different PRF was investigated in a further study. For this purpose, cuts were performed at a constant energy density with a variation of the pulse duration and pulse peak power. Pulse duration was kept constant (240 ns) with the effect that the pulse peak power decreased with increasing frequency (70 kHz: 13 kW, 102 kHz: 6 kW, 200 kHz: 2 kW, 291 kHz: 1 kW, 403 kHz: 0.7 kW, 490 kHz: 0.55 kW). The pulse duration was shortened (240–20 ns) to keep the pulse peak power quasi constant (70 kHz: 13 kW, 102/200/291 kHz: 10 kW, 403/490 kHz: 9 kW). The results for the chamfer width and the heat-affected zone derived from these experiments are shown in Figure 10. The PRF variation with constant pulse duration showed the same tendencies as in the previously presented models in Figures 8 and 9.

A reduction of the pulse length with a quasi-constant pulse peak power led to a larger chamfer width and HAZ, both at the anode and at the cathode (Figure 10). On the anode side, the reduction of the pulse length led to a significant enlargement of the heat-affected zone for the PRF 102 and 200 kHz. Due to the higher PRF and the high intensities, the plasma formed was of higher energy, leading to enhanced thermal stress of the electrode surface and, thus, potentially to a higher ablation. The reduction of the HAZ by the increased frequency of the ns laser pulses was thus determined largely by the low pulse energy. Only by a much greater reduction of the pulse length of less than 10 picoseconds, a cold cutting is possible [14].

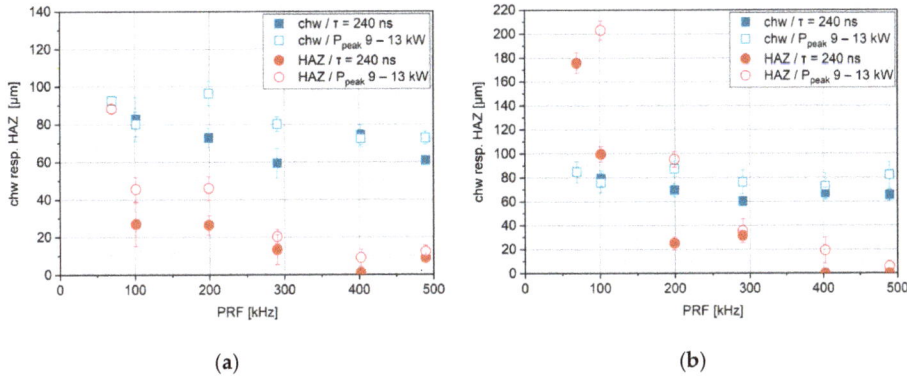

Figure 10. Influence of the pulse duration at different pulse repetition frequencies on the chamfer width and heat-affected zone of the anode (**a**) and the cathode (**b**) cutting edge.

The increased chamfer width for the anode and the cathode at higher pulse peak powers at higher PRF confirmed the assumption that was previously made on the anode and cathode model. In addition to the decrease in pulse peak power, the decrease in pulse energy at higher PRF led to a reduction of the chamfer width.

In the following experiment, the scalability of the cutting speed or the influence of the number of hits at constant energy density was investigated (Figure 11). The results for the anode showed that at constant energy density, the chamfer width increased at a reduced rate. A reason for this was the lower intensity and the correlated intensity distribution, as well as the lower energy of the pulse, which narrowed the profile of the material removal threshold. As a result, the geometric distance between a full cut and the ablation of the active material increased (see Figure 3.). In general, the cutting kerf, as well as the entire area in which material was removed, became smaller as a result. Because of the displacement of the removal thresholds, a larger chamfer width was produced at lower speeds for a specific energy density.

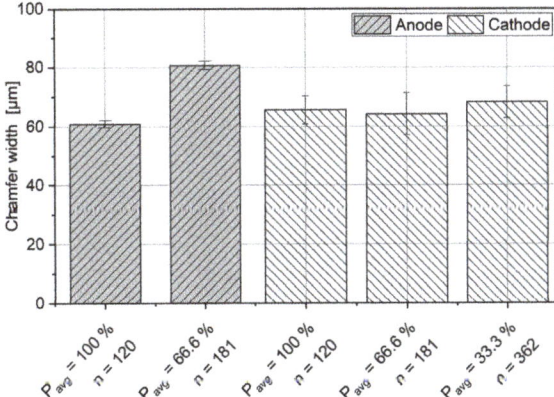

Figure 11. Influence of the intensity and the number of hits per area increment on the chamfer width at a constant energy density of 328 jcm^{-2}.

The fluence resulting from the reduction of the average power to 33.3% was still 1.15 jcm^{-2} and was thus above the limit for the formation of plasma for aluminum. Despite the high number of hits, it could be assumed that shielding effects were negligible as the intensity was much lower, and the larger plasma formation led to increased removal of the active material. When the average energy was reduced to one-third of the maximum power, no cut could be made through the copper collector at a frequency of 490 kHz for the energy density being studied. Due to the strong reduction of the pulse energy and peak pulse power, the beam could no longer be coupled because of the low absorption of the copper. When cutting copper, it is first heated by the radiation until it oxidizes [10]. As soon as the material oxidizes, the radiation can be much better coupled into the material, and only then leads to the sufficiently high absorption of the laser radiation for a complete cut.

Cathode investigations showed similar tendencies, which were much less pronounced. Compared to the anode, the cathode could be cut at 33% of the maximum average power, although the cathode's aluminum collector was twice as thick as the anode's copper collector. The results basically showed that the necessary energy density could be used as a second parameter to define a cut-through limit.

3.2. Influence of the Cutting Edge Characteristics and the Process Parameters on the Electrochemical Performance

Based on the knowledge gained from the experiments, a parameter study was developed to investigate the influence of the presented product and process properties on the electrochemical performance of the electrode, or the cell. In a first study, a cell was built with anodes and cathodes cut using the same laser and system parameters. The parameter configuration for this cell was referred to as V0 and served as a reference system for the variation of the parameter configuration. This parameter configuration was very reliable in terms of possible fluctuations in the focus position and the layer thickness. In a first limitation, it was examined whether the cutting edge on the anode side, or the cathode side, had a greater influence on the electrochemistry and, thus, the performance of the cell. In further experiments, only the cutting edge of the performance controlling electrode was examined. For the first experiments, cells with anodes (V1) and cathodes (V2) with a very large chamfer width were built. To produce this very large chamfer width, the electrodes were cut with a very high energy density and intensity. In the following, the parameters PRF (V3), V_c (V4), and τ (V5) were varied at a constant energy density at the electrode. Based on the parameter configurations shown in Table 1, it was possible to evaluate the previously presented cutting edges characteristics and process characteristics with regard to their influence on the electrochemical performance of the cell.

Table 1. Set laser parameters to investigate the impact on electrochemical performance.

Electrode	Parameter	Unit	V0	V1	V2	V3	V4	V5
Anode	V_c	mms^{-1}	300	50	300			
	PRF	kHz	70	70	70			
	τ	ns	240	240	240			
	ED	jcm^{-2}	328	1968	328			
	I_{Peak}	Wcm^{-2}	3.02×10^8	3.02×10^8	3.02×10^8			
	n_{line}	-	17	103	17			
Cathode	V_c	mms^{-1}	300	50	300	300	100	300
	PRF	kHz	70	70	70	490	490	490
	τ	ns	240	240	240	240	240	20
	ED	jcm^{-2}	328	1968	328	328	328	328
	I_{Peak}	Wcm^{-2}	3.02×10^8	3.02×10^8	3.02×10^8	0.17×10^8	0.1×10^8	2.09×10^8
	n_{line}	-	17	103	17	120	362	120

Cutting speed (V_c), Pulse repetition frequency (PRF), Pulse duration (τ), Energy density (ED), Intensity (I_{Peak}), Number of laser pulses per surface increment (n_{line}).

The evaluation of the electrochemical performance showed that the cutting edge of the cathode exerted the greater influence. Therefore, the influence of cut edge characteristics and process configurations at the cathode was investigated. In the following, the characteristics chamfer width and heat-affected zone have been shown for the examined parameter configurations V0 to V5.

The comparison of the features in Figure 12 showed that the total affected area consisting of the heat-affected zone and chamfer width was the largest for the reference parameter V0 for both the anode and the cathode. The largest chamfer width with almost identical characteristics showed the parameters V1 for the anode and V2 for the cathode. The further investigations with the parameters V3 to V5 showed, on the cathode side, the smallest influenced area with an average chamfer width of 70 μm to 80 μm.

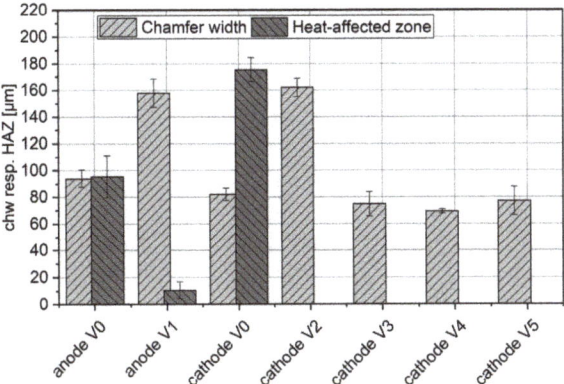

Figure 12. Response to the parameters, shown in Table 1.

The investigation of the influence of the presented cutting edge characteristics on the electrochemical performance was carried out by means of the cyclization routine defined in Section 2.4. The result of the diagnosis of the electrochemical performance is shown in the form of normalized cyclization curves in Figure 13. The cyclization curves showed that in comparison to the reference (V0), the cathode cutting edge (V2) exerted a significantly greater influence on the cycle stability than the anode (V1). In this case, the mean value for all cells, with the anode (V1), lied on the mean value of the reference cells (V0) with a similar standard deviation after 350 cycles. The cells with the cathode (V2) were far above this value and thus had significantly higher cycle stability. All tests to determine the influence of the chamfer width (V3–V5) on the cathode side showed that with low chamfer width, no improvement of the cycle stability could be achieved compared to V2. Since the cells V2 to V5 had higher cycle stability than the reference cells V0, the statement could be made that the heat-affected zone on the cathode side had a greater influence on the electrochemical stability of the cell than the chamfer width. Furthermore, the cycling curves of cells V3 and V5 were nearly identical after 350 cycles. From this, it could be deduced that the pulse duration or pulse peak power as process parameters did not exert a significant influence on the electrochemical performance. The cells with cathodes cut at constant line energy at reduced speed (V4) showed greater cycle stability after 350 cycles than the V3 and V5 cells. Despite the optimized process control, the cycle stability of the cells V4 was below that of V2.

The cells V3 to V5 showed a greater capacity drop than the cells from the series V2, although they had a small chamfer width and no heat-affected zone. This was an indication that in addition to the previously known and analyzed product features, another, previously unrecognized feature, influenced the electrochemical performance.

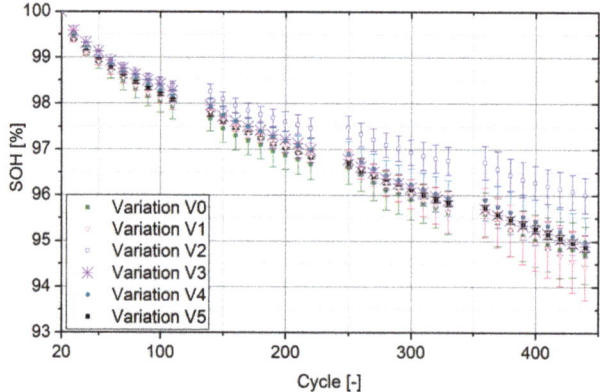

Figure 13. Capacity fading during the long-time cyclization of the investigated electrode/cell configuration.

3.3. Further Investigations of the Cutting Process and the Electrode Surface to Explain the Electrochemical Behavior

In addition to the physical characteristics of the cutting edge, it was found that the different parameters led to a different degree of flying sparks, as shown in Figure 14. This could be recorded by imaging techniques. The sparking could lead to contamination of the electrode and thus affect the electrochemical performance. The images showed that with increased pulse repetition frequency (V0 → V3), the distance of the spark flight increased by 2.5 times. This could be explained by the increased number of hits if the cutting speed remained the same. If the number of hits on the active material, as well as on the solid and molten collector, increased, the number and acceleration of the ablation products would increase too. A reduction of the cutting speed with constant line energy (V4) showed a strong reduction of the sparks. The lower level of the spark formation was due to the lower energy and intensity per pulse, as well as the reduced travel speed. The profile of the sparkling flight was very similar to the sparking profile at low pulse repetition frequency and traversing speed (V2). Since variation V2 was cut with a lower pulse repetition frequency than variation V4, this could lead to less contamination of the electrode and thus explain the impairment of the performance of V4 compared to V2.

In addition to the assessment of the flying sparks, the recordings could strengthen the assumptions made in the previous sections concerning the formation of the plasma. A qualitative comparison of the images in Figure 14 showed that V0, V2, and V5 showed a clearly purple-colored plasma formation, whereas, for the other parameters—V3 and V4, the plasma presumably lied below the ablation gas phase, and thus was significantly smaller. This also explained why at shorter pulse lengths or higher pulse peak powers, the material removal and the formation of the HAZ at a constant energy density was greater.

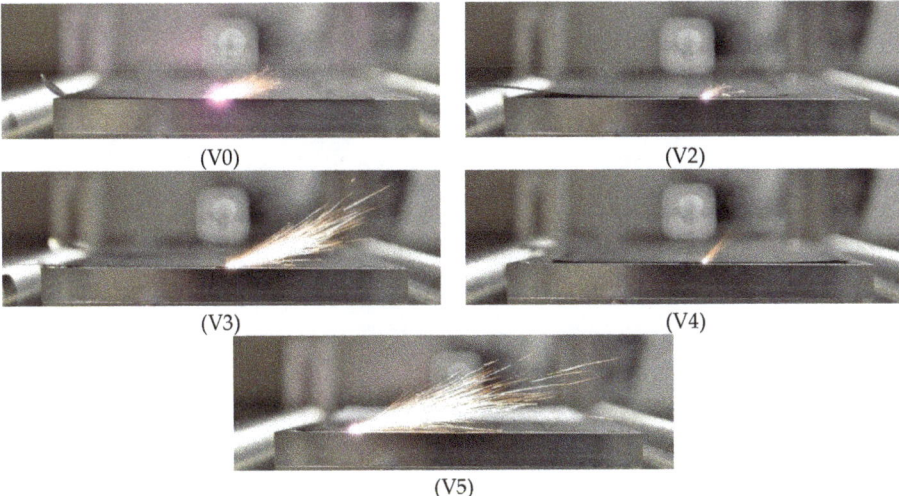

Figure 14. Optical process investigations regarding the flying sparks.

The change in spark travel might lead to a change in the degree of contamination of the electrodes and, thus, in addition to the cut edge quality, adversely affect the electrochemical performance. The contaminations of the electrode were analyzed by SEM/EDX images and are shown in Figure 15. Contamination products in the form of metal spatters could be identified on all electrodes tested. The reference cathode had the strongest metal spatter contaminations. The lowest contamination occurred with the parameter variations V2 and V4.

These results showed that fewer sparks formation led to less contamination of the electrode surface in the area of the cutting edge. Based on the results of the electrochemical diagnosis and the analysis of the cutting edges, it could be assumed that the contaminations on the cathode surface exerted a greater influence on the electrochemical performance than the heat-affected zone and the chamfer width.

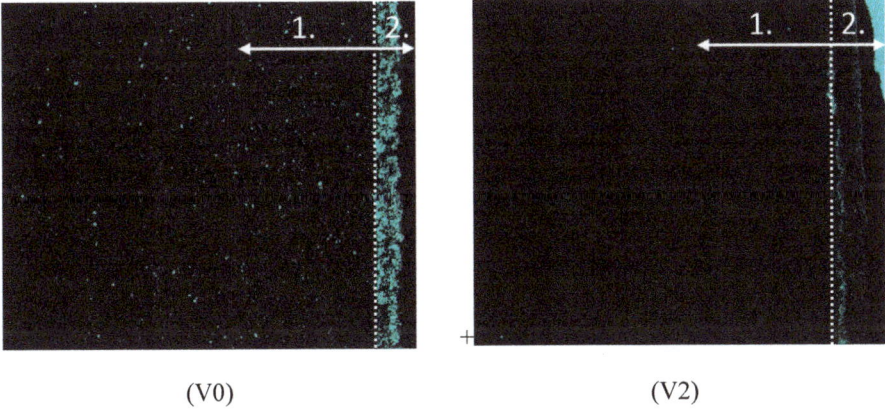

Figure 15. *Cont.*

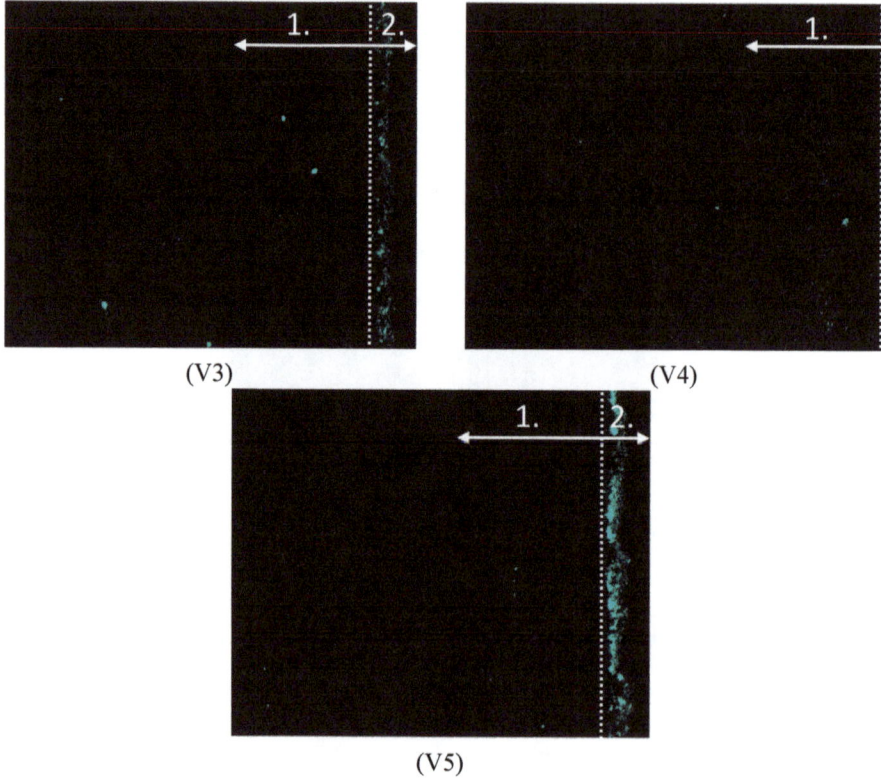

Figure 15. SEM/EDX analysis of the cathode surface next to the cutting edge: Contamination of the surface by molten aluminum splashes are marked turquoise: 1. Electrode surface, 2. Cut zone.

4. Conclusions

By means of a pulsed nanosecond fiber laser, electrodes could be cut without significantly affecting the cell cycle stability. The influence of the laser parameters or the cutting edge showed only a small influence on the examined variations. Considering a linear behavior, the parameter V1 led to a theoretical number of cycles of 1670 until the cells reached a state of health (SOH) of 80%. The best parameter V2 resulted in a cycle number of 1760. Since the capacity drop tended to decrease, it could be assumed that the cells would also reach higher cycles up to a SOH of 80%.

The examined characteristics—chamfer width and heat-affected zone—could be adjusted by means of the pulse duration, the pulse repetition frequency, and the cutting speed. In principle, the results showed that the chamfer width decreased with decreasing energy density or with increasing cutting speed. Furthermore, with a constant pulse duration and energy density, higher pulse repetition frequencies could result in a smaller chamfer width. This was probably due to the fact that both the decrease of the energy and the decrease of the intensity led to a geometric shift of the removal and the thermal load threshold. This shift was probably explained by the Gaussian intensity profile and the corresponding slope of the intensity in relation to the spot size, as well as by the intensity and energy-dependent formation of the plasma. The experiments with different pulse durations with the same pulse energy showed, for different pulse repetition frequencies and constant energy density, that the material removal and the thermal load, among the pure laser-material interaction, depended on the plasma or the intensity. The analysis of the size of the heat-affected zone revealed that it was essentially influenced by the pulse repetition frequencies. The cause of this dependency could be explained by the same assumptions that have been made previously for the formation of the chamfer width. Both the anode and the cathode showed similar tendencies. In general, the anode could be cut much faster than the cathode due to the double material thickness of the collector, so that the parameter space for the model development in terms of speed for the cathode turned out smaller than for the cathode.

Investigations to evaluate the influence of the chamfer width and the heat-affected zone on the electrochemical performance of large-sized multicompartment pouch cells showed that the HAZ had a greater influence than the chamfer width. Furthermore, the results showed that contamination products in the form of metal spatter influenced the electrochemical performance more than the width of the chamfer and probably also the HAZ. It could be shown by imaging techniques that with very low speeds and high intensities and pulse energies with a moderate number of hits, metal spatters could be reduced.

Furthermore, it was possible to reduce the melting spatters even for a high number of hits per surface increment with a low traversing speed, intensity, and pulse energy. The results showed, so far, that the formation of the metal spatters was a function of the number of hits per surface increment, the speed of travel, the intensity, and the energy per pulse.

5. Outlook

The results showed that the electrochemical performance of the electrodes, or the cells, was essentially affected by contamination products and less by the chamfer width or heat-affected zone investigated in this study. Based on this knowledge, it is imperative to find out which contaminations and concentrations are to be considered critical with regard to the electrochemical performance. Furthermore, the pulsed ns laser system should be used to investigate which parameters lead to which contaminations and whether the cw laser system offers process-inherent advantages with regard to the contamination of the electrode surface. On the basis of these further investigations, unknown new laser systems could be designed process-safe to fulfill the minimum requirement of four Sigma rejects.

Author Contributions: T.J. and M.W.K. wrote and edited the main parts of the paper, S.H. analyzed the REM/EDX data. L.H. analyzed the electrochemical data. W.H. analyzed the electrochemical data and administered the funding acquisition. K.D. administered the research project. All authors contributed to scientific discussions.

Funding: The results were generated in the project DaLion (03ET6089) funded by the Federal Ministry of Economic Affairs and Energy. The authors express their gratitude for the financial support by the Ministry and for the project management by the ProjektträgerJülich.

Acknowledgments: Hereby I would like to thank my student assistant Julien Essers for his technical support.

Conflicts of Interest: The authors declare no conflict of interest.

References

1. Pachauri, R.K.; Allen, M.R.; Barros, V.R.; Broome, J.; Cramer, W.; Christ, R.; Church, J.A.; Clarke, L.; Dahe, Q.; Dasgupta, P.; et al. *Climate Change 2014: Synthesis Report. Contribution of Working Groups, I. II and III to the Fifth Assessment Report of the Intergovernmental Panel on Climate Change*; IPCC, Ed.; IPCC: Geneva, Switzerland, 2014.
2. Trenberth, K.E.; Fasullo, J.T. Tracking Earth's Energy: From El Niño to Global Warming. *Surv. Geophys.* **2012**, *33*, 413–426. [CrossRef]
3. Hettesheimer, T. *Strategische Produktionsplanung in Jungen Märkten. Ein Systemdynamischer Ansatz zur Konzeption und Dynamischen Bewertung von Produktionsstrategien am Beispiel der Lithium-Ionen-Traktionsbatterie*; Fraunhofer Verlag: Stuttgart, Germany, 2018; ISBN 978 3 8396 1275 0.
4. Schüneman, J.-H. *Modell zur Bewertung der Herstellkosten von Lithiumionenbatteriezellen*; Sierke Verlag: Göttingen, Germany, 2015; ISBN 978-3-86844-704-0.
5. Choi, J.W.; Aurbach, D. Promise and reality of post-lithium-ion batteries with high energy densities. *Nat. Rev. Mater.* **2016**, *1*, 359. [CrossRef]
6. Ding, Y.; Cano, Z.P.; Yu, A.; Lu, J.; Chen, Z. Automotive Li-Ion Batteries: Current Status and Future Perspectives. *Electrochem. Energy Rev.* **2019**, *2*, 1–28. [CrossRef]

7. Albright, G. Cylindrical vs. Prismatic Cells: Life, Safety, Cost. Available online: http://www.batterypoweronline.com/images/Allcell.pdf (accessed on 3 October 2019).
8. Schröder, R.; Aydemir, M.; Seliger, G. Comparatively Assessing different Shapes of Lithium-ion Battery Cells. *Procedia Manuf.* **2017**, *8*, 104–111. [CrossRef]
9. Beuth Verlag GmbH. *DIN 8588:2013-08, Fertigungsverfahren Zerteilen_- Einordnung, Unterteilung, Begriffe*; Beuth Verlag GmbH: Berlin, Germany, 2013.
10. Jansen, T.; Kandula, M.W.; Blass, D.; Hartwig, S.; Haselrieder, W.; Dilger, K. Evaluation of the Separation Process for the Production of Electrode Sheets. *Energy Technol.* Available online: https://www.researchgate.net/publication/333333666_Evaluation_of_the_Separation_Process_for_the_Production_of_Electrode_Sheets (accessed on 10 October 2019).
11. Jansen, T.; Blass, D.; Hartwig, S.; Dilger, K. Processing of Advanced Battery Materials—Laser Cutting of Pure Lithium Metal Foils. *Batteries* **2018**, *4*, 37. [CrossRef]
12. Hoffmann, L.; Grathwol, J.-K.; Haselrieder, W.; Leithoff, R.; Jansen, T.; Dilger, K.; Dröder, K.; Kwade, A.; Kurrat, M. Capacity Distribution of Large Lithium-Ion Battery Pouch Cells in Context with Pilot Production Processes. *Energy Technol.* **2019**, *21*, 1900196. [CrossRef]
13. Kreling, S. *Laserstrahlung mit Unterschiedlicher Wellenlänge zur Klebvorbehandlung von CFK*; Shaker: Aachen, Germany, 2015; ISBN 9783844037685.
14. Bliedtner, M. *Lasermaterialbearbeitung*; Carl Hanser Verlag GMBH: München, Germany, 2013; ISBN 9783446421684.
15. Chichkov, B.N.; Momma, C.; Nolte, S.; Alvensleben, F.; Tünnermann, A. Femtosecond, picosecond and nanosecond laser ablation of solids. *Appl. Phys. A* **1996**, *63*, 109–115. [CrossRef]
16. Sundaram, S.K.; Mazur, E. Inducing and probing non-thermal transitions in semiconductors using femtosecond laser pulses. *Nat. Mater.* **2002**, *1*, 217–224. [CrossRef]
17. Sugioka, K.; Meunier, M.; Piqué, A. *Laser Precision Microfabrication*; Springer-Verlag: Heidelberg, Germany; New York, NY, USA, 2010; ISBN 978-3-642-10522-7.
18. Stafe, M.; Marcu, A.; Puscas, N. *Pulsed Laser Ablation of Solids*; Springer-Verlag: Berlin, Germany; Heidelberg GmbH & Co. KG: Heidelberg, Germany, 2013; ISBN 978-3-642-40977-6.
19. Luetke, M.; Franke, V.; Techel, A.; Himmer, T.; Klotzbach, U.; Wetzig, A.; Beyer, E. A Comparative Study on Cutting Electrodes for Batteries with Lasers. *Phys. Procedia* **2011**, *12*, 286–291. [CrossRef]
20. Lee, D. Investigation of Physical Phenomena and Cutting Efficiency for Laser Cutting on Anode for Li-Ion Batteries. *Appl. Sci.* **2018**, *8*, 266. [CrossRef]
21. Lee, D.; Oh, B.; Suk, J. The Effect of Compactness on Laser Cutting of Cathode for Lithium-Ion Batteries Using Continuous Fiber Laser. *Appl. Sci.* **2019**, *9*, 205. [CrossRef]
22. Lee, D.; Patwa, R.; Herfurth, H.; Mazumder, J. High speed remote laser cutting of electrodes for lithium-ion batteries: Anode. *J. Power Sources* **2013**, *240*, 368–380. [CrossRef]
23. Kronthaler, M.R.; Schloegl, F.; Kurfer, J.; Wiedenmann, R.; Zaeh, M.F.; Reinhart, G. Laser Cutting in the Production of Lithium Ion Cells. *Phys. Procedia* **2012**, *39*, 213–224. [CrossRef]
24. Lutey, A.H.A.; Fortunato, A.; Carmignato, S.; Ascari, A.; Liverani, E.; Guerrini, G. Quality and Productivity Considerations for Laser Cutting of LiFePO4 and LiNiMnCoO2 Battery Electrodes. *Procedia CIRP* **2016**, *42*, 433–438. [CrossRef]
25. Demir, A.G.; Previtali, B. Remote cutting of Li-ion battery electrodes with infrared and green ns-pulsed fibre lasers. *Int. J. Adv. Manuf. Technol.* **2014**, *75*, 1557–1568. [CrossRef]
26. Cabalín, L.M.; Laserna, J.J. Experimental determination of laser induced breakdown thresholds of metals under nanosecond Q-switched laser operation. *Spectrochim. Acta Part B At. Spectrosc.* **1998**, *53*, 723–730. [CrossRef]

© 2019 by the authors. Licensee MDPI, Basel, Switzerland. This article is an open access article distributed under the terms and conditions of the Creative Commons Attribution (CC BY) license (http://creativecommons.org/licenses/by/4.0/).

Article

Electrical Modelling and Investigation of Laser Beam Welded Joints for Lithium-Ion Batteries

Sören Hollatz [1,*], Sebastian Kremer [1,2], Cem Ünlübayir [2], Dirk Uwe Sauer [2,3,4,5], Alexander Olowinsky [1] and Arnold Gillner [1,6]

1. Fraunhofer Institute for Laser Technology ILT, Steinbachstr. 15, 52074 Aachen, Germany; sebastian.kremer@rwth-aachen.de (S.K.); alexander.olowinsky@ilt.fraunhofer.de (A.O.); arnold.gillner@ilt.fraunhofer.de (A.G.)
2. Chair for Electrochemical Energy Conversion and Storage Systems, Institute for Power Electronics and Electrical Drives (ISEA), RWTH Aachen University, Jägerstrasse 17-19, 52066 Aachen, Germany; cue@isea.rwth-aachen.de (C.Ü.); sr@isea.rwth-aachen.de (D.U.S.)
3. Institute for Power Generation and Storage Systems (PGS), E.ON ERC, RWTH Aachen University, Mathieustrasse 10, 52074 Aachen, Germany
4. Jülich Aachen Research Alliance, JARA-Energy, Templergraben 55, 52056 Aachen, Germany
5. Helmholtz Institute Münster (HI MS), IEK 12, Forschungszentrum Jülich, 52425 Jülich, Germany
6. Chair for Laser Technology, RWTH Aachen University, Steinbachstr. 15, 52074 Aachen, Germany
* Correspondence: soeren.hollatz@ilt.fraunhofer.de; Tel.: +49-241-8906-613

Received: 13 March 2020; Accepted: 10 April 2020; Published: 21 April 2020

Abstract: The growing electrification of vehicles and tools increases the demand for low resistance contacts. Today's batteries for electric vehicles consist of large quantities of single battery cells to reach the desired nominal voltage and energy. Each single cell needs a contacting of its cell terminals, which raises the necessity of an automated contacting process with low joint resistances to reduce the energy loss in the cell transitions. A capable joining process suitable for highly electrically conductive materials like copper or aluminium is the laser beam welding. This study contains the theoretical examination of the joint resistance and a simulation of the current flow dependent on the contacting welds' position in an overlap configuration. The results are verified by examinations of laser-welded joints in a test bench environment. The investigations are analysing the influence of the shape and position of the weld seams as well as the influence of the laser welding parameters. The investigation identifies a tendency for current to flow predominantly through a contact's edges. The use of a double weld seam with the largest possible distance greatly increases the joint's conductivity, by leveraging this tendency and implementing a parallel connection. A simplistic increase of welded contact area does not only have a significantly smaller effect on the overall conductivity, but can eventually also reduce it.

Keywords: resistance measurement; contact quality; laser beam welding; aluminium; copper; lithium-ion batteries; battery systems; spatial power modulation; single mode fibre laser

1. Introduction

Over the past years, the demand for large battery packs for electric vehicles (EV) has steadily increased with the ongoing electrification of the transportation sector and a growing demand for greater ranges. State of the art EV battery packs consist of a large quantity of cells connected in series to achieve the desired voltage level and in parallel in order to enable higher charge- and discharge-currents. For example, the EV Tesla Model S comprises of total count of over 7000 type 18,650 battery cells inside its battery pack [1]. A single defective connection can lead to failure or a reduction in performance. The quality of the joint has a decisive influence on the already discussed sustainability and safety of electric vehicles [2] Increased resistance at a welded joint causes more heat loss at this spot and leads to an increased electrical and thermal load on the individual cells, which in turn can lead to failure or accelerated aging. Laser beam welding is a promising technology to contact battery cells enabling automated, fast and precise production of conductive joints. In comparison to other conventional welding techniques, such as resistance spot welding, the laser beam welding has a reduced thermal energy input [3]. Compared to ultrasonic welding, the laser beam welding technique does not induce a mechanical force [4]. The resulting transition resistances are in the range of the basic material resistances. The overall performance of the battery pack is therefore improved by the reduction of the ohmic resistance of the joints and heat loss inside the battery cell.

Furthermore, laser beam welding produces a small heat-affected zone. In the context of production, laser beam welding is well suited to be integrated into almost fully automated production lines in the manufacturing process of battery packs and EVs. The joining of aluminium and copper is particularly challenging in laser welding as the metal pair forms intermetallic phases, which can yield lower weld qualities [5–7]. These phases can be identified in cross sections, see Figure 1. For the investigation, the different colours in the mixing zone (dark grey and yellow areas) give a first indication on the concentration of the metals. For further investigation an energy dispersive X-ray spectroscopy can be performed, but will not part of this study.

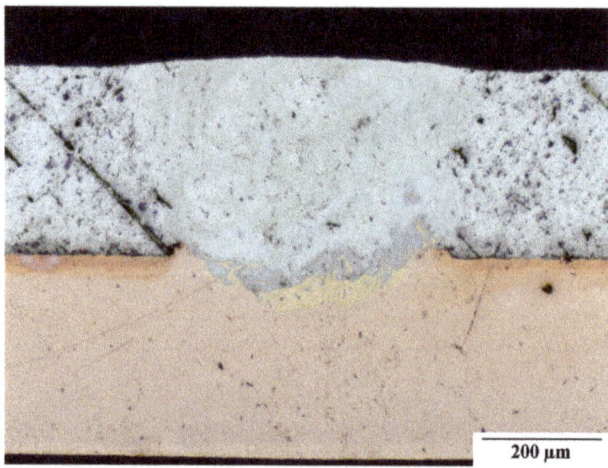

Figure 1. Cross section of a laser welded aluminium and copper joint (P = 294 W, v = 120 mm/s, A = 0.15 mm, f = 1000 Hz).

This paper showcases an evaluation of various laser welded joints for the connection of pouch cell terminals to the battery pack in an overlap configuration. The specimen design is related to pouch cells. Due to the focus on the connection quality, no functional cells are used for this investigation. First, an ohmic-resistance model for the joints is introduced. With the help of this model the current flow across the overlap transition is analysed. Lastly different geometries of welds were chosen and compared in terms of their conductivity.

2. State of the Art

2.1. Measurement of Electrical Resistance of Laser-Welded Joints

A current passing through a conductor encounters an electrical resistance, analogous to an opposing force by mechanical friction. This resistance is defined in Ohm's law as the proportion of voltage across and current through the same conductor. It is dependent on the specific electrical resistance of the conductor's material and its dimensions according to [8].

$$R = \rho \cdot \frac{l}{A} \tag{1}$$

R: resistance (Ω); ρ: specific resistivity ($\Omega \cdot m^2/mm$); l: conductor's length (mm) and A: conductor's cross-sectional area (m^2).

The specific electrical resistance is furthermore dependent on the material's temperature and exact chemical composition. Hence impurities can have an effect on the resistivity.

The measurement of an electrical resistance can be executed by a combined measurement of the voltage and current, as suggested by Ohm's law. A popular method for this combined measurement is the so called four-terminal sensing. The four-terminal sensing describes the introduction of a defined current through the conductor and a separated voltage measurement. By separating the sensing wires the measured voltage does not falsely include the voltage across the current carrying wires. As a voltage measurement usually has a high impedance the current through the voltmeter can be neglected for significantly lower measured resistances.

In view of laser-welded joints of battery contacts, the analysis of the electrical resistance might present a suitable indicator for the weld quality. This postulation is based on the effect of impurities in the material and the dimension of the actual contact area on the joints conductivity. However, solely the quantity of the resistance might result in incorrect results when comparing different laser-welded joints. For instance, a joint of worse quality can show a significantly higher conductivity due to a larger cross-sectional area of the conductor. To allow valid comparisons between joints of different dimensions, the electrical resistance must be further processed to yield the so-called contact quality index.

The contact quality index (CQI, or resistance equivalence factor) describes the proportion of the joint's resistance in respect to its base materials and dimensions [9,10]. To calculate the CQI of a lap joint an additional measurement of the material's resistances is necessary, besides determination of the actual joint resistance. Assuming a constant cross-sectional area of the conductors and constant measurement distances, these can be acquired as per [9,11] shown in Figure 2.

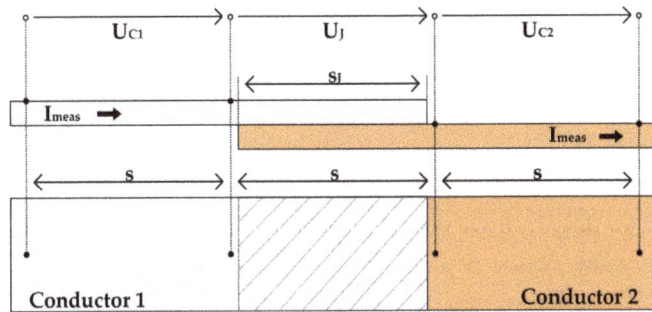

Figure 2. Schematic illustration for the calculation of the contact quality index (CQI).

The base resistance for the joint is derived by assuming a seamless, resistance-less joint from one material into the other over its length. The expected resistance for this geometrically optimal case would equal half the sum of the measured material's resistances, namely the average. The *CQI* can now be calculated by dividing the actual joint's resistance by the base joint's resistance. The equation is shown in (4).

$$R'_{Ci} = \frac{U_{Ci}}{I_{meas}} \text{ and } R_{Ci} = R'_{Ci} \cdot \frac{s_J}{s} \text{ for } i \in \{1,2\} \tag{2}$$

$$R'_J = \frac{U_J}{I_{meas}} \text{ and } R_J = R'_J - (R'_{C1} + R'_{C2}) \cdot \frac{s - s_J}{2 \cdot s} \tag{3}$$

$$CQI = \frac{2 \cdot R_J}{R_{C1} + R_{C2}} \tag{4}$$

R'_{Ci}: measured conductor resistance (Ω); R_{Ci}: conductor resistance for length s_J (Ω); R'_J: measured joint resistance (Ω); R_J: corrected joint resistance (Ω); U_{Ci}: voltage across conductor i (V); U_J: voltage across joint (V); I_{Meas}: measuring current (A); s: measuring distance (mm); s_j: joint distance (mm); CQI: contact quality index (1).

A CQI value of 1 can be interpreted as a joint of similar conductivity as the base materials; a value less than 1 indicates a higher, a value higher than 1 a lower conductivity. With the CQI joints of different materials and dimensions can now be compared for its joining method and effect on electrical resistivity.

2.2. Laser Beam Welding with Spatial Power Modulation

For the joining of materials with high thermal conductivity, laser beam welding is a suitable process. Using small focus diameters of a few 10 µm, the resulting high intensities are able to melt and vaporize the material to achieve deep and narrow weld seams. The process is defined with two process stages, the heat conduction welding and the deep penetration welding. For heat conduction welding the material is molten due to the absorption of the laser beam's energy on the surface. This process significantly depends on the absorptivity of the material. Significantly higher welding depths can be achieved by exceeding a characteristic intensity threshold using a deep penetration welding process. Therefore the material is vaporized and a keyhole is formed inside the molten pool. Multiple interactions of the laser beam inside this keyhole lead to an increase of the energy input resulting in a higher weld seam depth [12].

To manipulate the shape of the weld seam cross section and to stabilize the process during a deep penetration weld, a spatial power modulation can be used. Therefore the linear feed is superposed with a circular oscillation movement. The path is then characterized by an oscillation amplitude A, frequency f and feed rate v_f. In [13] a change from a v-shaped to a rectangular shaped weld seam cross section has been seen. Furthermore, an influence on the hardness, on the mixing of the materials in overlap configurations and on the roughness of the weld seam surface have been identified [13,14].

2.3. Laser Beam Welding of Electrical Contacts

For laser welding of electrical contacts, the contact resistance is the most important index, particularly indicated by the previously defined CQI. Therefore [15] has investigated similar material joints and reached a CQI of 0.55 for a copper connection and 0.57 for aluminium. In this case the joining partners have been connected by two parallel weld seams in an overlap configuration. By using two lines with the highest distance possible, the material in the overlap area is connected in parallel and shows a reduced transition resistance due to the higher current-carrying cross-section.

The investigation of [16] focuses on welding of dissimilar materials using a pulsed Nd:YAG laser. Weld seam depth and temperature gradient in the melt pool are controlled by a temporal power modulation. The investigations show a reduced mixing of the materials, higher process stability and higher seam quality.

Depending on the application, aluminium or copper is usually used to conduct electricity. Due to the reduced density, aluminium is used for lightweight applications, while copper with its higher conductivity is used when space is limited. Joining aluminium to copper, leads to numerous challenges. The differences in melting point, thermal conductivity and expansion cause tensions during the solidification, which can lead to cracks in the weld seam. Furthermore, the materials are soluble in the liquid state and form intermetallic phases inside the mixing zone of the weld seam. Besides the increase of hardness and crack sensitivity, the intermetallic phases increase the electrical resistance [5–7,17].

3. Electrical Equivalent Model for Joint

The investigation focused on laser-welded lithium ion pouch cells. The relatively broad contacts of these cells consist of aluminium and copper and offer a large contact area. Hence, in the joining of lithium ion pouch cells the overlap joint represents a suitable joining method. The joint in this investigation was aligned along the current direction, producing a straight junction with an expected homogenous current density in the conductors' cross-sectional areas (Figure 2).

The given lap joint can be modelled by using electrical equivalent circuit diagram from [4]. Each joining partner, M and m, was subdivided into equally sized stretch elements resistances. The partners were joined by bridge elements (indexed with J) representing the current-carrying interconnections. The resulting equivalent circuit is shown in Figure 3. The dimensions of each resistance in the equivalent circuit could be mapped according to regarded materials, joining methods and joint areas to represent the equivalent, real lap joint. All simulation results were based on basic electrical equations implemented in Python.

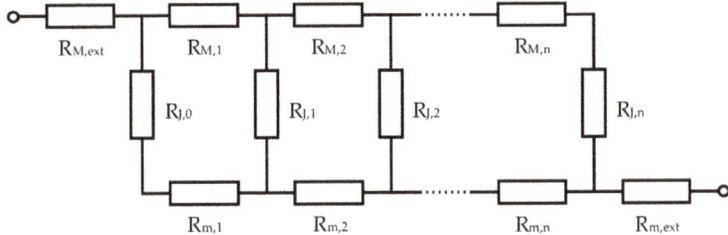

Figure 3. Model of the lap joint using an electrical equivalent circuit diagram.

The simulation of an electrical current through the equivalent diagram of a joint with homogenous resistance revealed a significant phenomenon. The current had an inherent tendency to flow along the edges of the joint, represented by the first and last bridge elements of index 0 and n. The tendency increased with decreasing bridge resistance in comparison to the joining partners' resistances (indexed with C). This phenomenon was explained by comparing the given equivalent circuit to a cascaded bridge circuit. With decreasing bridge resistance, the model approximated a cascaded bridge circuit, which by definition did not carry any current through the bridges when balanced (of equal resistance ratio). In fact, it is theoretically impossible for the current to be equal across all bridges in the equivalent circuit with $n > 1$ of a given joint, if no variance in the resistances is introduced. This can be proven mathematically by assuming equal bridge currents and overall equal resistances in the model ($n > 1$) and yielding a contradiction by calculating the overall joint resistance via circuit diagram simplification and the mesh current method.

For the simulation of laser-welded seams on the joint, the electrical values for the model were determined experimentally. Considering the limitations of the two-dimensional model, the bridge elements modelled a contact line across the whole conductor's width. The model was set to be of order $n = 110$, hence having 111 distinct bridge resistances (indices 0 through 110). The bridge elements $R_{J,i}$, representing either a weld or a purely frictional contact across the joint, could be determined by combining the measurements of a representative laser-welded joint, a purely frictional contact joint and a joint consisting solely of a laser welded seam. The produced bridge resistances show an improvement in conductivity for laser-welded bridges as compared to frictional contacts by a factor of 9800, 8750 and 178,000 for aluminium–copper, copper–copper and aluminium–aluminium joints respectively. The significantly higher factor in the pure aluminium joint can be explained by high resistances of its frictional contact due to surface oxidation [18]. A simulation of current through a joint with simplified resistances yields the proportional current through each bridge element $R_{J,i}$ presented in Figure 4 (model of order $n = 110$ model with $R_{M,i} = R_{m,i} = R_M/n$).

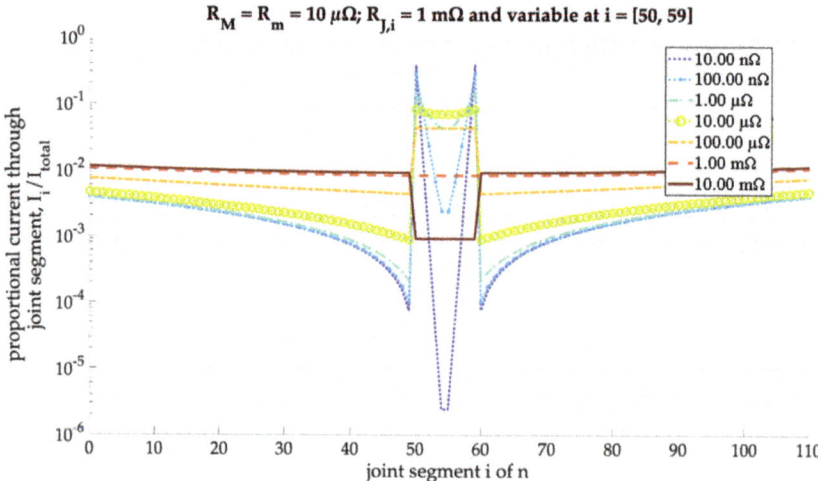

Figure 4. Simulation of current through a simplified model with a central weld seam.

The figure clearly shows the higher conductivity of the laser-weld, spanning over the central ten bridge elements. Additionally, the previously explained tendency of the current flowing through the joint's edges could be identified in the rising current rates towards the bridge indices of 0 and 110. Modifying the identical model to have the laser-welded bridges to be divided to the joint's edges, yields simulated current rates as shown in Figure 5.

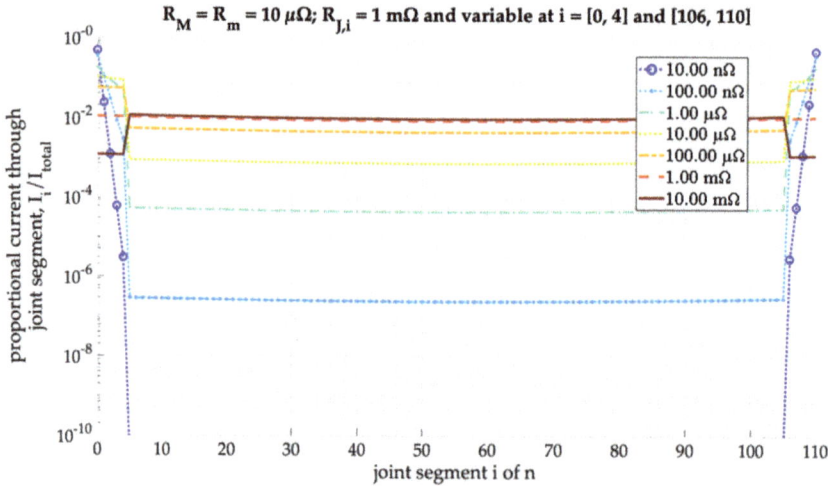

Figure 5. Simulation of current through a simplified model with weld seams on the joint's edges.

The first five and last five slots were now assigned to the laser-welded resistances and hence represented a joint of overall similar cross-sectional surface area to the central weld example. Due to the tendency of edge currents the overall carried current through the weld was significantly higher than in the example of the central weld; the current through the frictional contact was negligibly small. As a result, the implementation of a "double weld" not only effectively doubled the conductivity of the joint (by connecting the materials in parallel, increasing the conductor's cross-sectional area in direction of main current flow) but also complemented the current to predominantly flow through the laser-welded seams on the joints edges.

A variance of the resistances for the modelled materials introduced a shift in the bridges' current densities, whereas higher densities were found at the side of the lower conductive material (and accordingly the end of the material with higher conductivity). The distribution roughly resembled the rate of currents found in parallel resistances of different magnitude. Overall the simulations yield an understanding of current distribution in the overlap joint and its variance introduced by weld placement or different materials.

4. Metrological Investigation of Resistances of Laser Welded Joints

The experimental part of the investigation focused on the comparison of different welding characteristics and geometries on overlap joints. The equipment consisted of a laser welding machine, a micro-ohmmeter and a custom test bench. The laser machine was an IPG YLR 1000 SM, single-mode fibre laser with a maximum emission power rating of 1 kW. The ohmmeter was a LoRe precision micro-ohmmeter from Werner Industrielle Elektronik and had a resolution of 1 nΩ in low measurement ranges starting at 10 nΩ. The custom test bench was designed to measure overlap joints in a manner to obtain both the joints resistance and CQI in respect to the materials, while retaining a maximum standard deviation of ±45 µm in the measurement tips' placement.

The investigation included the survey of different welding parameters and of different weld geometries altogether. The specimens were fabricated from aluminium Al99.5 and copper Cu-ETP metal strips of dimensions 20 mm width, 85 mm length and 0.3 mm strength. The specimen geometry and material were based on a connection of pouch cell batteries. As per Figure 6 the joints overlap s_J was dimensioned to be of 10.5 mm length.

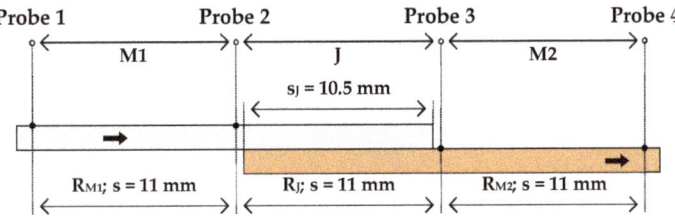

Figure 6. Schematic set-up of measuring the specimen.

The measuring sections span over s = 11 mm; the deviation between measurement and actual joint distance (s and s_J respectively) was eliminated mathematically post measurement. The produced joints included Al–Al, Cu–Cu and Al on Cu joining; Cu on Al joints were averted due to high instabilities in the welding process. The differences in material properties and the occurrence of intermetallic phases led to weld defects such as cracks.

The investigation of different welding parameters was conducted on the geometry of a central weld. Variances were introduced in respect to weld length across the joint, weld width along the joint and for the case of Al–Cu joints the induction of resistive intermetallic phases by altering the laser's power. The joint's resistance progressively increased with a reduction of weld length (Figure 7).

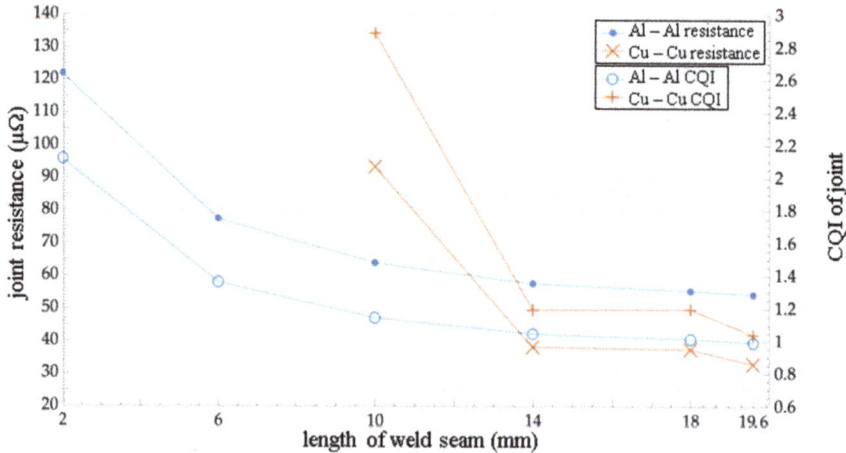

Figure 7. Resistance and *CQI* of the central weld joints with varying length.

Although the variation of weld widths introduced instabilities in the welding process, an overall slight resistance reduction was also measured with the increase of weld width. The intentional introduction of intermetallic phases was to test the measurability of such. The measured resistances did indeed show a dependency to the introduced laser power and hint a local minimum for optimal parametrization (Figure 8).

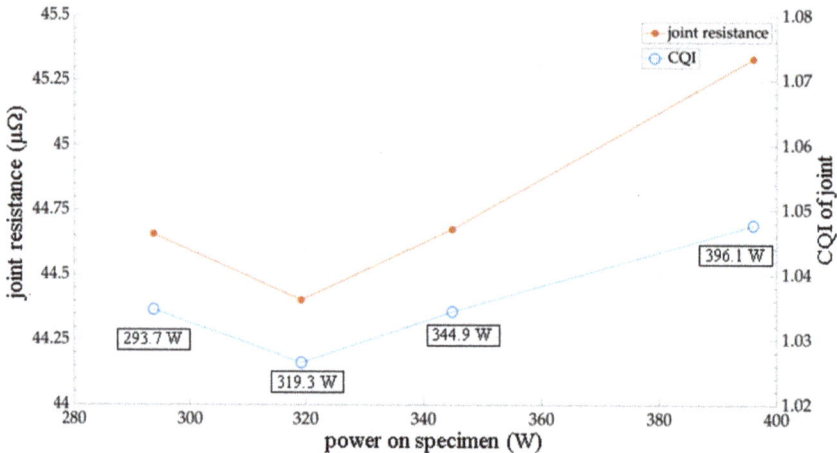

Figure 8. Resistance and *CQI* of central weld joints with varying laser powers and hence intermetallic phases ($v_f = 100$ mm/s, $A = 0.2$ mm, $f = 1000$ Hz).

The weld geometries and their measured CQIs are presented in Table 1. They were categorized by central, long-side and double geometries; additionally, asymmetrical geometries for Al–Cu joints were examined. The values for each configuration indicate the *CQI* for each material combination Al–Al (A), Cu–Cu (C) and Al–Cu (M). The long-sided and the sawtooth welds were only tested with a copper–copper connection as they were not expected to yield significantly different results across material combinations. The double weld with pattern and the asymmetrical configurations were likewise only implemented as an aluminium–copper connection.

Table 1. CQIs of different weld geometries for Al–Al (A), Cu–Cu (C) and Al–Cu (M). Laser parameters: amplitudes $A_A = A_C = A_M = 0.2$ mm unless specified; frequencies $f_A = f_C = f_M = 1$ kHz; feed rates $v_{f,A} = v_{f,C} = v_{f,M} = 100$ mm/s unless specified and power (on specimen) $P_A = 243$ W, $P_C = 498$ W, $P_M = 294$ W.

Joint	Central	Long-Side	Double	Asymmetrical
	Single	Single	Double	Shifted double
A	0.9924	-	0.5250	-
C	1.0010 [1]	1.1831 [3]	0.5220 [3]	-
M	1.0051 [2]	-	0.5321	0.5809
	Close double	Double	Double with pattern	Shrunk double
A	0.9713	-	-	-
C	1.2271 [3]	1.17781 [3]	-	-
M	0.9824	-	0.5436	0.5550 [4]
	Sawtooth			Arrow
A	-			-
C	0.9946			-
M	-			0.9414

[1] $A = 0.1$ mm, $v_f = 120$ mm/s. [2] $A = 0.15$ mm and $v_f = 120$ mm/s. [3] Process produced unexpectedly high resistances. Hypothesis: Higher thermal impact affected material deformations and hence elevated frictional contact resistances.
[4] Values after correction for inhomogeneous current flow. Raw measurements show higher resistance for single and lower resistance for double long-side welds.

As presented in the table, a single central weld could reach sufficient contact quality with respect to the reference materials. All material combinations could achieve values close to a CQI of 1, emulating the conductance of the materials. All long-sided joints show a higher CQI compared to the single central line. This may result due to inhomogeneous current densities, not utilizing the full extent of the available material as shown in Figure 9 below. The inhomogeneous current flow could be corrected by recalculating the inner two measurement point. Those measurement points were expected to be affected by the current flow as presented in Figure 8. The recalculation was done by taking the measured total resistance between the outer measurement points and subtracting the expected resistances of the metals on both sides. The expected resistances were calculated by averaging the respective measurements of the remaining, unaffected specimens. The resulting substitute resistances were expected to deliver more resembling and comparable estimates in cases of inhomogeneous current flow.

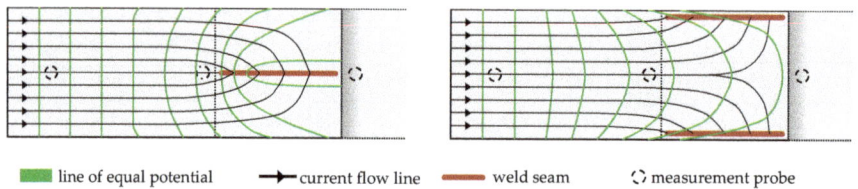

Figure 9. Schematic of proposed inhomogeneous current densities affecting voltage measurements.

By using double welds positioned on the joint's edges, the *CQI* was reduced to the near theoretical optimum value of 0.5. With this configuration, the available material was utilized as a parallel connection, effectively doubling the conducting cross sectional area. As seen with the double weld with pattern, a further increase of the connection area by an additional weld seam did not lead to a further reduction of the *CQI* for an aluminium copper connection.

Due to the materials' differing resistivity, the current was predominantly through the more conductive material, in this case copper. To determine the influence of the different material properties and the weld seam geometry, three asymmetrically configurations were investigated. By shifting the weld seam closer to the copper edge, an increase of the *CQI* was measured. A slight increase was also seen with a reduced weld seam length on the aluminium edge. By using an arrow geometry, the *CQI* was increased significantly due to the use of just one weld seam instead of two, but was still lower compared to a single central weld seam.

5. Discussion

During the investigation the model could be verified to represent the single and double joints, when initialized with representative data. However, it cannot represent more complex joints, where the conductor's cross-sectional areas did not predominantly carry strictly perpendicular currents. For the simulation of such joints a more complex three-dimensional model is required. The phenomenon of current distributions within the joint, as seen in the model, provides implications for the manufacturing of battery joints.

For applying a laser welding process in electrical applications, the results lead to following design guide lines. To achieve a *CQI* of 1, it is sufficient to apply a single weld seam along the whole width of the joining partners. Requirement is a stable welding process for contacting the joining partners. The width of the weld seam, rectangular to the current flow direction did not have a significant influence (compare central slim double line in Table 1). The measured values indicate a slight *CQI* reduction for the similar aluminium and the aluminium copper connection.

By greatly increasing the distance between the two weld seams the *CQI* was reduced to nearly 0.5 leading to the assumption that the position of the weld seams had a greater effect on the resistance than increasing the connection width and accordingly seam area with additional weld seams. That concludes that weld seams should be positioned as wide apart as possible to use the parallel connection of the two conductors. This measurement result is further supported by the simulation, which identified a predominant current flow through the joint's edges.

The identified dependency of the laser power on the resistance, leads to the assumption that an increased weld depth is not improving the *CQI*. The measurements showed higher values with increased laser power and therefore weld depths. The reason for this behaviour might be an increased mixing of the copper and aluminium, which leads to an increased occurrence of intermetallic phases. These phases inhibit the current flow in the weld seam and lead to an increase of the measured resistances. Furthermore, besides affecting the weld seam's resistance, it can supposedly reduce the materials' conductivity. As seen with the joint "Double with pattern" from Table 1, the addition of a pattern to the normal double joint increases the *CQI*. The introduction of further intermetallic phases deals a greater effect on conductivity, than the increase of welded area, in the joint's central area, less significant to the bridge-currents.

6. Conclusions

To reduce the electric loss in the connection of battery cells for electric vehicles, the joining process and the resulting transition resistance are essential. By introducing a model representing the joints' partial resistances, the current flow through the connection could be investigated. As a consequence, various joint geometries were investigated using a laser welding process to leverage and examine the observed edge current phenomenon. Naturally the current density along the material edges of the overlap, which are perpendicular to the current flow, was predominantly higher compared to the inner contact area. By measuring a *CQI* for each proposed connection, the influence of different weld seam geometries could be identified. Using double welds close to the edges of the overlapped materials yielded the investigations minimum *CQI* of about 0.52 and should hence be considered for manufactural purposes. The sheer increase of the welded contact area should be critically assessed, as it had a significantly smaller influence to the *CQI* and could even reduce the joint's conductivity.

Author Contributions: Idea, study design and laser welding were done by S.H. The development of the measurement system and measurements have been performed by S.K. The electrical modelling and theoretical background was contributed by C.Ü. A.O. contributed to the laser welding process and study design. A.G. and D.U.S. contributed to all parts. All authors have read and agreed to the published version of the manuscript.

Funding: This research received no external funding.

Conflicts of Interest: The authors declare no conflict of interest.

References

1. Shi, W.; Hu, X.; Jin, C.; Jiang, J.; Zhang, Y.; Yip, T. Effects of imbalanced currents on large-format LiFePO4/graphite batteries systems connected in parallel. *J. Power Sour.* **2016**, *313*, 198–204. [CrossRef]
2. Mauger, A.; Julien, C.M. Critical review on lithium-ion batteries: Are they safe? Sustainable? *Ionics* **2017**, *23*, 1933–1947. [CrossRef]
3. Katayama, S. Introduction: Fundamentals of laser welding. In *Handbook of Laser Welding Technologies*; Woodhead Publishing: Cambridge, UK, 2013; pp. 3–17.
4. Brand, M.J.; Schmidt, P.A.; Zaeh, M.F.; Jossen, A. Welding techniques for battery cells and resulting elextrical contact resistances. *J. Energy Storage* **2015**, *1*, 7–14. [CrossRef]
5. Poprawe (Hrsg.), R. *Tailored Light 2—Laser Application Technology*; Springer: Heidelberg, Germany, 2011.
6. Bliedtner, J.; Müller, H.; Barz, A. *Lasermaterialbearbeitung*; Carl Hanser: München, Germany, 2013.
7. Standfuß, J.; Rath, W.; Valentin, M.; Falldorf, H. Laserschweißen von Mischverbindungen. *Laser Tech. J.* **2011**, *8*, 24–26. [CrossRef]
8. Frielingsdorf, H.; Lintermann, F.J. *Elektrotechnik—Allgemeine Grundbildung*, 2nd ed.; Stam: Köln, Germany, 1993.
9. Schmalen, P.; Plapper, P. Resistance Measurement of Laser Welded Dissimilar Al/Cu Joints. *JLMN-J. Laser Micro Nanoeng.* **2017**, *2*, 189–194.
10. Schmidt, J.P. *Verfahren zur Charakterisierung und Modellierung von Lithium-Ionen-Zellen*; KIT Scientific Publishing: Karlsruhe, Germany, 2013.
11. Schmidt, P.A.; Schweier, M.; Zaeh, M.F. Joining of Lithium-Ion Batteries Using Laser Beam Welding: Electrical Losses of Welded Aluminum and Copper Joints. In Proceedings of the 31st International Congress on Applications of Lasers & Electro-Optics, Anaheim, CA, USA, 23–27 September 2012; p. 915.
12. Hügel, H.; Graf, T. *Laser in der Fertigung*; Vieweg + Teubner: Wiesbaden, Germany, 2009.
13. Schmitt, F. *Laserstrahl Mikroschweißen mit Strahlquellen hoher Brillanz und örtlicher Leistungsmodulation*; RWTH Aachen University: Aachen, Germany, 2012.
14. Häusler, A.; Mehlmann, B.; Olowinsky, A.; Gillner, A.; Poprawe, R. Efficient Copper Microwelding with Fibre Lasers using Spatial Power Modulation. In *Lasers in Enginneering 36*; Old City Publishing: London, UK, 2017; pp. 133–146.
15. Schmidt, P.A. *Laserstrahlschweißen elektrischer Kontakte von Lithium-Ionen-Batterien in Elektro- und Hybridfahrzeugen*; Technischen Universität München: München, Germany, 2015.

16. Gedicke, J.; Olowinsky, A.; Artal, J.; Gillner, A. Influence of temporal and spatial laser power modulation on melt pool dynamics. In Proceedings of the 26th International Congress on Applications of Lasers & Electro-Optics, Orlando, FL, USA, 29 October–1 November 2007; pp. 816–822.
17. Mys, I.; Schmidt, M.; Eßer, G.; Geiger, M. Verfahren zum Schweißen Artungleicher Metallischer Fügepartner, Insbesondere von Aluminium-Kupfer-Verbindungsstellen. Deutschland Bayern Patent Application No. DE102004009651B4, 9 October 2008.
18. Mercier, D.; Mandrillon, V.; Volpi, F.; Verdier, M.; Bréchet, Y. Quantitative evolution of electrical contact resistance between Aluminum thin films. In Proceedings of the 2012 IEEE 58th Holm Conference, Portland, OR, USA, 23–26 September 2012.

© 2020 by the authors. Licensee MDPI, Basel, Switzerland. This article is an open access article distributed under the terms and conditions of the Creative Commons Attribution (CC BY) license (http://creativecommons.org/licenses/by/4.0/).

Article

Cell Replacement Strategies for Lithium Ion Battery Packs

Nenad G. Nenadic *,†, **Thomas A. Trabold** † and **Michael G. Thurston** †

Rochester Institute of Technology, Rochester, NY 14523, USA; tatasp@rit.edu (T.A.T.); mgtasp@rit.edu (M.G.T.)
* Correspondence: nxnasp@rit.edu; Tel.: +1-585-662-8250
† These authors contributed equally to this work.

Received: 13 May 2020; Accepted: 17 July 2020; Published: 23 July 2020

Abstract: The economic value of high-capacity battery systems, being used in a wide variety of automotive and energy storage applications, is strongly affected by the duration of their service lifetime. Because many battery systems now feature a very large number of individual cells, it is necessary to understand how cell-to-cell interactions can affect durability, and how to best replace poorly performing cells to extend the lifetime of the entire battery pack. This paper first examines the baseline results of aging individual cells, then aging of cells in a representative 3S3P battery pack, and compares them to the results of repaired packs. The baseline results indicate nearly the same rate of capacity fade for single cells and those aged in a pack; however, the capacity variation due to a few degrees changes in room temperature ($\simeq \pm 3$ °C) is significant ($\simeq \pm 1.5\%$ of capacity of new cell) compared to the percent change of capacity over the battery life cycle in primary applications ($\simeq$20–30%). The cell replacement strategies investigation considers two scenarios: early life failure, where one cell in a pack fails prematurely, and building a pack from used cells for less demanding applications. Early life failure replacement found that, despite mismatches in impedance and capacity, a new cell can perform adequately within a pack of moderately aged cells. The second scenario for reuse of lithium ion battery packs examines the problem of assembling a pack for less-demanding applications from a set of aged cells, which exhibit more variation in capacity and impedance than their new counterparts. The cells used in the aging comparison part of the study were deeply discharged, recovered, assembled in a new pack, and cycled. We discuss the criteria for selecting the aged cells for building a secondary pack and compare the performance and coulombic efficiency of the secondary pack to the pack built from new cells and the repaired pack. The pack that employed aged cells performed well, but its efficiency was reduced.

Keywords: capacity fade; secondary applications; end-of-life; cell balancing; temperature effects

1. Introduction

Large lithium-ion battery packs are emerging in both vehicular and stationary energy storage applications, with rapidly increasing market penetration expected in the coming decades. The extent of battery system commercialization in both vehicle and renewable energy applications will depend upon the environmental and economic benefits that can be realized relative to incumbent technologies and other advanced mobility and energy technologies, such as fuel cells [1]. The effective cost of battery systems can be reduced by amortizing the cost over longer usage cycles. Two ways to extend the usage cycle of battery systems are (1) to extend the life of cells and packs in the original application, and (2) to reuse cells for other applications. For example, several studies have indicated that the cost of plug-in hybrid vehicle battery packs may be offset by repurposing vehicle batteries in grid support systems [2], and some automotive original equipment manufacturers (OEMs) are actively pursuing this option with energy technology companies [3,4]. For vehicle applications, Marano et al. [5] built a general, high-level model and, based on conservative assumptions, estimated that vehicles equipped with lithium ion batteries can last up to 10 years and provide the equivalent of 150,000 miles of travel. For battery packs that have failures or significant capacity loss prior to reaching the expected life-cycle, some means of recapitalization of the cell value is important to the overall cost to benefit ratio. Although many studies have addressed fundamental degradation modes of common battery electrode materials, these studies are often conducted at the "button cell" or single-cell 18650 scale where effects of assembly, packaging, and integration are not fully comprehended. The importance of acquiring a detailed understanding of cell aging, individually and in packs, has been recognized previously, and much recent research has focused on techniques for battery health monitoring and prognostics of battery packs in electric vehicles (e.g., review articles by [6–8]). Designing and implementing strategies for first identifying and then isolating failures of individual cells within a pack is challenging, although some potential methods have been proposed [9–13]. Recently, Li et al. [14] proposed three categories of approaches for multicell state estimation:

- treating the battery pack as a single cell of high voltage and capacity;
- applying single-cell *state-of-charge* (SOC) estimation methods to every cell in a pack, but this approach is computationally intensive and cumbersome for practical application;
- quantifying individual cell SOC by analyzing variations in open circuit voltage and internal resistance.

The difficulty in assessing and comparing many of these advanced battery system-level monitoring approaches is that direct, in-situ data from electric vehicles or storage systems are not readily available in the open literature. Therefore, many researchers have relied on experimental and modeling studies that start with simpler multi-cell systems and then attempt to extrapolate these findings to more complex, commercial-scale systems. Dubarry et al. have reported on cell aging and the degradation mechanisms of a composite positive electrode [15,16]. Understanding the origins of cell variations can be used for building more robust packs [17]. Moreover, it is well known that multi-cell (pack) aging behavior can be quite different from that associated with single cells, due to the need for cell balancing and thermal management, among other effects [18,19]. Thus, it is important to first fully characterize aging behavior at the individual cell level as a function of the pertinent operating parameters and for different electrode materials [20]. The cell characterization can be used for accelerated estimation of remaining capacity and state of charge [21].

In the current research program, after quantifying the aging of individual LiCo 18650 cells at a statistically significant level, the evaluation process was systematically extended to small packs which represent small-scale versions of larger commercial battery systems. Pack-level testing was intended to gain insight into a variety of practical issues associated with commercial battery systems. The selected pack was a 3×3 cell arrangement (three cells are connected in series to form a string and then three strings are connected in parallel, i.e., 3S3P configuration), with its associated charging and discharging processes, and enabled comparison of aging of cells in the pack versus individual cell aging. The replacement strategies considered two scenarios.

The first scenario, the replacement of an early life failure, addresses an important open question for maintenance of battery packs. The traditional approach in pack maintenance is to replace all cells at once to control the mismatches. This approach is clearly untenable for very large battery packs. Even for packs built in a hierarchical fashion, where cells are first assembled into sub-modules, which, in turn, form larger modules, this replacement philosophy does not work because replacement of a single cell in a module would require replacement of all the cells in the module, and, by extension of this approach, all the sub-modules, etc. Replacement of all cells as a result of an early-life failure in a large pack is clearly not economically viable; therefore, an alternative strategy needs to be established. One strategy for minimizing imbalance and premature aging in this scenario is to maintain an inventory of cells aged to different levels of capacity fade and to select the appropriately aged cell, or cells, to effect the repair. The experimental results reported here have been obtained on a small pack, which is a module that could be used in larger packs. Since larger packs are built hierarchically, where modules are often treated as larger cells, the conclusions of this study should provide important insight into the behavior of larger packs as well.

The second scenario addresses the problem of secondary uses for the cells in a less demanding application after the end of useful life in a higher-performance application. These cells, while no longer suitable for the original applications, may be deemed adequate for less demanding applications. For example, studies have indicated that the cost of plug-in hybrid vehicle battery packs may be reduced by repurposing vehicle batteries in grid support systems because modules would be sold to the secondary user and the primary user would not have to assume the processing cost associated with safe disposal [2,22,23]. For example, Schneider et al. [24], who developed methods for assessment and reuse of nickel metal hydride (NMH) cells, found that, on average, about 37% of discarded cells have sufficient remaining capacity for reuse. Lih et al. [25] identified technological challenges and analyzed secondary uses of lithium ion batteries from an economic point of view. The potential of lithium ion batteries, after they serve their useful life, for grid applications has been considered by Kamath of the Electric Power Research Institute (EPRI) [26], who hypothesized that cost per production volume may be lower than lead-acid batteries. Neubauer et al. have performed a techno-economic analysis of vehicle batteries for secondary uses [27–29]. They concluded that an uninterruptible power supply (UPS) system based on used lithium ion batteries has the potential to be more cost effective than lead-acid batteries, with superior longevity, specific energy, and energy density. The considerable potential for secondary applications has been widely recognized. The present study examines empirically practical problems of maintenance and rebuilding of packs using a small 3 × 3 pack as the platform.

2. Methodology

The objective of the empirical study was twofold: to compare aging of lithium ion cells individually and in small packs and to test investigate cell rebuilding packs in the context of two case studies. This objective was executed by three sets of tests: the first set compared aging of single 1860 LiCo cells to their aging in small packs, the second set examined performance of a used pack after one of the cell was replaced by a new cell and the third set examined scenarios of rebuilding packs from two packs that were considered "failed". The details of individual set of tests are described in subsequent sub-sections. The main metrics were capacity fade, impedance changes, and coulombic efficiency.

The study employed two test stands: one for single-cell testing and the other for battery pack evaluations. The single-cell apparatus was a Maccor 4600 battery test system, used for initial cell characterization, pre-aging of individual cells, and periodic monitoring of the cells subjected to pack-level cycling. Additional details on the test procedure are provided in the supplementary material.

The test stand for pack testing is shown in Figure 1a. The packs consisted of nine cells in 3S3P configuration: three strings of three serially connected cells were connected in parallel, as shown in Figure 1b. The pack employed passive cell balancing, based on the commercial, off-the-shelf balancing circuit. A diagram of the test stand is depicted in Figure 1c. More details on cell balancing are provided in the supplementary material.

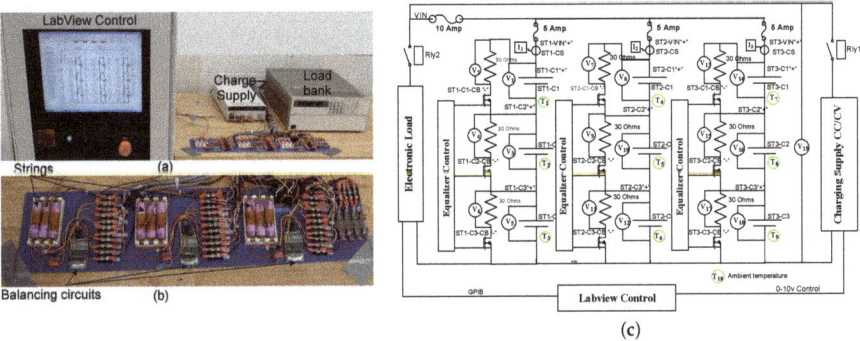

Figure 1. 3S3P pack (**a**) LabVIEW-controlled fixture; (**b**) enlarged view of the pack; (**c**) schematic.

2.1. Individual and Pack Aging of 18650 LiCo Cells

For this study, cell aging was limited to full charge-discharge cycles of battery cells. A typical full charge-discharge cycle, shown in Figure 2a, consisted of discharge at 2.33 A ($1Q_n$) and the best C-rate (I_c = 1.63 A, or $0.7Q_n$) – constant voltage (at V_c = 4.2 V) charge, separated by 20-min rest periods. Q_n denotes nominal capacity of a new cell. The tests were conducted at ambient temperature, but the temperature was monitored in all tests. The motivation for operating both cells and pack at ambient temperature came from many practical pack implementation, which do not control the ambient temperature. The cells were operated in the same environment as the packs. The cells in a pack (see Figure 1) were aged essentially in the same way, as illustrated in Figure 2b.

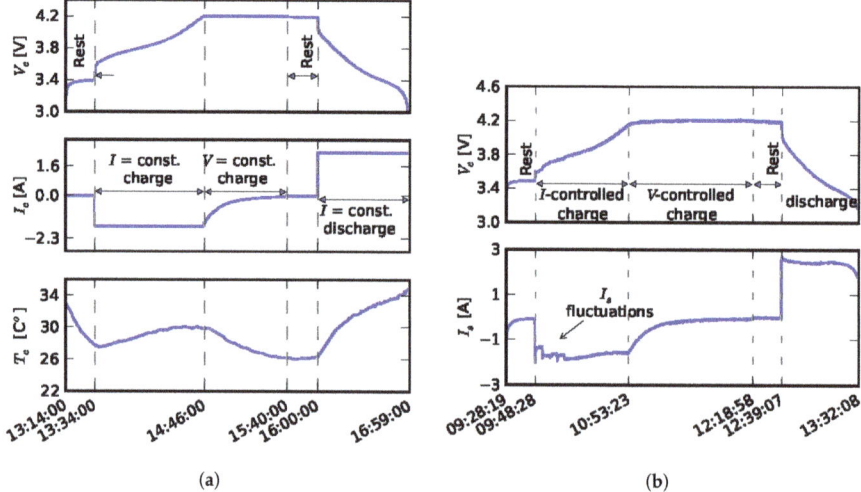

Figure 2. Waveforms associated with charge-discharge cycle (**a**) single-cell testing; (**b**) pack testing.

To compare pack aging of cells with individual cell aging, we aged sixteen cells on the Maccor single-cell tester and nine cells in a pack. Figure 3a shows the capacity fade of individually aged cells vs. number of cycles, measured at discharge. Each of the sixteen cells was aged until its capacity faded to 90% of new (i.e., 10% capacity fade). The bottom plot of Figure 3a displays a histogram of number of cycles that led to 10% capacity fade. The average rate of capacity fade for the group of sixteen cells was computed from all the measurement points using the least mean squares, and the result (−0.11%/cycle) is indicated in the plot. As shown in Figure 3b, capacity fade of cells aged in a pack had a very similar degradation rate (−0.1%/cycle).

Note that the capacity fade profiles are not strictly monotonic but generally display multiple local maxima and minima. These variations are largely due to sensitivity of capacity to relatively small variations in the ambient temperature, as can be seen in Figure 4. The cell temperature at the end of discharge (the red dashed trace of the bottom of Figure 4a) follows the ambient temperature (the orange trace of the bottom of Figure 4a). Moreover, very strong linear correlation coefficient of 0.99 was observed between the change in temperature at the end of discharge over two subsequent cycles ΔT_{ed} and the change in capacity at the end of discharge Q_{ed} over two subsequent cycles, as seen in Figure 4b. These results strongly suggest that temperature management is very important in practical applications for which it is not reasonable to maintain the ambient temperature at a fixed value. However, controlling the range of this variability is very important. The temperature variation between cycles can be perceived as noise. From Figure 4b, the change in capacity was 0.46%/°C. Recall that the degradation rate described above was 0.11%/cycle. Thus, it is seen that the "error" in capacity due to change in ambient temperature of only 1 °C was more than four times larger than normal cycle-to-cycle degradation. Temperature variation may affect internal resistance, which, in turn, affects the terminal voltage and effective SOC. Figure 4c zooms into temporal variation of the capacity for an individual cell. The inset shows a histogram of the coulombic efficiency, defined as the ratio of charge at the end of discharge and the end of charge, $\eta = Q_{cd}/Q_{cc}$. It is important to note that coulombic efficiency was used in this study and not the energy efficiency (the ratio of total energy during discharge and charge). In many places in the text below, we state this explicitly, by referring to it as coulombic efficiency. Sometimes, we refer to it simply as efficiency, but this paper considers only coulombic efficiency. The histogram shows that, within a single cycle, the efficiency can be even higher than 100%. This means that capacity measured during charge (EOC) is sometimes lower than capacity measured during discharge (EOD). Of course, this efficiency could not be sustained over many cycles; however, the average efficiency was very high over a range of 40 cycles (99.85% average efficiency over cycles 20 to 60).

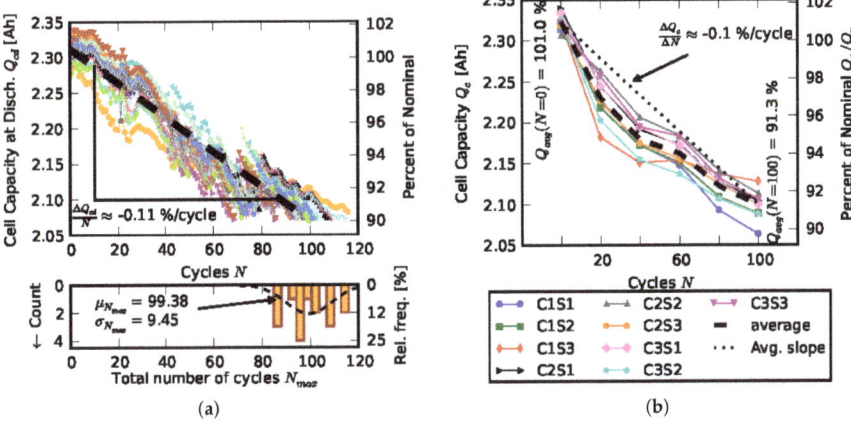

Figure 3. (a) capacity fade (100–90%) as function of number of cycles; (b) capacity fade of nine cells within a pack over 100 cycles.

A common health indicator of battery aging is impedance [13,30–35]. Figure 5a shows the average impedance spectrogram of new cells, and of the same cells after they have been aged to 90% of their nominal capacity. Several key frequencies are indicated by arrows. The standard frequency used for health indication is f = 1 kHz. For this lithium ion chemistry, the resistance at f = 1 kHz corresponds to the high-frequency intercept with the real axis. This resistance, denoted by $\Re\{\underline{Z}_{Batt}\}$, is approximately equal to the ohmic resistance of the battery [36]. $\underline{Z}_{Batt}$ has a convenient equivalent circuit representation and is relatively easy to extract [37], but it is not the most sensitive parameter for indirect monitoring of cell aging. Figure 5b shows the change of the real part of the impedance at 1 kHz for sixteen cells as they are aged to 90% of their new capacity, with the colored markers indicating individual measurements, and dashed lines indicating the shape of the fitted normal distributions. The real part of the impedance at f = 1 kHz is the standard metric for the impedance; the impedance spectra of Figure 5a indicates that, for these cells, larger impedance change occurred at lower frequencies, in the [0.1, 1] Hz range. This indication is further confirmed in Figure 5c, which illustrates the change in the real part of the impedance for the same group of sixteen cells at f = 1.0 Hz. The absolute change of the mean resistance, averaged over sixteen cells, at f = 1.0 Hz was 4.12 mΩ (7.6%) compared to a 0.64 mΩ change in resistance (1.6%) at f = 1.0 kHz. The disadvantage of measuring impedance at f = 1 Hz, at the corner of Warburg region, is that measurements take longer.

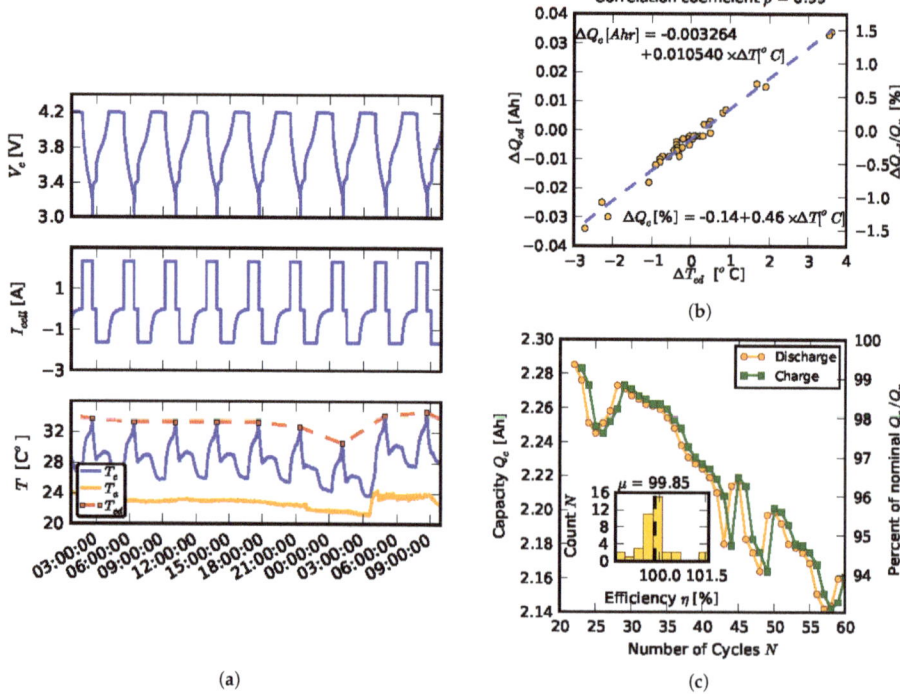

Figure 4. (a) characteristic waveforms for a single cell: voltage V_c, current I_c, temperature T_c, and ambient temperature T_a; (b) scatter plot of change in capacity at discharge in subsequent cycles ΔQ_{ed} vs. temperature of the cell at the end of discharge ΔT; (c) more details on capacity variation of a single cell over time.

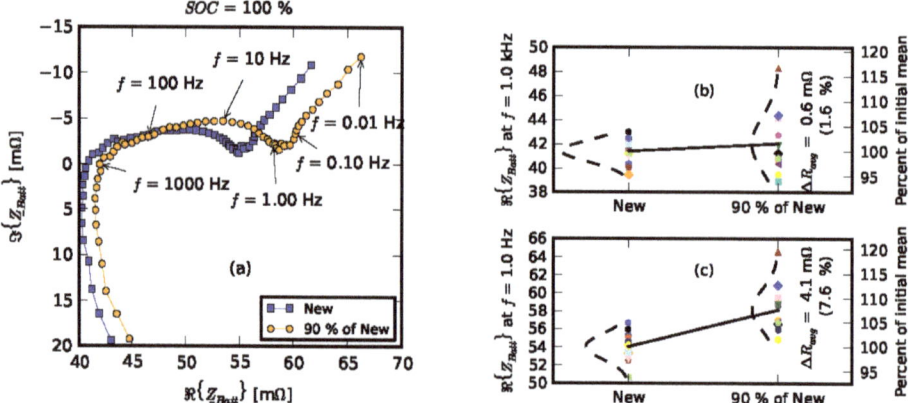

Figure 5. Impedance changes during single-cell aging. (a) average impedance spectra; (b) real part change for individual cells at $f = 1$ kHz; (c) real part change for individual cells at $f = 1$ Hz.

The impedance spectra of pack aging data are given in Figure 6: Figure 6a shows the evolution of averaged spectra, Figure 6b shows the evolution of distributions of impedance real parts of individual cells at $f = 1$ kHz, and Figure 6c shows the evolution of real part of impedance at $f = 1$ Hz.

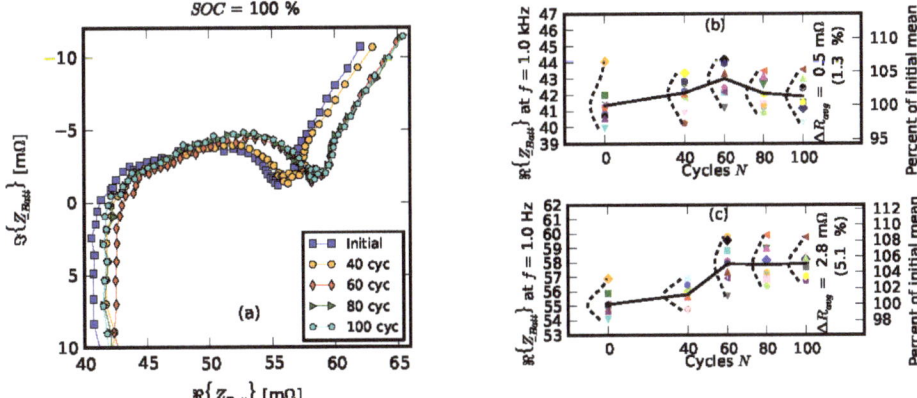

Figure 6. Impedance changes during pack cell aging. (**a**) average impedance spectra; (**b**) real part change for individual cells at $f = 1$ kHz; (**c**) real part change for individual cells at $f = 1$ Hz.

2.2. Scenario 1: Early Life Failure

The first scenario represents the case when one cell in a pack fails early with respect to the expected life of the pack. To simulate this situation, we pre-aged cells in the individual pack tester to 90% of their initial capacity. In demanding applications, such as electric vehicles, the cells are considered usable only when the capacity is higher than 80% of the nominal capacity. Therefore, for these applications, cells aged down to 90% of their nominal capacity are less than 50% of their useful life because cell degradation is nonlinear and typically slows down (see Figure 7).

Nine individually-aged cells were formed into a pack (3S3P), where the cells were matched based on their capacity and impedance. Then, after a few cycles, one of the aged cells was replaced by a new cell. Figure 8 shows capacity of individual cells first aged separately until their capacity faded to 90% of nominal, and then assembled into a pack. Individual cell capacities are considered here as the key metric because they exclude effects of other pack components, such as equalization circuit. The traces are labeled by the location of the cell in the pack CiSj where i denotes the cell within a string and j denotes the string. For example, C1S2 is the first cell in the second string. The cells were always placed in the same location throughout the testing, except in the case of the C1S1 site, where the replacement took place. The original cell is denoted by C1S1o and the replacement cell by C1S1r.

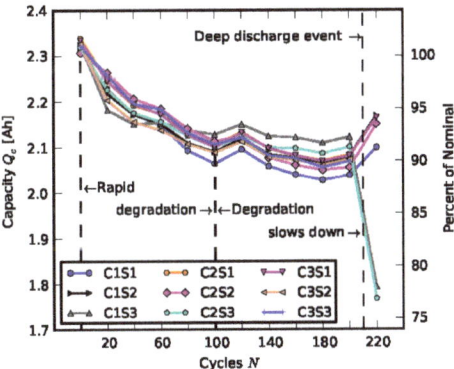

Figure 7. Individual capacities of the baseline pack comprised of new cells.

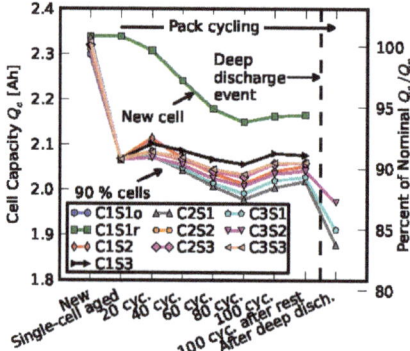

Figure 8. Capacities of individual cells in an early life failure scenario where one cell C1S1 is replaced with a new cell after 10% capacity fade. Traces are labeled by the site of the cell in the pack.

Comparing capacity fades of the cells of the pack, comprised of eight aged and one new cell to the capacity fades of the reference pack, plotted in Figure 7, shows that a new cell nicely coexists with the aged cells and did not introduce any obvious increase in the overall rate of capacity fade. The slope of the capacity fade of the new cell is higher than that of the aged cells, but this behavior is consistent with cell aging in general. Capacities of the new cells in Figure 7 fade faster over the course of the first 100 cycles and the relative rate of capacity degradation slows down.

2.3. Scenario 2: Rebuilding a Pack from Two Failed Packs

The pack with eight pre-aged cells and one new was behaving well for one hundred cycles before it was subjected to deep discharge to generate cells for the second scenario. A deep discharge event is a very severe case of cell degradation and it was induced here to simulate a harsh case of field failure because future integrators of packs for secondary applications may not have access to the usage history of cells in their primary applications.

The conventional knowledge in battery integration systems has been that new cells should never be mixed with old cells. Before conducting this experiment, it was suggested by some domain experts that the pre-aging cells for replacement may be required for reliable operation of the repaired pack. Our results suggest that the potentially expensive proposition of maintaining an inventory of pre-aged cells may not be necessary.

To create the second scenario for cell replacement strategy, two 3S3P packs were first subjected to deep discharges, as shown in Figures 7 and 8. The deep discharge events caused what can be considered to be major pack failures, and the cells recovered from these deeply discharged packs are good candidates for simulating cell repurposing processes. To simulate a failure in the battery management system, the cells were left overnight to discharge through a set of resistors used for cell balancing, allowing the terminal voltages to drop considerably below the minimum value required by the cell manufacturer. This scenario intended to mimic one of the worst-case practical situations because the operational history of cells in primary application, including failures, is not available to the integrator of the packs based on used cells. After the deep discharges, the individual cells were recovered by charging them at low current (100 mA). This process was able to recover twelve out of the original eighteen cells. The remaining six cells had their current interrupt devices triggered which rendered them unusable for the Scenario 2 experiments. After the recovery, the cell capacity was measured on the single-cell tester. Table 1 shows the capacities of the recovered cells.

Table 1. Capacity of the recovered cells.

Cell ID	Cell Capacity Q_c (Ah)
6	1.83
8	1.85
14	2.11
16	2.11
17	2.06
19	2.13
20	max→ 2.14
21	2.12
23	1.73
24	1.69
25	min → 1.58
27	2.02
μ	1.95
σ	0.20

Table 2 shows the real part of the impedance of the surviving cells at frequencies of 0.1, 1 and 1000 Hz. In the first part of the study, we found that resistance at 1 Hz showed more sensitivity to aging than the typically used value at 1 kHz.

Table 2. Resistances of the recovered cells.

Cell ID	$R_{f=0.1\,Hz}(m\Omega)$	$R_{f=1\,Hz}(m\Omega)$	$R_{f=1\,kHz}(m\Omega)$
6	88.18	77.98	49.01
8	90.24	80.41	max→52.23
14	64.85	62.12	44.00
16	72.07	70.09	51.55
17	72.42	69.55	50.27
19	64.47	61.74	43.56
20	min→62.61	min→60.64	44.74
21	64.15	61.54	43.56
23	max → 113.70	max→86.66	min→43.41
24	99.37	79.64	49.14
25	105.02	83.75	49.15
27	69.98	65.81	45.65
μ	80.59	71.66	47.19
σ	17.96	9.56	3.36

The relationship between the capacity and real part of the impedance at 1 Hz has a high correlation coefficient of $\rho = -0.92$ (Figure 9). The high correlation between resistance and capacity was expected (see e.g., [38]). While we recognize that there are several degradation mechanisms in lithium ion cells (including degradation of active material, impedance rise by formation of solid-electrolyte interphase layer, lithium inventory loss by side reactions, and loss of carbon as conductive additive from the cathode [39]), the impedance change was the dominant feature that was readily detectable in our phenomenological approach. It is reasonable to suggest that some of the scatter is due to measurement of the impedance spectra. The ability to reasonably assess cell condition from impedance is very important for rebuilding a pack because the capacity measurement is a considerably longer process. The real part of the resistance at $f = 1$ Hz can be measured within seconds, whereas determining the capacity at 1 Q_n may take several hours, depending on cell capacity and the rest period. More accuracy in capacity estimation can be achieved by taking more data and applying considerably more computation [2].

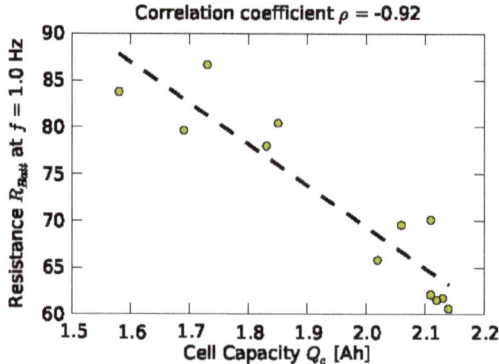

Figure 9. Capacities of individual cells in early life failure scenario.

As stated above, negative correlation between remaining capacity and impedance increase has been observed (e.g., [38]). An equivalent circuit model, such as double-exponential model [37], depicted in Figure 10, provides an intuitive, phenomenological way to provide simple, first-order interpretation. The model consists of two electrical ports: hidden and terminal. The state of the hidden port, model by capacitance, and $\dot{Q}$ denotes the rate of charge. It contains a current-dependent current source, which is simply unity multiplied by the cell current, cell capacity C_{cell}, and self-discharge resistance R_{sd}. The terminal port has two directly measurable quantities, viz. battery terminal cell voltage V_c and cell current I_c. The circuit components of the terminal port, open-circuit voltage V_{oc}, R_{cell}, R_1, C_1, R_2, and C_2 are not directly observable and have to be inferred [21]. In addition, they are functions of charge level Q. At low frequencies, the impedance approaches its real (purely resistive component) component $\underline{Z}_{Batt} \to \Re\{\underline{Z}_{Batt}\} = R_{cell} + R_1 + R_2$ as $f \to 0$. During the discharge at a constant current, the impedance is approximately purely resistive. As this resistive part increases, so does the effective voltage drop across it, which effectively reduces the terminal cell voltage V_c, whose level is used to limit the charge and discharge, based upon manufacturer's specifications. Thus, increase in $R_{cell} + R_1 + R_2$ effectively reduces discharge capacity. The impedance at 1 Hz strongly depends on state of charge [40], which was accounted for by measuring impedance at the same level of SOC. However, temperature also significantly affects the low-frequency impedance. This aspect was only accounted for in a statistical manner, by comparing the two sets of distributions, which seemed to be separated significantly.

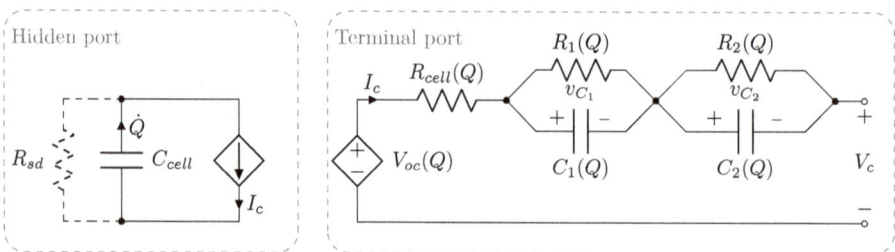

Figure 10. Equivalent circuit model (adapted from [37]).

There are many ways to arrange the 12 recovered cells into a 9-cell pack. One approach is to maximize the capacity of the pack. This arrangement is achieved simply by first grouping the cells in descending order with respect to their capacity and then populating the pack across the strings starting from left to right in the first pass (where "left-to-right" signifies arbitrary signed direction perpendicular to the direction along the strings), and then continuing from right to left in the second pass, etc. For the case at hand, only three such passes are required. The resulting pack configuration is shown in Table 3.

Table 3. Cell and string capacities (in (Ah)) for Option 1) that maximizes total pack capacity.

Configuration	String 1	String 2	String 3
C_1	2.14	2.13	2.12
C_2	2.06	2.11	2.11
C_3	2.02	1.85	1.83
Total	6.22	6.09	6.06

An alternative approach is to arrange the cells in a manner that equalizes their string capacities. This arrangement minimizes string-to-string equalization, which, in turn, minimizes the losses during rest periods. While the previous arrangement only requires sorting, the string equalization is slightly more demanding. The arrangement shown in Table 4 was arrived at by employing an optimization procedure, which was in this case implemented in Python.

Table 4. Cell and string capacities (in (Ah)) for Option 2) that best match capacity among the three strings.

Configuration	String 1	String 2	String 3
C_1	1.83	2.11	2.06
C_2	1.85	2.12	1.73
C_3	1.85	1.58	2.02
Total	5.81	5.81	5.81

The second option, attractive from the efficiency viewpoint, proved less reliable for the pack. It turned out that the safety features embedded in our pack design were less tolerant to mismatches within a string then string-to-string mismatches. The cells with considerably lower capacity and higher resistance were difficult to balance. Thus, maximizing the capacity of the pack was found simpler, and more robust within our pack implementation.

3. Results and Analysis

While there are several important metrics for battery packs, this study focused on coulombic efficiency because it is strongly affected by mismatches among cells in packs built for secondary applications. We also consider temperature effects and the effect of cell balancing scheme, viz. passive vs. active cell balancing.

3.1. Efficiency Comparison

The pack performance is assessed with respect to its overall efficiency. Figure 11 shows a composite plot for pack efficiency. The top subplot shows the capacity at the end of discharge Q_{pd} and capacity at the end of charge Q_{pc}. Their ratio, pack coulombic efficiency η_p, defined as

$$\eta = \frac{Q_{pd}}{Q_{pc}} \quad (1)$$

and expressed in percent, is plotted on the bottom subplot. It is important to note that, while the pack was charged using constant current and constant voltage conditions, the discharge was conducted only in constant current condition. This approach is consistent with the single-cell charge-discharge profile and is reasonable from the pragmatic viewpoint of the user. The histogram of the efficiency is plotted in the right subplot in the horizontal direction, where the y-axis of the histogram is scaled to match the y-axis of the efficiency plot.

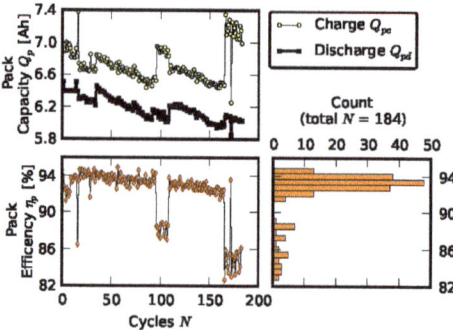

Figure 11. Coulombic efficiency of a pack during cycling (new cells).

The main mode of the distribution of the coulombic efficiency of a pack comprised of all new cells is approximately 93%. As mentioned above, the pack cycling was interrupted periodically to take capacity measurements of individual cells and to record impedance spectra. Sometimes, after the test was resumed, the pack may have operated at somewhat different global capacity. This explains the occasional large step-shaped drops in efficiency. In addition to these abrupt drops, one can see that the overall efficiency degraded slowly, at approximately the same rate as the cell capacity. The efficiency of individual cells was about 99% on average (see Section 2.1, Figure 4c). The additional energy loss was attributed to cell balancing.

Figure 12 shows the efficiency of the pack built for Scenario 1 (early life failure study). Comparing this figure to Figure 11, it appears that the efficiency corresponding to the dominant mode of the distribution was very comparable to that of the pack comprised of new cells.

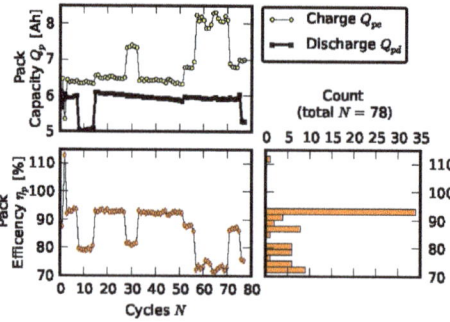

Figure 12. Coulombic efficiency of a pack during cycling (Scenario 1).

However, for the Scenario 1 pack, we observed more frequent abrupt drops in efficiency. One of the main indicators of cell aging is impedance increase [33]. Spectra of average impedance associated with the aged early life failure pack are displayed in Figure 13 in the usual way, with the x-axis being the real part of the impedance, $R = \Re\{\underline{Z}_{Batt}\}$, and the y-axis being the negative imaginary part of the impedance, $Y = -\Im\{\underline{Z}_{Batt}\}$. Both R and Y are expressed in mΩ, with the frequency ranging from 10 mHz to 10 kHz. The plotted impedance traces are averages of nine individual-cell impedance measurements. There is a total of ten sets of measurements:

- *Initial* signifies the measurements on new cells.
- *Few cycles* signifies the impedance of nine cells after they were initially run in the pack for three cycles.
- *10% fade* signifies the measurements on the cell after they were individually aged on the single-cell tester to 90% of their nominal capacity.

- *Replacement* signifies resistance measurements after one cell was replaced with a new cell; *+20 cycles*, *+40 cycles*, *+60 cycles*, *+80 cycles*, and *+100 cycles* signify measurements as the "repaired" pack was operated for 100 cycles.
- *Recovered cells* signify the impedance of a few cells that survived the deep discharge event.

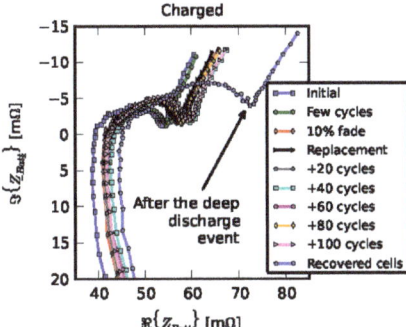

Figure 13. Average impedance spectra of a pack comprised of eight aged cells and one new (Scenario 1).

At a high level, the shifts in the impedance spectra were larger initially. During the aging phase (+20 cycles through +100 cycles), the shifts were relatively small. Larger shifts occur at the points that correspond to lower frequencies. While $R[f = 1 \text{ kHz}]$ changed significantly after the deep discharge event on average, two of the six surviving cells were only moderately affected. This observation suggests that even after relatively violent failures, a subset of cells from a pack may retain nearly the same characteristics as before the failure.

Figure 14 shows distribution of the real part of the impedance measurements at $f = 1$ Hz. Here, the history of the testing is shown in the form of the x-axis tick marks. The measurements of individual cells are labeled with unique markers. Nine individual cell measurements correspond to each x-axis tick mark. The solid line connects the means of the nine cells, while the dashed curves represent the fitted normal distributions. While average resistance increased significantly after the deep discharge event, two cells were barely affected, as noted above.

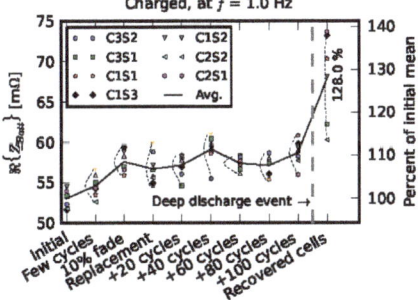

Figure 14. Real part of the impedance at $f = 1$ Hz at different stages of aging of early life failure pack (Scenario 1).

The coulombic efficiency of the rebuilt pack (Scenario 2) is shown in Figure 15. This pack clearly operates at lower efficiency than the new pack (Figure 11) and the early life failure pack (Figure 12). Its dominant mode was at 85% efficiency. The reduced efficiency was due to larger mismatches among the comprising cells which required more balancing because passive cell balancing circuits dissipate imbalanced charge on a resistor. This result suggests that an active cell-balancing scheme may have potential for the pack based on repurposed cells because the return on the investment for these packs is faster. We return to this point at the end of the section.

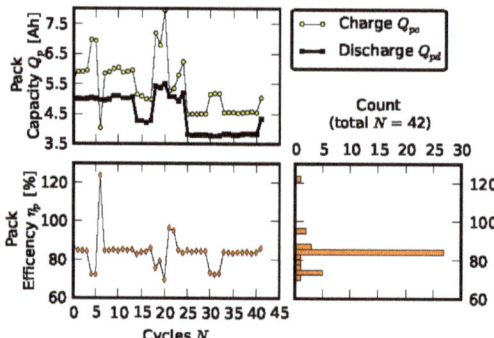

Figure 15. Coulombic efficiency of a pack during cycling (Scenario 2).

Figure 16 compares the histograms of coulombicic efficiency of the three packs: the pack composed of new cells, the pack with an early failure, and the pack rebuilt from used cells. The dominant modes of the pack composed of new cells and the pack with an early failure largely overlap, and the dominant mode associated with the "rebuilt pack" is noticeably lower.

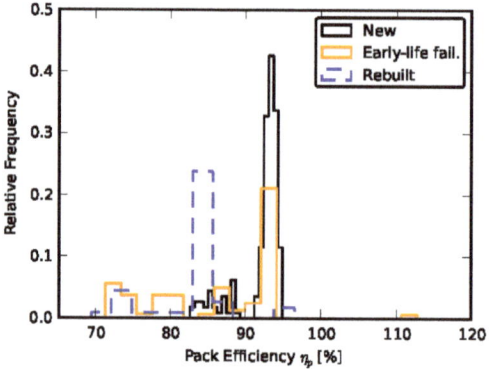

Figure 16. Overlaid coulombic efficiencies for the three packs.

3.2. Temperature Effects

Local cell temperature variations and overall pack heating are important concerns in pack design. Within the confines of this study, we examined how temperature of individual cells increased during pack operation.

Figure 17 shows the distribution of temperature differences between individual cells and ambient temperature for the three packs. All test temperature measurements are included. There were no significant pack-to-pack differences in heating of individual cells, but the distribution of temperature differences of the rebuilt pack was slightly wider than the other two temperature distributions. The objective of the time domain plot on the bottom is to show that the "tail" of the distribution was not associated with the last cycles. The plot aligns the last event where ΔT exceeded 10 °C for each of the three packs. The x-axis is temperature difference ΔT and was scaled to be the same as that of the histogram above; the y-axis is time. The peak corresponds to the end of discharge. The maximum ΔT of the bottom plot is considerably smaller than the maximum ΔT of the histogram. Thus, the tails of the histogram did not correspond to the aging, but to random variation. An alternative view of the data are provided in the supplementary file.

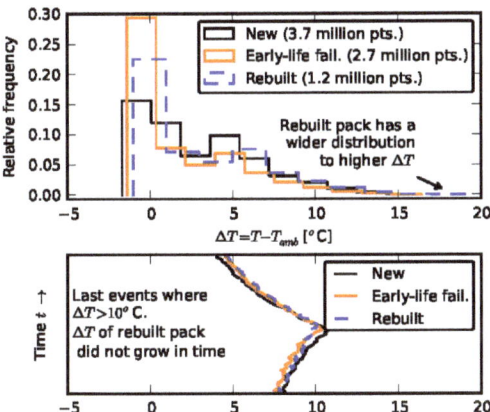

Figure 17. Comparison of heating of aging batteries.

3.3. Effects of Balancing Scheme

As stated above, the mismatched cells required more balancing, and increased balancing directly translated into reduced efficiency. Some of the lost efficiency could be recovered by employing active cell balancing for the repurposed packs. To demonstrate this empirically, we created a pack consisting of a single string of three new serially connected cells and cycled this simple pack employing either passive or active balancing. The passive balancing scheme was the same as that employed for the test described Section 2.1, with the details provided in the supplementary material.

The active balancing circuits employed a capacitor and solid state switches, as illustrated in Figure 18. The logic of the switches, denoted by symbols φ_1–φ_3, connected only one of the cells in parallel with the capacitor. The cells took turns with respect to their connection to the capacitor in a circularly cyclical manner, as shown by the sketched waveforms of φ_1–φ_3.

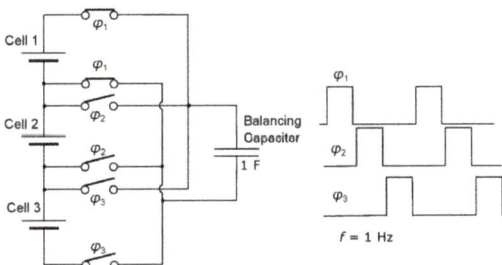

Figure 18. Circuit for active balancing of a 3S1P pack.

The data are shown in Figure 19a where total pack current waveforms are given in the top subplot and total charge associated with charge (bottom subplots) for a few charge/discharge cycles of the 3S1P pack while passive balancing scheme was used. The red and green colors were used to indicate charge and discharge cycles, while the blue color signifies the rest period. The efficiencies of individual cycles were denoted near the discharge points. The pack with passive cell balancing had 92.5% efficiency. Figure 19b shows the same information as Figure 19a when the pack employed active cell balancing.

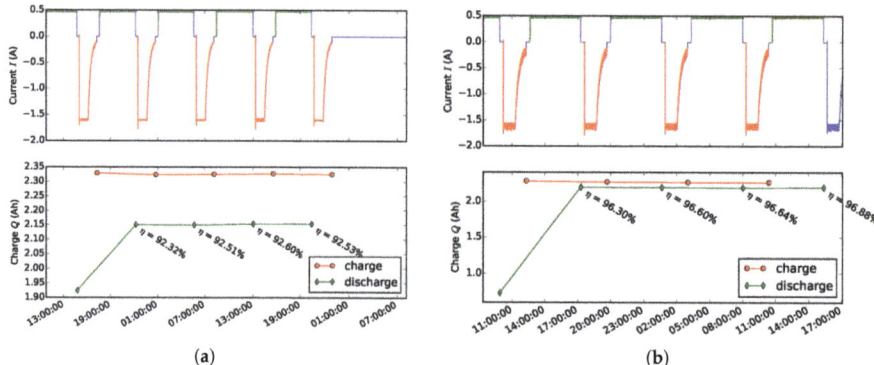

Figure 19. Comparison of cell balancing for a 3S1P (**a**) passive; (**b**) active.

The pack using the same cells and active balancing scheme had 96.6% efficiency, a 4.1% improvement over the passive balancing. In both cases, we ignored the first estimate because it would be based on a partial cycle. Because the average efficiency of new cells is almost 99.8% (see Section 2.1, Figure 4c), the improvement due to active balancing recovered more than half of the reduced efficiency, as illustrated in Figure 20.

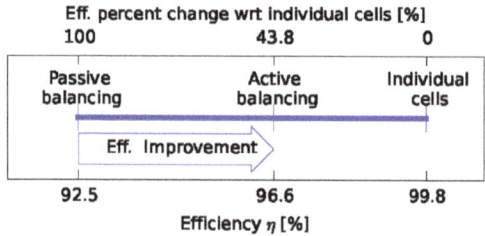

Figure 20. Efficiency of passive balancing, active balancing, and individual cells.

The remaining losses were due to the equalization process which required additional current flows between the cells and the capacitor and the resistive imperfections of real solid-state switches (their on-state resistance), real equalizing capacitor (its equivalent series resistance), and the cells' internal resistance. For example, during a charge cycle of an individual cell test, the current only flows into the cell, whereas during a charge cycle of the cell in a string, the current mostly flowed into the cell, but small amounts of current often flowed out of the cell during its connection to the capacitor. The balancing currents were completely dissipated if passive balancing circuit was used, and only partially dissipated in parasitic resistance of real circuit components in the active balancing circuit.

4. Conclusions

This study examined fundamental properties of aging lithium ion battery cells. We described the testing methodology and established that the cells in a carefully designed pack aged at the same rate as when they are individually aged. The careful design consisted of matching overall capacities of serial connections to minimize charge exchanges between strings and the associated dissipation, and also to match impedances within a serial connection to minimize equalization with its associated losses. We then considered two common scenarios in cell replacements: replacing a prematurely failed cell and building a new pack from cells of the damaged packs for less demanding applications. Less demanding applications considered here are those that can tolerate capacity of less than 70%. Stationary applications, e.g., microgrid storage, where the overall pack weight is not a limiting factor, are considered good examples of less demanding applications that can be source their cells and modules from high-demanding applications (e.g., transportation).

The first scenario is important for maintenance of existing packs and modules, especially large packs. We found that, at least for demanding applications where the pack operates until its cell drop to about 80% of the original capacity, a new cell coexisted well within a pack of aged cells. Therefore, within these applications, there may not be a need for a potentially expensive process of pre-aging cells.

The second scenario addressed the problem of repurposing of used cells in less demanding applications. To examine a severe case of a pack failure, we deeply discharged two packs and assembled a pack from the surviving cells. The deep discharge event was so severe that 6 of the original 18 cells were permanently damaged. Two approaches to rebuilding the packs were considered, but only matching the cells within a string worked robustly in our pack. After the pack was rebuilt in this fashion, more than forty cycles were successfully completed. Comparing the modes of the distribution, it was found that the overall pack efficiency of the rebuilt pack was about 8% lower than that of pack comprised of new cells. However, the heating of individual cells within the three packs was compared and no significant differences were found. We finally showed that increased losses due to larger mismatches can be partially compensated by employing active instead of passive cell balancing. The balancing scheme may be particularly critical for secondary applications where cells cannot be matched as well as in the primary applications, and where the initial mismatches are expected to grow further over time.

Lessons learned from this study will be employed in our future work of assessing opportunities for repurposing larger packs, where secondary applications may have specific use profiles distinctly different from the primary applications. For example, this information could be used in considering reuse of vehicle batteries in grid applications, such as peak shaving and smoothing of solar-generated power vs. their reuse in much smaller systems such as power tools, or toys. The future work may also consider other criteria, including economic and environmental performance.

Supplementary Materials: The following are available online at http://www.mdpi.com/2313-0105/6/3/39/s1, Figure S1: High-level view of the research program; Figure S2: Initial characterization and single-cell testing; Figure S3: Histogram of cell weights; Figure S4: Characteristic waveforms of a typical cycle; Figure S5:Impedance spectroscopy. (a) Average spectra. (b) Repeatability and potentiostat comparison; Figure S6: Schematic diagram of the 3S3P battery pack; Figure S7: (a) LabVIEW-controlled battery pack test stand. (b) Enlarged view of the pack; Figure S8:Voltage and current waveforms for a cell during pack cycling; Figure S9: Current distribution of three strings over four cycles; Figure S10:Voltage distribution of three cells within string 1 over four cycles; Figure S11: Cell balancing for one of the strings. Pack design employed 106 Astro "Blinky" from AstroFlight Inc. as the commercial cell balancer; Figure S12: Voltage distribution of three cells within string 1 over four cycles for active cell balancing; Figure S13:(a) Representative capacity of individual cells diverging in time (18650, lithium cobaltoxide chemistry). (b) Scatter plot of capacity fade vs. percent recovery; Figure S14: Capacity fade of eight lithium cobalt cells; Figure S15: Thermal image of the pack; Figure S16: Comparison of heating of lithium ion batteries; Table S1.Effect of T_a on T_{ed}; Algorithm S1: Charge cycle.

Author Contributions: Conceptualization, N.G.N., T.A.T., and M.G.T.; methodology, N.G.N. and T.A.T.; software, N.G.N.; validation, N.G.N., T.A.T., and M.G.T.; formal analysis, N.G.N. and T.A.T.; investigation, N.G.N., T.A.T., and M.G.T.; data curation, N.G.N.; writing—original draft preparation, N.G.N.; writing—review and editing, N.G.N., T.A.T., and M.G.T.; visualization, N.G.N.; supervision, M.G.T.; project administration, M.G.T.; funding acquisition, M.G.T. All authors have read and agreed to the published version of the manuscript.

Funding: This work was made possible by the Office of Naval Research under Award No. N0004-07-1-0823.

Disclaimer: Any opinions, findings, and conclusions or recommendations expressed in this material are those of the author(s) and do not necessarily reflect the views of the Office of Naval Research.

Acknowledgments: We gratefully acknowledge the help of our colleagues from Rochester Institute of Technology: Scott Dewey for his assistance in pack assembly and instrument setup, Robert Kosty for his assistance in collecting data and debugging the fixtures, and Joseph Wodenscheck and Art Dee for skillful LabVIEW implementation of the charge cycle and discharge cycle on the battery pack test stand.

Conflicts of Interest: The authors declare no conflict of interest. The funders had no role in the design of the study; in the collection, analyses, or interpretation of data; in the writing of the manuscript, or in the decision to publish the results.

References

1. Wagner, F.T.; Lakshmanan, B.; Mathias, M.F. Electrochemistry and the future of the automobile. *J. Phys. Chem. Lett.* **2010**, *1*, 2204–2219. [CrossRef]
2. Williams, B.D.; Lipman, T.E. Strategy for Overcoming Cost Hurdles of Plug-In-Hybrid Battery in California. *Transp. Res. Rec. J. Transp. Res. Board* **2010**, *2191*, 59–66. [CrossRef]
3. Ralston, M.; Nigro, N. *Plug-in Electric Vehicles: Literature Review*; Pew Center on Global Climate Change: Arlington, VA, USA, 2011. Available online: https://www.c2es.org/site/assets/uploads/2011/07/plug-in-electric-vehicles-literature-review.pdf (accessed on 22 July 2020).
4. Witkin, J. A Second Life for the Electric Car Battery. 2011. Available online: https://green.blogs.nytimes.com/2011/04/27/a-second-life-for-the-electric-car-battery/ (accessed on 22 July 2020).
5. Marano, V.; Onori, S.; Guezennec, Y.; Rizzoni, G.; Madella, N. Lithium-ion batteries life estimation for plug-in hybrid electric vehicles. In Proceedings of the 2009 IEEE Vehicle Power and Propulsion Conference, Dearborn, MI, USA, 7–10 September 2009; pp. 536–543. [CrossRef]
6. Waag, W.; Fleischer, C.; Sauer, D.U. Critical review of the methods for monitoring of lithium-ion batteries in electric and hybrid vehicles. *J. Power Sources* **2014**, *258*, 321–339. [CrossRef]
7. Rezvanizaniani, S.M.; Liu, Z.; Chen, Y.; Lee, J. Review and recent advances in battery health monitoring and prognostics technologies for electric vehicle (EV) safety and mobility. *J. Power Sources* **2014**, *256*, 110–124. [CrossRef]
8. Lu, L.; Han, X.; Li, J.; Hua, J.; Ouyang, M. A review on the key issues for lithium-ion battery management in electric vehicles. *J. Power Sources* **2013**, *226*, 272–288. [CrossRef]
9. Roscher, M.A.; Bohlen, O.S.; Sauer, D.U. Reliable state estimation of multicell lithium-ion battery systems. *IEEE Trans. Energy Convers.* **2011**, *26*, 737–743. [CrossRef]
10. Kim, G.H.; Smith, K.; Ireland, J.; Pesaran, A. Fail-safe design for large capacity lithium-ion battery systems. *J. Power Sources* **2012**, *210*, 243–253. [CrossRef]
11. Offer, G.J.; Yufit, V.; Howey, D.A.; Wu, B.; Brandon, N.P. Module design and fault diagnosis in electric vehicle batteries. *J. Power Sources* **2012**, *206*, 383–392. [CrossRef]
12. Paul, S.; Diegelmann, C.; Kabza, H.; Tillmetz, W. Analysis of ageing inhomogeneities in lithium-ion battery systems. *J. Power Sources* **2013**, *239*, 642–650. [CrossRef]
13. Zheng, Y.; Han, X.; Lu, L.; Li, J.; Ouyang, M. Lithium ion battery pack power fade fault identification based on Shannon entropy in electric vehicles. *J. Power Sources* **2013**, *223*, 136–146. [CrossRef]
14. Li, J.; Barillas, J.K.; Guenther, C.; Danzer, M.A. Multicell state estimation using variation based sequential Monte Carlo filter for automotive battery packs. *J. Power Sources* **2015**, *277*, 95–103. [CrossRef]
15. Dubarry, M.; Truchot, C.; Cugnet, M.; Liaw, B.Y.; Gering, K.; Sazhin, S.; Jamison, D.; Michelbacher, C. Evaluation of commercial lithium-ion cells based on composite positive electrode for plug-in hybrid electric vehicle applications. Part I: Initial characterizations. *J. Power Sources* **2011**, *196*, 10328–10335. [CrossRef]
16. Dubarry, M.; Truchot, C.; Liaw, B.Y.; Gering, K.; Sazhin, S.; Jamison, D.; Michelbacher, C. Evaluation of commercial lithium-ion cells based on composite positive electrode for plug-in hybrid electric vehicle applications. Part II: Degradation mechanism under 2C cycle aging. *J. Power Sources* **2011**, *196*, 10336–10343. [CrossRef]
17. Dubarry, M.; Vuillaume, N.; Liaw, B.Y. Origins and accommodation of cell variations in Li-ion battery pack modeling. *Int. J. Energy Res.* **2009**, *34*, 216–231. [CrossRef]
18. Gering, K.L.; Sazhin, S.V.; Jamison, D.K.; Michelbacher, C.J.; Liaw, B.Y.; Dubarry, M.; Cugnet, M. Investigation of path dependence in commercial lithium-ion cells chosen for plug-in hybrid vehicle duty cycle protocols. *J. Power Sources* **2011**, *196*, 3395–3403. [CrossRef]
19. Rao, Z.; Wang, S. A review of power battery thermal energy management. *Renew. Sustain. Energy Rev.* **2011**, *15*, 4554–4571. [CrossRef]
20. Sarre, G.; Blanchard, P.; Broussely, M. Aging of lithium-ion batteries. *J. Power Sources* **2004**, *127*, 65–71. [CrossRef]
21. Nenadic, N.G.; Bussey, H.E.; Ardis, P.A.; Thurston, M.G. Estimation of State-of-Charge and Capacity of Used Lithium-Ion Cells. *Int. J. Progn. Health Manag.* **2014**, *5*, 12.

22. Dudek, K.; Lane, R. Synergy between electrified vehicle and community energy storage batteries and markets: Super session 4. Energy storage. In Proceedings of the Power and Energy Society General Meeting, Detroit, MI, USA, 24–28 July 2011; p. 1.
23. Williams, D.M.; Gole, A.M.; Wachal, R.W. Repurposed battery for energy storage in applications of renewable energy for grid applications. In Proceedings of the 2011 24th Canadian Conference on Electrical and Computer Engineering (CCECE), Niagara Falls, ON, Canada, 8–11 May 2011; pp. 1446–01450.
24. Schneider, E.L.; Kindlein, W., Jr.; Souza, S.; Malfatti, C.F. Assessment and reuse of secondary batteries cells. *J. Power Sources* **2009**, *189*, 1264–1269. [CrossRef]
25. Lih, W.C.; Jieh-Hwang, Y.; Fa-Hwa, S.; Yu-Min, L. Second Use of Retired Lithium-ion Battery Packs from Electric Vehicles: Technological Challenges, Cost Analysis and Optimal Business Model. In Proceedings of the 2012 International Symposium on Computer, Consumer and Control (IS3C), Taichung, Taiwan, 4–6 June 2012; pp. 381–384. [CrossRef]
26. Kamath, H. Lithium Ion Batteries in Utility Applications. In Proceedings of the 27th International Battery Seminar & Exhibit, Fort Lauderdale, FL, USA, 15–18 March 2010.
27. Neubauer, J.; Pesaran, A.; Howell, D. *Secondary Use of PHEV and EV Batteries: Opportunities & Challenges (Presentation)*; No. NREL/PR-540-48872; National Renewable Energy Lab (NREL): Golden, CO, USA, 2010. Available online: https://www.nrel.gov/docs/fy10osti/48872.pdf (accessed on 22 July 2020)
28. Neubauer, J.; Pesaran, A. *PHEV/EV Li-Ion Battery Second-Use Project*; National Renewable Energy Laboratory: Golden, CO, USA, 2010. Available online: https://www.nrel.gov/docs/fy10osti/48018.pdf (accessed on 22 July 2020).
29. Neubauer, J.; Pesaran, A.; Williams, B.; Ferry, M.; Eyer, J. *Techno-Economic Analysis of PEV Battery Second Use: Repurposed-Battery Selling Price and Commercial and Industrial End-User Value*; Report; National Renewable Energy Laboratory (NREL): Golden, CO, USA, 2012.
30. Hawkins, J.M. Some field experience with battery impedance measurement as a useful maintenance tool. In Proceedings of the 16th International Telecommunications Energy Conference, INTELEC '94, Vancouver, BC, Canada, 30 October–3 November 1994; pp. 263–269. [CrossRef]
31. Lasia, A. Electrochemical impedance spectroscopy and its applications. *Mod. Asp. Electrochem.* **1999**, *32*, 143–248.
32. Amine, K.; Chen, C.H.; Liu, J.; Hammond, M.; Jansen, A.; Dees, D.; Bloom, I.; Vissers, D.; Henriksen, G. Factors responsible for impedance rise in high power lithium ion batteries. *J. Power Sources* **2001**, *97–98*, 684–687. [CrossRef]
33. Broussely, M.; Biensan, P.; Bonhomme, F.; Blanchard, P.; Herreyre, S.; Nechev, K.; Staniewicz, R.J. Main aging mechanisms in Li ion batteries. *J. Power Sources* **2005**, *146*, 90–96. [CrossRef]
34. Dees, D.; Gunen, E.; Abraham, D.; Jansen, A.; Prakash, J. Alternating Current Impedance Electrochemical Modeling of Lithium-Ion Positive Electrodes. *J. Electrochem. Soc.* **2005**, *152*, A1409–A1417. [CrossRef]
35. Tröltzsch, U.; Kanoun, O.; Tränkler, H.R. Characterizing aging effects of lithium ion batteries by impedance spectroscopy. *Electrochim. Acta* **2006**, *51*, 1664–1672. [CrossRef]
36. Li, J.; Murphy, E.; Winnick, J.; Kohl, P.A. The effects of pulse charging on cycling characteristics of commercial lithium-ion batteries. *J. Power Sources* **2001**, *102*, 302–309. [CrossRef]
37. Chen, M.; Rincon-Mora, G.A. Accurate electrical battery model capable of predicting runtime and IV performance. *IEEE Trans. Energy Convers.* **2006**, *21*, 504–511. [CrossRef]
38. Saha, B.; Goebel, K. Uncertainty Management for Diagnostics and Prognostics of Batteries using Bayesian Techniques. In Proceedings of the 2008 IEEE Aerospace Conference, Big Sky, MT, USA, 1–8 March 2008; pp. 1–8.
39. Shim, J.; Kostecki, R.; Richardson, T.; Song, X.; Striebel, K.A. Electrochemical analysis for cycle performance and capacity fading of a lithium-ion battery cycled at elevated temperature. *J. Power Sources* **2002**, *112*, 222–230. [CrossRef]
40. Waag, W.; Käbitz, S.; Sauer, D.U. Experimental investigation of the lithium-ion battery impedance characteristic at various conditions and aging states and its influence on the application. *Appl. Energy* **2013**, *102*, 885–897. [CrossRef]

© 2020 by the authors. Licensee MDPI, Basel, Switzerland. This article is an open access article distributed under the terms and conditions of the Creative Commons Attribution (CC BY) license (http://creativecommons.org/licenses/by/4.0/).

Article

Multi-Physics Equivalent Circuit Models for a Cooling System of a Lithium Ion Battery Pack

Takumi Yamanaka *, Daiki Kihara, Yoichi Takagishi and Tatsuya Yamaue

Kobelco Research Institute Inc., Kobe 6512271, Japan; kihara.daiki@kki.kobelco.com (D.K.);
takagishi.yoichi@kki.kobelco.com (Y.T.); yamaue.tatsuya@kki.kobelco.com (T.Y.)
* Correspondence: yamanaka.takumi@kki.kobelco.com; Tel.: +81-78-992-5976

Received: 10 July 2020; Accepted: 27 August 2020; Published: 29 August 2020

Abstract: Lithium (Li)-ion battery thermal management systems play an important role in electric vehicles because the performance and lifespan of the batteries are affected by the battery temperature. This study proposes a framework to establish equivalent circuit models (ECMs) that can reproduce the multi-physics phenomenon of Li-ion battery packs, which includes liquid cooling systems with a unified method. We also demonstrate its utility by establishing an ECM of the thermal management systems of the actual battery packs. Experiments simulating the liquid cooling of a battery pack are performed, and a three-dimensional (3D) model is established. The 3D model reproduces the heat generated by the battery and the heat transfer to the coolant. The results of the 3D model agree well with the experimental data. Further, the relationship between the flow rate and pressure drop or between the flow rate and heat transfer coefficients is predicted with the 3D model, and the data are used for the ECM, which is established using MATLAB Simulink. This investigation confirmed that the ECM's accuracy is as high as the 3D model even though its computational costs are 96% lower than the 3D model.

Keywords: equivalent circuit models; Li-ion battery packs; thermal management systems; electric vehicles

1. Introduction

Recently, electric vehicles (EVs) have gained popularity as transportation vehicles [1,2]. Lithium ion (Li-ion) batteries are widely installed in EVs because of their high-power density, high energy density, long lifetime, and less self-discharge [1]. Since Li-ion batteries are usually utilized as main energy sources, management systems for batteries are essential for the development of EVs. In particular, battery thermal management systems (BTMSs) are important because the battery temperature during an operation affects the performance, lifespan, and safety of the batteries [1].

Many cooling methods are used for BTMSs, and the coolant materials can be mainly categorized into three types: air, liquid and phase change materials (PCMs). Air cooling is one of the most commonly used methods. Although it is simple and its advantage is the weight of the refrigerant, it is not suitable for a large capacity battery pack because air has a low thermal conductivity and heat capacity [3]. The PCM has a great advantage in terms of keeping the temperature of the batteries uniform or preventing thermal runaway. Nevertheless, for the long-term operation, the method is limited due to PCM full melting [4]. Liquid cooling is more efficient and compact, and it is widely used by several EV manufacturers, such as Tesla and General Motors (GM) [4]. Although water, mineral oil, or mixture ethylene glycol and water is adopted as the liquid coolant materials, the mixture of ethylene glycol and water is normally used by the EV manufacturers [1]. This is because the mixture has a lower melting point than 0 °C, and it is suitable for operating even in cold environments. Because the direct liquid cooling is concerned about the electrical short, indirect cooling using the thermal conductive media is often adopted. As with the media, tubes, cold plates with channels or jacket are used [1]. Since the heat pipe has a highly efficient thermal conductivity, BTMSs which adopt the heat pipe as the media are currently being studied. Nevertheless, there is room for improvement such as reducing the system complexity [3]. In this work, since it is intended for general-purpose modeling, we will target the BTMS of the liquid cooling using a mixture of ethylene glycol and water, which is often used in popular EVs.

Numerical simulations are useful for predicting and evaluating the impact when varying the design, and they are expected to reduce the study period and cost. Numerical simulations are used in many reports for BTMSs, and the method can be categorized into two types: detailed models and equivalent circuit models (ECMs). Detailed models [5–13] usually target two-dimensional (2D) or three-dimensional (3D) thermal phenomena in battery cells or modules with finite element methods (FEMs) or finite volume methods (FVMs). Y. Chung et al. [5] performed a thermal analysis with a 3D model to improve the cooling design of pouch battery packs. Siruvuri et al. [6] also modeled a battery pack and the cooling channels and optimized the design to reduce the peak temperature by reversing the direction of the water flow in one of the channels. Although these detailed models have merits for grasping the mechanism and predicting the results more reliably, these are not suitable for predicting the large-scale phenomenon such as the whole battery pack because the computational costs are higher.

Because ECMs [3,4,14–17] have low computational costs, these models can be adopted for the whole battery pack and for controlling the interaction among multiple parts with electronic control units (ECUs) during the drive operation. The developed ECMs are useful considering the development of all the BTMSs installed in the ECUs. By having a uniform environment, some programs, such as MATLAB Simulink® (Natick, MA, USA), can be useful for establishing the ECMs. Y. Gan et al. [3] developed a thermal ECM for the heat pipe-based thermal management system for a battery module, and they validated the model with experimental data. M. Shen et al. [4] constructed a system simulation model by using a refrigerant-based battery thermal management system, and they improved the system's performance.

Since ECMs have more abstract modeling, it is more difficult to confirm intuitively that the model assumption is satisfied with the target phenomenon. In most reports for ECMs, some important parameters that are represented by the heat transfer coefficients or the pressure drops are determined with some theoretical equations. However, the structural design of the flow channels has become more complex, to improve the efficiency. In addition, it may not be clear that conventional theoretical equations can be generally adopted for the flow in channels. Moreover, because these equations sometimes have a range of applicable values or fitting parameters when taking into account the various impacts, such as geometric factors, the model design is more complex, and it may cause the model to contain potential mistakes.

In this study, we propose a framework in which the ECMs can be established with a unified method by using detailed 3D models. We also present a model for the thermal–electrical coupled ECM of the liquid cooling phenomenon for the battery pack, which includes the experimental methods, and the model is validated with experimental data.

2. Methodology for the Test Bench and the Detailed 3D Model

2.1. Electrical ECM Modeling with the Experimental Data

The battery pack for the liquid cooling experiment is constructed with five commercial Li-ion battery modules in which two pouch-type Li polymer cells are connected in series. The cell size is 36 mm × 125 mm × 6.5 mm. In this study, it is defined that the open-circuit voltage (OCV) of a cell is 3.0 V when the state of charge (SOC) is 0%, and the OCV of a cell is 4.2 V when the SOC is 100%. In addition, it is confirmed that the charging and discharging capacities are 3.0 Ah in the OCV range between 6.0 and 8.4 V for the module.

The experiment to evaluate the discharge properties of a module was performed based on the following procedures:

(i) Keeping the module at a set environmental temperature in the thermostatic chamber.
(ii) Charging the module following the constant current (CC)-the constant voltage (CV) procedure to 8.4 V with a 0.2C current rate.
(iii) Rest period of 60 min.
(iv) Discharging the module following the CC procedure with a set current rate until the SOC decreases by 10%.
(v) Rest period of 60 min.
(vi) Performing operations (iv) and (v) until the cell voltage reaches 6.0 V.

The combinations of the environmental temperature and current rate are shown in Table 1. The experiment for the charge properties was also performed with similar procedures. Then, the temporal behavior of the electrical potential per SOC 10% during the intermittent discharge or the charge under each condition were obtained, respectively. In the model, the SOC is calculated as follows.

$$SOC = SOC_0 + \int_0^t (I/C_c)dt, \qquad (1)$$

where SOC_0 is the initial state of the charge; t is the time; C_c is the capacity, which is equal to 3 Ah; and I is the electric current, in which a positive value represents the charge and a negative value indicates a discharge.

Table 1. The combinations of the experimental condition in order to evaluate the battery module's charge and discharge properties.

Environmental Temperature	Current Rate		
	1C (3A)	2C (6A)	3C (9A)
0 °C	Performed	-	-
20 °C	Performed	Performed	Performed
40 °C	Performed	-	-

For the electrical ECM for the batteries, the model shown in Figure 1a is adopted. The internal resistance is calculated from the experimental data as follows.

$$R = (V - OCV)/I, \qquad (2)$$

where R is the internal resistance per cell, V is the voltage per cell, and OCV is the open-circuit voltage. Because two cells in the module are assumed to be in the same state, V is half of the module voltage. Further, OCV is calculated by the cubic Hermitian interpolation polynomial with the voltage immediately before the CC discharge during the discharge experiment, as shown in Figure 2n.

In the electrical ECM, the internal resistance during the CC charge or discharge is calculated theoretically as follows.

$$R = [1 - \exp\{-t'/(R_2 C)\}]R_2 + R_1, \qquad (3)$$

where t' is the time after the charge or discharge, and R_1, R_2, and C are the resistance or capacitance components as shown in Figure 1a. R_1, R_2, and C are assumed be dependent on the temperature, current, and SOC, and they are optimized by fitting these parameters for the experimental data, which is calculated by applying Equation (2). The values of R_1, R_2, and C can be calculated by applying a linear interpolation or extrapolation with the database of the relationship between these parameters and the temperature, current, and SOC. The contours of the relationship are indicated in Figure 2a–m. From these figures, it is considered that the resistance R_1 and R_2 have a negative correlation with the temperature. This is because the electrochemical reaction degree and the Li-ion conductivity increase with a higher temperature. It is also considered that R_1 and R_2 have a negative correlation with the current rate. This is because the nonlinear relationship between the current and overpotential is indicated with the Butler–Volmer equation.

The heat generation of batteries is calculated by the following formula.

$$Q_{gen} = Q_{irr} + Q_{rev} = I(V - OCV) + IT\partial(OCV)/\partial T = I^2R + IT\Delta S/F, \quad (4)$$

where: Q_{gen} is the total heat generation, Q_{irr} is the irreversible heat generation due to the heat from the internal resistance, Q_{rev} is the reversible heat generation due to the entropy change, T is the temperature, ΔS is the entropy change, and F is the Faraday constant. By applying Equation (4), the entropy change is calculated by $F\partial(OCV)/\partial T$ and the results are shown in Figure 2o as the plot points. The data of the entropy change almost monotonically increases with the SOC, and the tendency and values are similar to the batteries that consist of the $LiCoO_2$ cathode and the graphite anode [18,19]. However, the values from SOC = 50% to SOC 80% vary strongly, and the tendency is different from the measurement in some reports [18,19]. We estimate that the tendency is affected by the measurement error because of the lack of experimental data. In this model, from the perspective that simple modeling is preferred, a linear regression is adopted for the entropy change when the SOC is less than 80% with the least squares methods. In addition, the entropy change when the SOC is more than 80% is not taken into account, which is illustrated in Figure 2o.

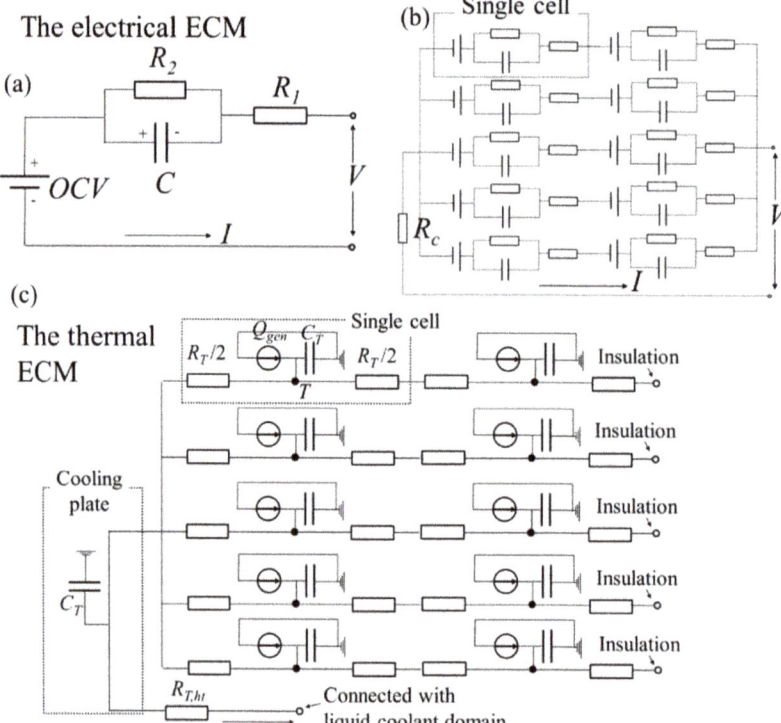

Figure 1. The schematic views of (**a**) the electrical equivalent circuit model (ECM) for a single cell, (**b**) the electrical ECM for the whole battery pack, and (**c**) the thermal ECM for the whole battery pack.

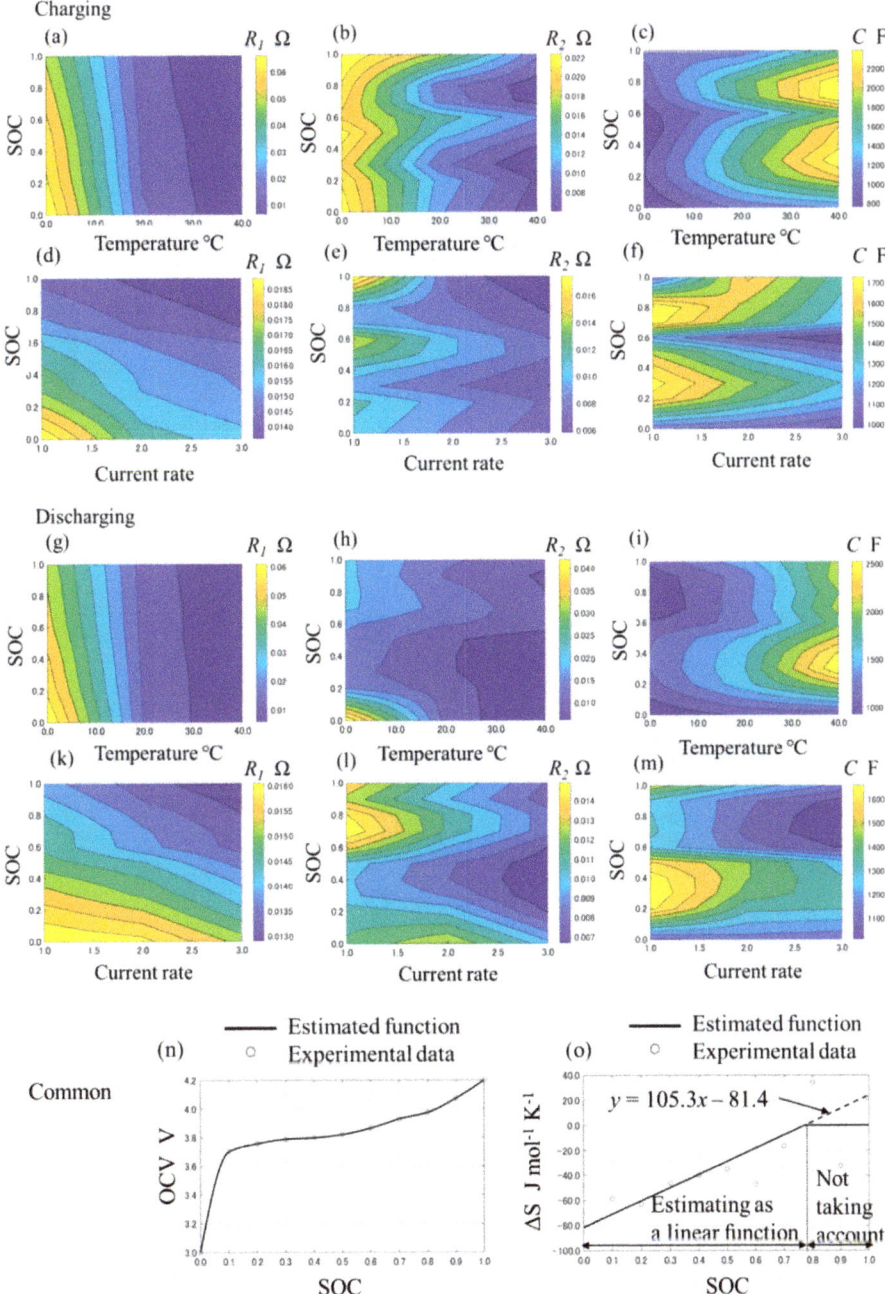

Figure 2. The relationship between the variable parameters and the arguments in the electrical ECM. (**a**–**f**) show the relationship between R_1, R_2, C and temperature, current, state of change (SOC) at charge, (**g**–**m**) show it at discharge, (**n**) shows the relationship between SOC and open-circuit voltage *(OCV)*, and (**o**) shows the relationship between SOC and ΔS.

2.2. Testing of the Liquid Cooling for a Battery Pack

The schematic view of the test bench is shown in Figure 3. The test bench is constructed roughly with five battery modules, a cooling plate, some tubes for the flow channels, and a chiller. Although the constitution of the test bench is simpler than the actual systems for the EVs, we consider that the test bench has roughly the same component as the actual systems, and it is sufficient to explain the usefulness for our framework. Five battery modules are connected in parallel, and the terminals are connected with the device for the battery management and the direct current (DC) power supply. We defined these batteries as a battery pack whose capacity is 15 Ah, and its operating voltage is from 6.0 V to 8.4 V. These modules are set on the cooling plate, which has a meandering flow channel. Liquid coolant is circulated among the chiller and the flow channel via the tubes, and the temperature of the liquid coolant in the chiller is maintained at 25 °C. As the liquid coolant, we adopt long life coolant, which is mainly a mixture of ethylene glycol and water. The batteries and cooling plate are covered with heat insulators in order to reduce the impact of the heat transfer from other than the liquid coolant. Therefore, most of the heat radiation is due to the outflow of the liquid coolant. The battery pack is pressed down at about 100 Pa to reduce the impact on the contact resistance between the batteries and the cooling plate. The temperature for the top of the battery modules and the liquid coolant flowing in the tubes is measured by using thermocouples. The volume flow rate of the liquid coolant through the cooling plate flow channel is measured with an electromagnetic flow meter.

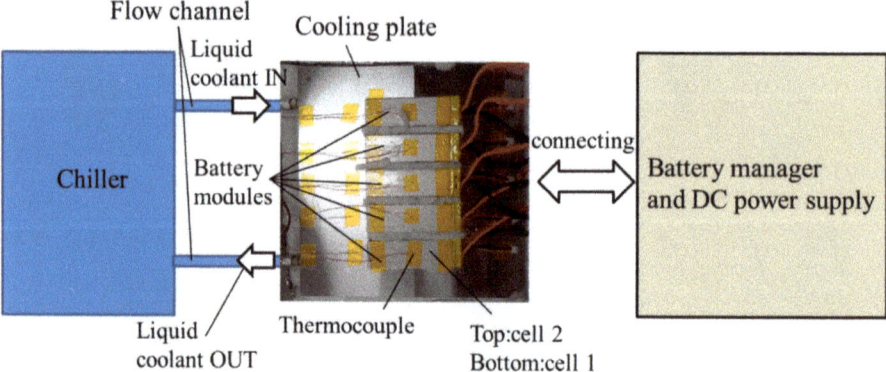

Figure 3. The schematic view of the test bench.

The charging and discharging for the battery pack were performed with the following procedures:

(i) Discharging of the battery following the CC-CV procedure to 6.0 V with the 0.2C current rate.
(ii) Rest period of 60 min.
(iii) Charging the battery to 8.4 V with a 2C current rate.
(iv) Discharging the battery to 6.0 V with a 2C current rate.
(v) Performing two cycles of operations (iii) and (iv).

The tests are performed twice under conditions for the chiller setting in which the flow rate is 0.0 l·min^{-1}, i.e., not flowing, or 1.0 l·min^{-1}. In this work, the case of not flowing is called case 1, and the case of 1.0 l·min^{-1} is called case 2.

2.3. Constructing the Detailed 3D Model

The detailed 3D model for the tests was developed. The overview of the calculation meshes and the geometry of the flow channel in the cooling plate are illustrated in Figure 4. The shapes are modeled by measuring the actual things. The prism meshes are adopted into the flow channel close to the wall of the cooling plate. In addition, they have two layers whose thickness is less than 1 mm to resolve the temperature boundary layers. The others are the tetrahedral meshes, and the total number of elements is 992,623. It is confirmed that the calculation results did not vary with more refined meshes. For the same reason, the relative tolerance of the progress for the next time step is set as 1.0×10^{-3}.

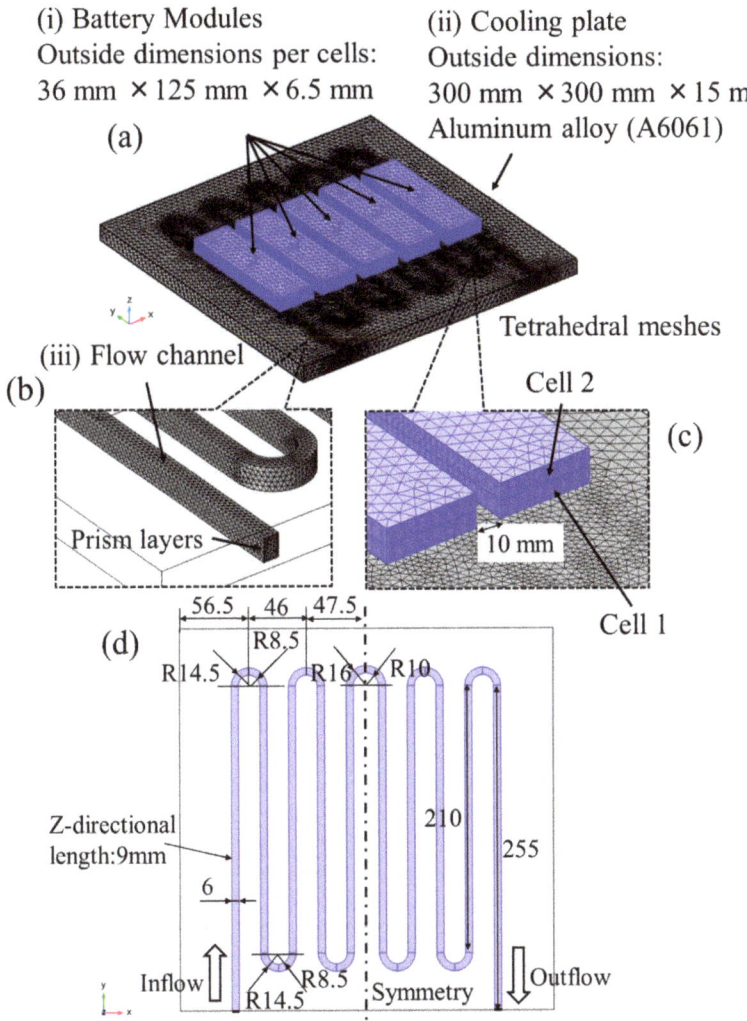

Figure 4. The overview of the calculational meshes and the geometry of the flow channel in the cooling plate. (**a**) shows the whole overview of the calculational meshes, (**b**,**c**) show the detail view of the flow channel and the cells, and (**d**) shows the geometry of the flow channel in the cooling plate.

The flow field and pressure field in the liquid coolant domain are presented as the equations of the mass continuity and the Navier–Stokes equations for incompressible flow, which are described as follows.

$$\nabla \cdot \mathbf{u} = 0, \tag{5}$$

$$\rho(\mathbf{u} \cdot \nabla)\mathbf{u} = \nabla \cdot \left[p\mathbf{I} + \mu \left\{ \nabla \mathbf{u} + (\nabla \mathbf{u})^T \right\} \right], \tag{6}$$

where **u** is the flow velocity vector, ρ is the density, p is the pressure, **I** is the identity matrix, and μ is the kinematic viscosity. In this model, the flow is assumed to be a steady laminar flow because the Reynolds number is low when the flow rate is less than 1.0 l/min. The heat transfer phenomenon in the whole domain is presented as the heat transfer equation, which is described below.

$$\rho C_p \partial T / \partial t + \rho C_p \mathbf{u} \cdot \nabla T + \nabla \cdot (-\mathbf{k} \nabla T) = Q_{gen} / V_L, \tag{7}$$

where C_p is the specific heat at a constant pressure, **k** is the thermal conductivity vector, and V_L is the volume of a cell. The flow velocity vector **u** in the liquid coolant domain can have a value other than **0**, and **u** = **0** in the other domain. Q_{gen} in the domains other than the batteries is 0. In this model, the heat generation is assumed to be uniform per the battery cell, and the electrical ECM is shown in Figure 1b, which is calculated per time-step to evaluate Q_{gen} in Equation (4). Note that the symbol R_c is the electrical contact resistance due to the connection of the conductors, and it is estimated to be 7.8 mΩ. The thermal contact resistance between the cooling plate and the liquid coolant is not taken into account because pressing the battery pack is supposed to reduce it sufficiently. This system of equations is discretized by the FEM and it is solved by using initial and boundary conditions. These numerical calculations are conducted by using COMSOL Multiphysics® ver. 5.4. Table 2 lists the physical properties that use this model.

Table 2. The physical properties using the simulations. Superscript "a" indicates the assumed value.

Item	Unit	Cell (Homogeneous Body)	Cooling Plate	Liquid Coolant (Long Life Coolant)
Material		Composite	A6061	Ethylene glycol 46wt% aqueous solution
Density	kg·m^{-3}	2000 [a]	2700 [20]	1054 [20]
Specific heat	J·kg^{-1}·K^{-1}	800 [a]	896 [20]	3412 [20]
Thermal conductivity	W·m^{-1}·K^{-1}	Parallel direction 30 [a] Vertical direction 0.5 [a]	180 [20]	0.43 [20]
Viscosity	Pa·s			3.12 × 10^{-3} [20]

The charging or discharging conditions for the battery pack are the same as the experiment. For the inlet boundary of the liquid coolant domain, the flow rate is 0.0 l·min^{-1} and the temperature is 23.5 °C under case 1. Meanwhile, the flow rate is 0.862 l·min^{-1} and the temperature is 25.2 °C under case 2. These values are given by averaging the measured data. The pressure is set as 0 Pa for the outlet boundary under case 2. Under case 1, the outlet boundary condition is assumed to be the same as the inlet boundary condition. The outside boundaries except the flow outlet are insulated thermally.

The assumptions made for the model are as follows.

(1) The flow of the liquid coolant is assumed to be a steady laminar flow. Therefore, the turbulence models are not adopted.
(2) The heat generation is assumed to be uniform per the battery cell.
(3) The inlet flow conditions are set by the measured data.
(4) If there is no flow, the outlet boundary condition is assumed to be the same as the inlet boundary condition.
(5) The heat loss to the outside is not taken into account except due to the flow outlet.

2.4. The Thermal–Electrical Coupled ECM

The thermal ECM for the battery pack cooling system was constructed based on the detailed 3D model. The schematic view is presented in Figure 1c. The thermal resistance, heat capacities, and heat generation are linked within the ECM. The temperature of each component, except for the liquid coolant domain, is calculated by the energy balance equation.

$$C_T \partial T/\partial t = Q_{link} + Q_{gen}, \tag{8}$$

where C_T is the heat capacity that is calculated by $\rho C_p V_L$, and Q_{link} is the heat transfer into or from the linked components, which can be determined from Equation (9).

$$Q_{link} = \Delta T/R_T, \tag{9}$$

where R_T is the thermal resistance, and ΔT is the temperature difference between the target component and the linked components. In the liquid coolant domain, the temperature can be determined by applying Equation (10).

$$C_T \partial T/\partial t = Q_{ht} + Q_f = \Delta T/R_{T,ht} + \dot{m} C_p \Delta T_f, \tag{10}$$

where Q_{ht} is the heat transfer that is exchanged between the cooling plate and the liquid coolant domain, Q_f is the heat due to the liquid coolant flow through the flow channel, $R_{T,ht}$ presents the liquid–solid heat transfer as the thermal resistance, $\dot{m}$ is the mass flow rate, and ΔT_f is the liquid temperature difference between the inlet and the outlet. $R_{T,ht}$ is described as follows.

$$R_{T,ht} = 1/hA, \tag{11}$$

where h is the heat transfer coefficient, and A is the liquid–solid cross-sectional area.

From the 3D model's result, the temperature distribution in the cells has a large gradient in the z-direction and it is in the cooling plate, which is not large. Therefore, z-directional thermal links are taken into account for the cell, and one component is taken into account for the cooling plate. The parameter values are calculated from Table 2 and the geometry of the 3D model. For example, the cell thermal resistance R_T is calculated by $l_T/(A_T k_z)$ where l_T is the length of the heat pass, A_T is the cross-sectional area, and k_z is the z-directional thermal conductivity. The thermal–electrical coupled ECM is constructed by using MATLAB Simulink® R2019a/Simscape, and the schematic views are illustrated in Figure 5.

The heat transfer coefficient h is estimated by the results of the 3D model. The calculations are performed with the above 3D model, in which the uniform temperature of 26 °C was set to the liquid–solid cross-sectional boundary. The flow and temperature field were calculated only for the liquid coolant domain. Further, the temperature on the inlet boundary is 25 °C and the results are given under the various flow rate conditions where the range of the flow rate is from 0 l/min to 1 l/min. The heat transfer coefficient can be determined as follows.

$$h = \iint q_n dA / (T_s - T_{f,inlet}) A, \tag{12}$$

where q_n is the normal directional heat flux on the liquid–solid cross-section; T_s is the solid temperature, which is equal to 26 °C; and $T_{f,inlet}$ is the inlet liquid temperature, which is equal to 25 °C. In the ECM, the values that were calculated by the cubic Hermitian interpolation polynomial are adopted.

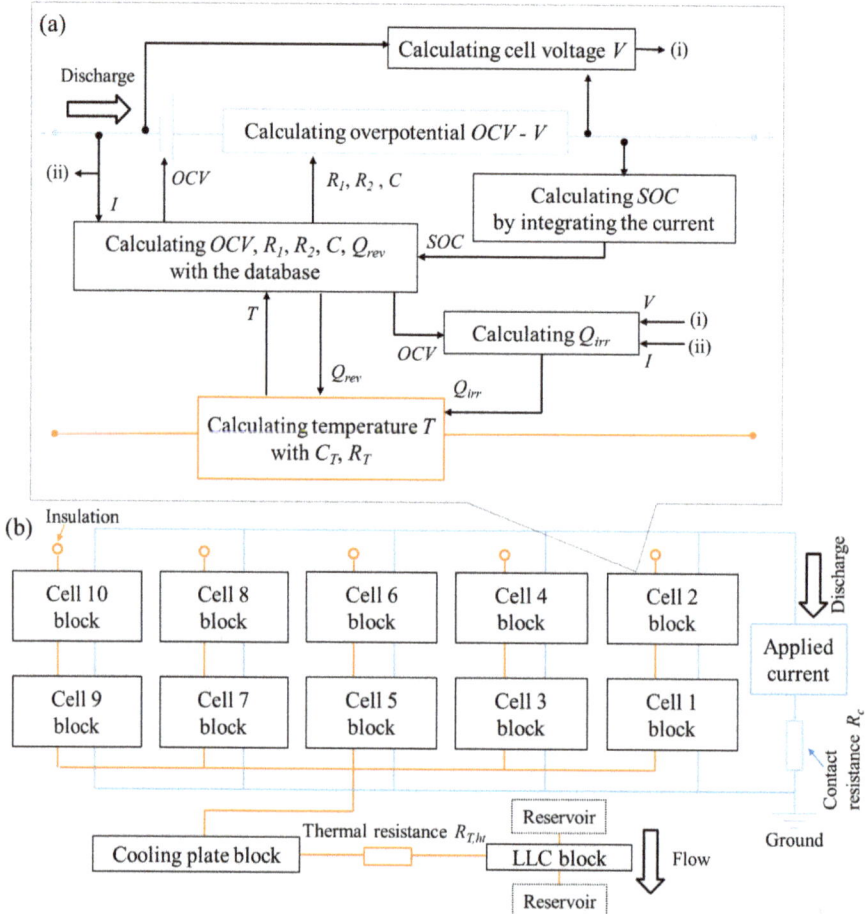

Figure 5. The schematic views of the electrical-thermal coupled ECM for (**a**) the single cell and (**b**) the whole battery pack. Note that the blue lines indicate the electrical links and the orange lines indicate the thermal links.

3. Results and Discussion

3.1. The Test Bench and the 3D Detail Model

The temperature distributions after 300 s and 2700 s under case 2 are shown in Figure 6. Large temperature gradients in the Z direction in the cells are confirmed after 300 s and 2700 s although the temperature gradients in the X and Y direction are small. Moreover, after 300 s, the temperature in the cells is lower than the cooling plate because of the endothermic reactions of the entropy change during charging. On the other hand, after 2700 s, the temperature in the cells is higher, and it reaches approximately 36 °C because of heat generation from the Joule heat and the exothermic reactions of the entropy change during discharge. The temperature for the outflow is approximately 0.5 °C higher than the inflow. The time histories of the total voltage and temperature for cases 1 and 2 are shown in Figure 7. From Figure 7a,c, under both cases, the voltage of the battery pack increases from 6.0 to 8.4 V during charging, and it decreases from 8.4 to 6.0 V during discharging. It is confirmed that the values agree with the experimental data. From Figure 7b,d, the temperature on the surface of cell 2 decreases immediately after the starting charge because of the endothermic reactions, and it increases significantly after it starts charging again. The difference between case 1 (without flow) and case 2 (with flow) per cycle is smaller than the temperature rise and drop due to the heat generation or the endothermic reactions because the applied current is high rate (2C). The temperature drop under case 1 is considered to be mainly due to the endothermic reactions and heat transfer to the cooling plate which has high heat capacity and high thermal conductivity. In the experimental data of the first cycle, the total temperature rise on the surface of cell 2 is 11.6 °C (initial: 23.9 °C, and final: 35.5 °C) under case 1 and 10.1 °C (initial: 24.9 °C, and final: 35.0 °C) under case 2. Therefore, approximately 1.5 °C is the impact of the flow. Especially when the temperature becomes high, the difference between the 3D model and the measured data increases slightly, and the maximum error is approximately 2 °C. This is mainly because heat dissipation, such as the heat flow via air or the pipes that are connected with the cooling plate, is not taken into account in the 3D model. In this work, because the set flow rate is relatively low, the heat paths other than the liquid cooling are not sufficiently small to ignore its effects. Nevertheless, the values during charging agree well with the measured data, and the difference does not increase per cycle under the two experimental conditions. In addition, the heat capacities and heat generation are supposed to be valid. From Figure 7e, the tendency of the temperature at the outlet agrees with the experimental data. This fact indicates that the heat dissipation due to the liquid coolant flow is similar between the 3D model and the experiment. From the above, it can be determined that the 3D model is validated sufficiently with the experiment.

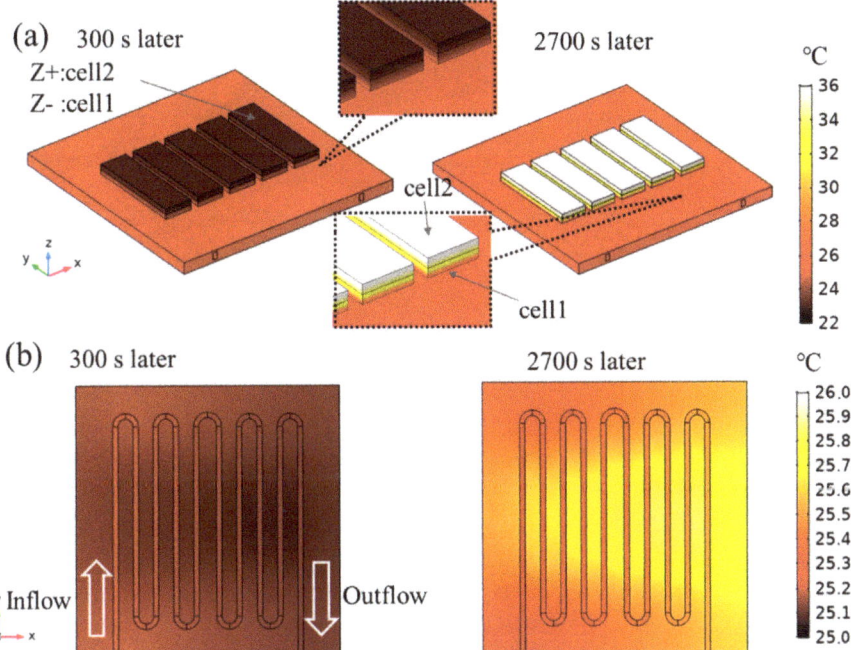

Figure 6. (**a**) The temperature distributions on the surfaces and (**b**) it in the z-directional center cross-section of the flow channel after 300 s and 2700 s in the 3D model under case 2.

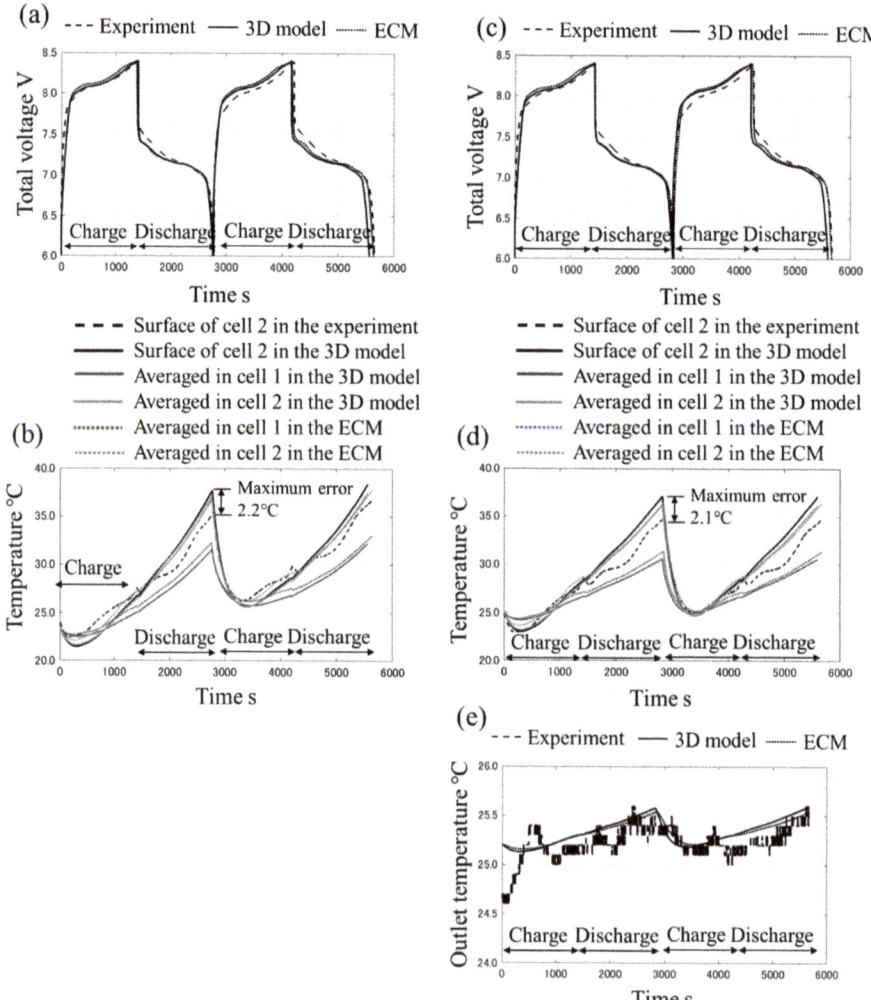

Figure 7. The time histories of the total voltage under (**a**) case 1 and (**c**) case 2, the time histories of averaged temperature and surface temperature under (**b**) case 1 and (**d**) case 2, and (**e**) the time histories of outlet temperature.

Under case 2, the calculation time of the model is 140 min performed by using two Intel® Xeon® CPU E5-2699 v3 @ 2.30 GHz processors with a memory of 128 GB, and the calculation cost may not be small enough. Therefore, it may need more effort or more powerful computational power to calculate a huge scale phenomenon such as battery packs with tens of battery cells, which is often installed for EVs.

3.2. The Thermal–Electrical Coupled ECM

According to Figure 7, it is clear that the results of the ECM agree well with the 3D model. From this, the ECM is supposed to be a surrogate model for the 3D model because the heat coefficient is determined with its data and the links of the thermal ECM are also determined based on its temperature distribution results. Table 3 shows the summary of computational environment and calculation costs of our models. Under case 2, the calculation time of the ECM is 5.0 min, which is 96% lower than the 140 min of the 3D model.

Table 3. Computational environment and calculation costs of our models.

Computational Environment		
CPUs	Intel® (Santa Clara, CA, USA) Xeon® (Santa Clara, CA, USA) CPU E5-2699 v3 @ 2.30 GHz	
Memory size	128 GB	
Calculation costs	Case 1	Case 2
3D detail model calculated with 4 parallel number	138 min	140 min
ECM	5.0 min	5.0 min

In the ECM, the value calculated by the 3D detail model is adopted as the liquid–solid heat transfer coefficient, shown in Figure 8a. In comparison, the parameters are also calculated by applying conventional theoretical equations. The heat transfer coefficient can be determined by applying Equation (13).

$$h = k_f Nu/D_h, \tag{13}$$

where k_f is the liquid thermal conductivity, D_h is the hydraulic diameter of the flow channel, and Nu is the Nusselt number, which is known to be constant if the flow is a fully developed laminar flow in a straight flow channel and the heat fluxes or temperature are uniform [21]. As demonstrated in the literature [21], if the duct is a rectangular cross-section and the temperature is uniform, Nu is 3.08 when the aspect ratio is 1.43, and Nu is 3.39 when the aspect ratio is 2.0. The aspect ratio of the flow channel is 1.5, and Nu is 3.12 based on linear interpolation. According to Figure 8a, the results of the 3D model increase linearly with the flow rate, and the tendency is different from the above theoretical results. When the flow rate is less than 0.2 l·min^{-1}, the heat transfer coefficient is less than the line where $Nu = 3.12$. This is because the prerequisite of the above theoretical equation is not satisfied. It is assumed that the axial conduction can be negligible [21] under the conditions where the heat diffusion into the inlet boundary is the main phenomenon and the axial conduction is not negligible. When the flow rate is more than 0.2 l·min^{-1}, there is an increase in the flow rate. This is because the heat exchange efficiency is higher than a straight flow channel due to the flow bias as shown in Figure 8d. More consideration for this effect is described in Appendix A in detail.

The pressure drop Δp in the 3D model can be calculated as follows.

$$\Delta p = \bar{p}_{inlet} - \bar{p}_{outlet}, \tag{14}$$

where $\bar{p}_{inlet}$ is the mean pressure in the inlet boundary, and $\bar{p}_{outlet}$ is the mean pressure in the outlet boundary. The pressure drop in a straight flow channel can be theoretically calculated by applying the Darcy–Weisbach equation as demonstrated in Equation (15).

$$\Delta p = f \rho u_m^2 L / 2D_h, \tag{15}$$

where L is the total length of the flow channel, f is the Darcy friction factor, and u_m is the cross-sectional mean flow velocity in the flow channel. There are formulations to calculate f, which have been proposed by S.W. Churchill [22] and V. Bellos et al. [23] and so on. Nevertheless, if only laminar flow is considered, the Darcy friction factor f can be calculated as follows [24,25].

$$f = 64/Re = 64/(\rho u_m D_h/\mu), \tag{16}$$

where *Re* is the Reynolds number. Moreover, the impact of the curved geometry of the flow channel can be taken into account by applying Equation (17) [24,25].

$$\Delta p = f \rho u_m^2 L / 2D_h + K_b \rho u_m^2 / 2,$$ (17)

where K_b is the bend loss coefficient, which is dependent on the bend angle and the ratio between the bend radius and the hydraulic diameter. By having a bending angle of 180 degrees [24], K_b = 0.413 when the bend radius is 11.5 mm and K_b = 0.401 when the bend radius is 13.0 mm.

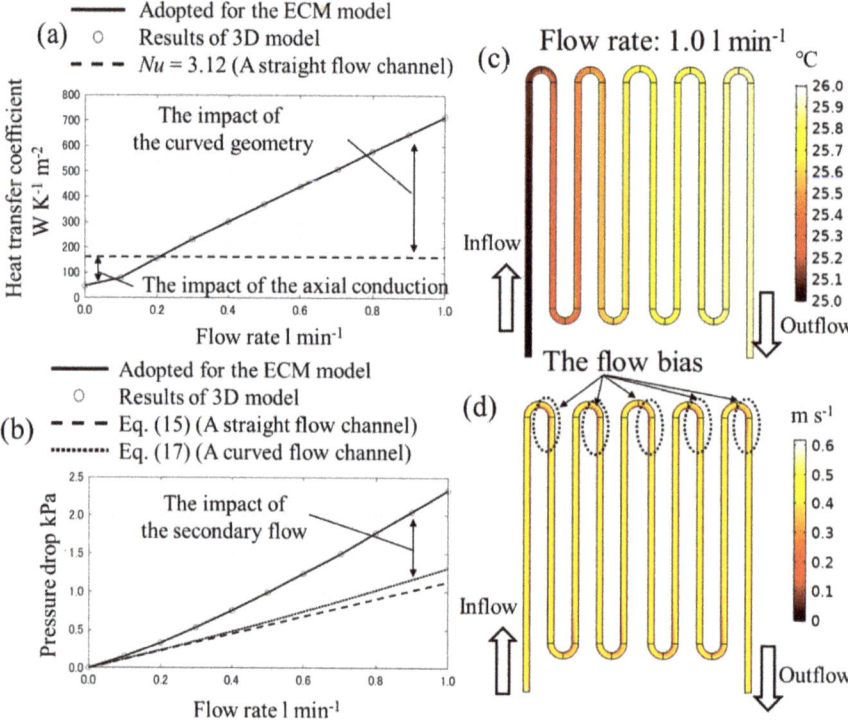

Figure 8. The relationship (**a**) between the flow rate and the heat transfer coefficient, (**b**) it between the flow rate and the pressure drop, (**c**) the temperature distribution and (**d**) the velocity magnitude distribution in the z-directional center cross-section of the flow channel when the flow rate is 1.0 l·min^{-1}.

According to Figure 8b, it is confirmed that the gradient of the 3D model is higher by applying Equations (15) and (17). This is because of the secondary flow's difference between the rectangular cross-section and the circular cross-section. Note that the evaluation for K_b is performed based on the circular cross-section flow. In the rectangular cross-section, the corner obstructs the secondary flow and it may increase the pressure loss.

It needs to be emphasized that the theoretical equations can be modified to take into account the presented impacts and it is adopted for the ECM. Nevertheless, in our framework, these processes are not necessary except for the validation of the 3D model, and the modeling consideration may become simpler. Although only laminar flow model is used in this demonstration for our framework, the turbulence model can be adopted for the 3D models.

4. Conclusions

In this study, we describe a framework in which a thermal–electrical coupled ECM for a cooling system of a battery pack is constructed with a detailed 3D model. It is confirmed that the ECM has the same accuracy as the 3D model, which is validated through experimental results. The conclusions are described as follows:

(1) To match the ECM's results with the 3D model, the liquid–solid heat transfer coefficient and the links of the thermal ECM are determined with the 3D model's results.
(2) The ECM's accuracy is as high as the 3D model even though its computational costs are 96% lower than the 3D model.
(3) In terms of the 3D model's disadvantages, there may be a cost for constructing and validating the 3D model, and it may need more effort or more powerful computational power to calculate a huge scale phenomenon such as battery packs with tens of battery cells, which are often installed for EVs.
(4) The tendency of the liquid–solid heat transfer coefficient and the pressure drop do not agree well with the 3D results and some theoretical equations, since this phenomenon does not satisfy some prerequisites for the theoretical equations. Although theoretical equations can be determined and adopted for the ECM, the 3D model is advantageous because of its simplicity and certainty.

Author Contributions: Conceptualization, T.Y. (Takumi Yamanaka) and T.Y. (Tatsuya Yamaue); methodology, T.Y. (Takumi Yamanaka), D.K. and Y.T.; software, T.Y. (Takumi Yamanaka); validation, T.Y. and D.K.; formal analysis, T.Y.; investigation, T.Y.; resources, T.Y.; data curation, T.Y.; writing—original draft preparation, T.Y.; writing—review and editing, T.Y., Y.T. and T.Y. (Tatsuya Yamaue); visualization, T.Y. (Takumi Yamanaka); supervision, T.Y. (Tatsuya Yamaue); project administration, T.Y.; funding acquisition, T.Y. All authors have read and agreed to the published version of the manuscript.

Funding: This research received no external funding.

Conflicts of Interest: The authors declare no conflict of interest.

Nomenclature

A	Liquid–solid cross-sectional area, m^2
A_T	cross-sectional area of the heat pass, m^2
a	variable
b	variable
C	capacitance component in the electrical ECM, F
C_c	capacity of a battery cell, Ah
C_p	specific heat at a constant pressure, J·kg^{-1}·K^{-1}
C_T	heat capacity component in the thermal ECM, J·K^{-1}
D_h	hydraulic diameter of the flow channel, m^2
f	Darcy friction factor
F	Faraday constant, 96485 C·mol^{-1}
h	heat transfer coefficient, W·K^{-1}·m^{-2}
I	electric current, A
$\mathbf{I}$	identity matrix
k	thermal conductivity vector, W·m^{-1}·K^{-1}
k_f	liquid thermal conductivity, W·m^{-1}·K^{-1}
K_b	bend loss coefficient
L	total length of the flow channel, m
l_T	length of the heat pass, m
$\dot{m}$	mass flow rate, kg·s^{-1}
Nu	Nusselt number
OCV	open-circuit voltage, V

p	pressure, Pa
$\bar{p}_{inlet}$	mean pressure in the inlet boundary, Pa
$\bar{p}_{outlet}$	mean pressure in the outlet boundary, Pa
Q_f	heat due to the liquid coolant flow through the flow channel, W
Q_{gen}	total heat generation of a battery cell, W
Q_{ht}	heat transfer exchanged between the cooling plate and the liquid coolant domain, W
Q_{irr}	irreversible heat generation of a battery cell, W
Q_{link}	heat transfer into or from the linked components in the thermal ECM, W
Q_{rev}	reversible heat generation of a battery cell, W
q_n	normal directional heat flux on the liquid–solid cross-section, W·m^{-2}
R	internal resistance of a battery cell, Ω
R_1	resistance component in the electrical ECM, Ω
R_2	resistance component in the electrical ECM, Ω
R_c	resistance component in the electrical ECM, Ω
R_T	thermal resistance component in the thermal ECM, K·W^{-1}
$R_{T,ht}$	thermal resistance of the liquid–solid heat transfer in the thermal ECM, K·W^{-1}
Re	Reynolds number
S	entropy, J·K^{-1}
SOC	state of charge
SOC_0	initial state of charge
t	time, s
t'	time after charge or discharge, s
T	temperature, °C
T_f	liquid temperature, °C
$T_{f,inlet}$	inlet liquid temperature, °C
T_s	solid temperature, °C
$\mathbf{u}$	flow velocity vector, m·s^{-1}
u_m	cross-sectional mean flow velocity in the flow channel, m·s^{-1}
V	voltage, V
V_L	volume of a battery cell, m^3
Greek	
ε	effective roughness, m
μ	kinematic viscosity, Pa·s
ρ	density, kg·m^{-3}

Appendix A

In the appendix, more consideration for the heat transfer coefficients is described. According to Figure 8a, the heat transfer coefficients increase linearly with the flow rate. As described in Section 2.4, the calculation performed with the 3D model, in which the uniform temperature of 26 °C was set to the liquid–solid cross-sectional boundary, and the temperature on the inlet boundary is 25 °C. The results are given under the various flow rate conditions where the range of the flow rate is from 0 l/min to 1 l/min. The heat transfer coefficients are calculated by the results and Equation (12).

According to the right side of Equation (12), the normal directional heat flux on the liquid–solid cross-section q_n is the only component which is dependent on the flow rate. Figure A1b shows the normal directional heat flux profile under the various flow rate. On the curve line the values increase linearly with the flow rate though the values do not vary on the straight line. It is considered that this is because of the variation of temperature due to the flow bias in the curve channel. From Figure A1d,f, in the curve channel the variation of temperature and velocity in the vicinity of the boundary is larger. On the other hand, from Figure A1c,e, in the straight channel the variation of them is smaller. Because the normal directional heat flux is influenced significantly from the temperature in the vicinity of the boundary, the fact can be the main factor to increase the normal directional heat flux and the heat transfer coefficients.

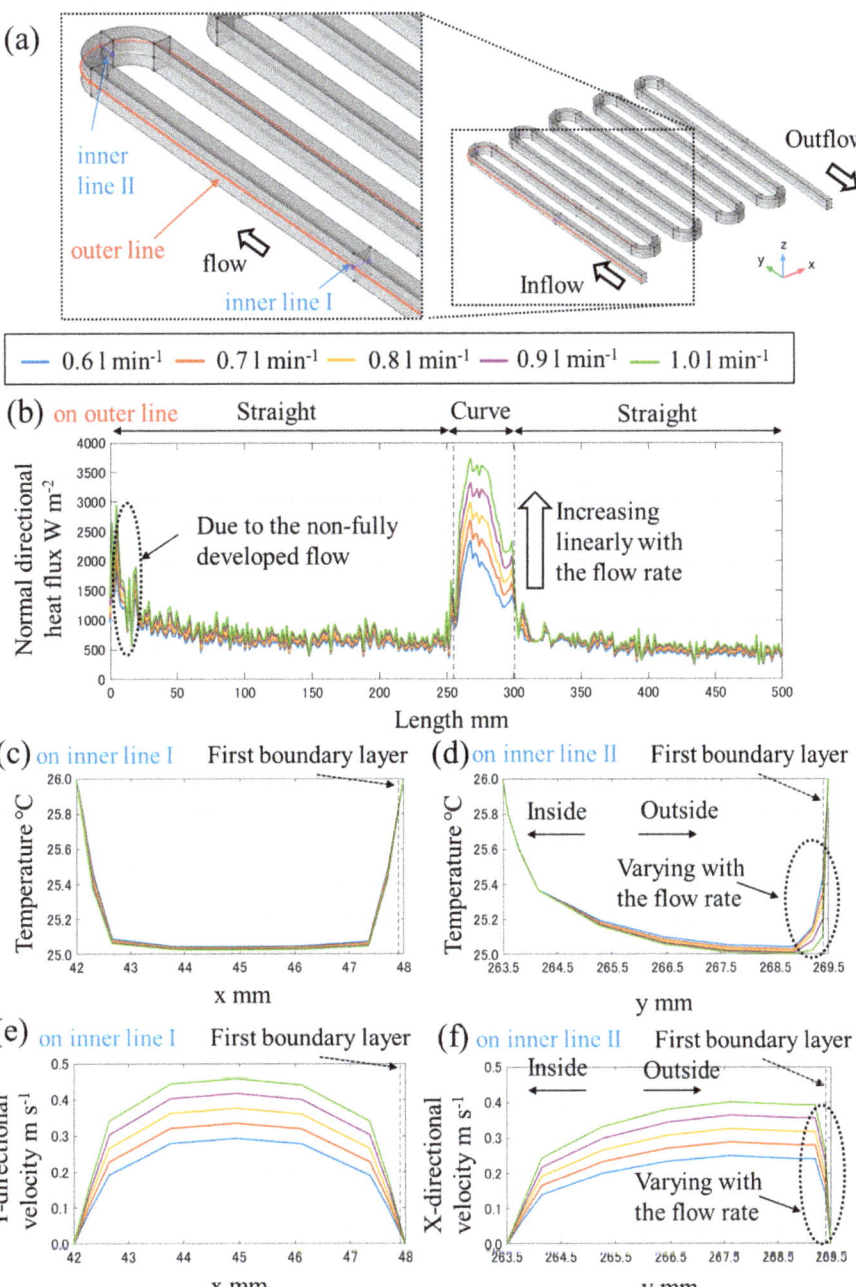

Figure A1. (a) The overview of the data output lines, and under the various flow rate (b) the normal directional heat flux profile on the outer line, the temperature profile on (c) the inner line I and (d) the inner line II, and the velocity profile on (e) the inner line I and (f) the inner line II.

References

1. Xia, G.; Cao, L.; Bi, G. A review on battery thermal management in electric vehicle application. *J. Power Sources* **2017**, *367*, 90–105. [CrossRef]
2. Raihan, A.; Siddique, M.; Mahmud, S.; Heyst, B.V. A comprehensive review on a passive (phase change materials) and an active (thermoelectric cooler) battery thermal management system and their limitations. *J. Power Sources* **2018**, *401*, 224–237.
3. Gan, Y.; Wang, J.; Liang, J.; Huang, Z.; Hu, M. Development of thermal equivalent circuit model of heat pipe-based thermal management system for a battery module with cylindrical cells. *Appl. Therm. Eng.* **2020**, *164*, 114523. [CrossRef]
4. Shen, M.; Gao, Q. System simulation on refrigerant-based battery thermal management technology for electric vehicles. *Energy Convers. Manag.* **2020**, *203*, 112176. [CrossRef]
5. Chung, Y.; Kim, M.S. Thermal analysis and pack level design of battery thermal management system with liquid cooling for electric vehicles. *Energy Convers. Manag.* **2019**, *196*, 105–116. [CrossRef]
6. Siruvuri, S.V.; Budarapu, P.R. Studies on thermal management of Lithium-ion battery pack using water as the cooling fluid. *J. Energy Storage* **2020**, *29*, 101377. [CrossRef]
7. Zhang, Z.; Wei, K. Experimental and numerical study of a passive thermal management system using flat heat pipes for lithium-ion batteries. *Appl. Therm. Eng.* **2020**, *166*, 114660. [CrossRef]
8. Severino, B.; Gana, F.; Palma-Behnke, R.; Estevez, P.A.; Calderon-Munoz, W.R.; Orchard, M.E.; Reyes, J.; Cortes, M. Multi-objective optimal design of lithium-ion battery packs based on evolutionary algorithm. *J. Power Sources* **2014**, *267*, 288–299. [CrossRef]
9. Li, M.; Liu, Y.; Wang, X.; Zhang, J. Modeling and optimization of an enhanced battery thermal management system in electric vehicles. *Front. Mech. Eng.* **2019**, *14*, 65–75. [CrossRef]
10. Lu, Z.; Meng, X.Z.; Wei, L.C.; Hu, W.Y.; Zhang, L.Y.; Jin, L.W. Thermal management of densely-packed EV battery with forced air cooling strategies. *Energy Procedia* **2016**, *88*, 682–688. [CrossRef]
11. Li, W.; Zhuang, X.; Xu, X. Numerical study of a novel battery thermal management system for a prismatic Li-ion battery module. *Energy Procedia* **2019**, *158*, 4441–4446. [CrossRef]
12. Jilte, R.D.; Kumar, R. Numerical investigation on cooling performance of Li-ion battery thermal management system at high galvanostatic discharge. *Eng. Sci. Technol. Int. J.* **2018**, *21*, 957–969. [CrossRef]
13. Tang, W.; Xu, X.; Ding, H.; Guo, Y.; Liu, J.; Wang, H. Sensitivity analysis of the battery thermal management system with a reciprocating cooling strategy combined with a flat heat pipe. *ACS Omega* **2020**, *5*, 8258–8267.
14. Qin, D.; Li, J.; Wang, T.; Zhang, D. Modeling and simulating a battery for an electric vehicle based on modelica. *Automot. Innov.* **2019**, *2*, 169–177. [CrossRef]
15. Wei, C.; Hofman, T.; Caarls, E.I.; van Iperen, R. Zone model predictive control for battery thermal management including battery aging and brake energy recovery in electrified powertrains. *IFAC PapersOnLine* **2019**, *52*, 303–308. [CrossRef]
16. Madani, S.S.; Schaltz, E.; Kaer, S.K. An electrical equivalent circuit model of a lithium titanate oxide battery. *Batteries* **2019**, *5*, 31. [CrossRef]
17. Brand, J.; Zhang, Z.; Agarwal, R.K. Extraction of battery parameters of the equivalent circuit model using a multi-objective genetic algorithm. *J. Power Sources* **2014**, *247*, 729–737. [CrossRef]
18. Viswanathan, V.V.; Choi, D.; Wang, D.; Xu, W.; Towne, S.; Williford, R.E.; Zhang, J.; Liu, J.; Yang, Z. Effect of entropy change of lithium intercalation in cathodes and anodes on Li-ion battery thermal management. *J. Power Sources* **2010**, *195*, 3720–3729. [CrossRef]
19. Takano, K.; Saito, Y.; Kanari, K.; Nozaki, K.; Kato, K.; Negishi, A.; Kato, T. Entropy change in lithium ion cells on charge and discharge. *J. Appl. Electrochem.* **2002**, *32*, 251–258. [CrossRef]
20. The Japan Society of Mechanical Engineers (Ed.) *JSME Data Book: Heat Transfer*, 5th ed.; The Japan Society of Mechanical Engineers: Tokyo, Japan, 2017; pp. 284–332.
21. Incropera, F.P.; Dewitt, D.P.; Bergman, T.L.; Lavine, A.S. *Fundamentals of Heat and Mass Transfer*, 6th ed.; John Wiley & Sons: New York, NY, USA, 2007; pp. 519–520.
22. Churchill, S.W. Friction-factor equiation spans all fluid-flow regimes. *Chem. Eng.* **1977**, *84*, 91–92.
23. Bellos, V.; Nalbantis, I.; Tsakiris, G. Friction modeling of flood flow simulations. *J. Hydraul. Eng.* **2018**, *144*, 04018073. [CrossRef]

24. Kitto, J.B.; Stultz, S.C. *Steam: Its Generation and Use*, 41th ed.; Babcock and Wilcox Company: Akron, OH, USA, 2005; pp. 99–102.
25. White, F.M. *Fluid Mechanics*, 7th ed.; McGraw-Hill: New York, NY, USA, 1999; pp. 389–391.

 © 2020 by the authors. Licensee MDPI, Basel, Switzerland. This article is an open access article distributed under the terms and conditions of the Creative Commons Attribution (CC BY) license (http://creativecommons.org/licenses/by/4.0/).

Case Report

Future Portable Li-Ion Cells' Recycling Challenges in Poland

Agnieszka Sobianowska-Turek * and Weronika Urbańska

Section of Waste Technology and Land Remediation, Faculty of Environmental Engineering, Wroclaw University of Technology, 27 Wybrzeże Wyspiańskiego St, 50-370 Wrocław, Poland; weronika.urbanska@pwr.edu.pl
* Correspondence: agnieszka.sobianowska-turek@pwr.edu.pl; Tel.: +48-663-104-614

Received: 7 November 2019; Accepted: 9 December 2019; Published: 12 December 2019

Abstract: The paper presents the market of portable lithium-ion batteries in the European Union (EU) with particular emphasis on the stream of used Li-ion cells in Poland by 2030. In addition, the article draws attention to the fact that, despite a decade of efforts in Poland, it has not been possible to create an effective management system for waste batteries and accumulators that would include waste management (collection and selective sorting), waste disposal (a properly selected mechanical method) and component recovery technology for reuse (pyrometallurgical and/or hydrometallurgical methods). This paper also brings attention to the fact that this EU country with 38 million people does not have in its area a recycling process for used cells of the first type of zinc-carbon, zinc-manganese or zinc-air, as well as the secondary type of nickel-hydride and lithium-ion, which in the stream of chemical waste energy sources will be growing from year to year.

Keywords: spent batteries and accumulators; Li-ion cells; legislation; recycling

1. Introduction

Lithium-ion chemical energy sources (Li-ion) dominate the market for secondary type batteries (accumulators); almost all mobile phones and laptops are powered by lithium cells [1,2]. In addition, the growing market for electric and hybrid cars [3] is a new industry generating demand for Li-ion batteries. Lithium-ion batteries contain several of valuable metals such as cobalt, copper, lithium, nickel, manganese, aluminium and iron [4–6]. Cobalt is one of the less common metals in the Earth's crust, hence its market value is high at 80,491$/MT [7], and its recovery is profitable [8]. Lithium recovery is essential for the development of electric car production [9]; the current price is 16,500$/MT [7]. Mass production of vehicles powered by lithium-ion batteries will increase demand for lithium, so its price will increase and recovery will be profitable [10–12]. Another argument in favour of recycling batteries is the need to protect the environment from pollution by heavy metals or complex organic substances contained in Li-ion batteries [13,14].

It is also very important that, despite a decade of efforts in Poland, it has not been possible to create an effective management system for waste batteries and accumulators that should include waste management (collection and selective sorting), waste disposal (a properly selected mechanical method) and component recovery technology for reuse (pyrometallurgical and/or hydrometallurgical methods). The fact that this European Union country with a population of 38 million does not have in its area a recycling process for used cells of the first type of zinc-carbon, zinc-manganese or zinc-air, as well as the secondary type of nickel-hydride and lithium-ion, which in the stream of chemical waste energy sources will be growing from year to year, means there is a potential benefit from a future solution.

2. Market for Portable Batteries and Accumulators in the European Union (EU) and Poland

The total mass of portable batteries and accumulators placed on the EU market between 2009 and 2017 is estimated at 1,921,000 MT and collected at 694,000 MT [15]. Data for 2017, not yet collected from all member states of the EU, show that in 2016 215,000 MT portable cells were introduced and 94,000 MT (~43.72%) were collected. The largest mass of portable chemical energy sources was introduced to the EU market by Germany 45,511 MT (~21.2%), the UK 38,659 MT (~18.0%), France 29,491 MT (~13.7%) and Italy 25,197 MT (~11.7%). Poland is ranked fifth with a weight of 12,585 MT, which accounted for about 5.9% of all portable cells on the European market two years ago. Unfortunately, there are no data available on portable lithium-ion batteries and accumulators passing through the EU market. According to data published on the Accurec Recycling GmbH website [16], around 70,000 MT of Li-ion cells for different applications will be introduced to the EU market in 2020, while only slightly more than 8000 MT of this type of waste will be recycled. Nevertheless, it is estimated that the global market for lithium cells will be worth USD 93.1 billion in 2025 [17] and that around 50,000 MT of such waste will be recycled in the member states by this year [18].

In Poland, pursuant to Article 72 of the Batteries and Accumulators Act 2009 [19], the Chief Inspector of Environmental Protection prepares an annual report on the state of management of batteries and accumulators and their waste. The report contains data on the number of companies placing batteries and accumulators on the market as well as on the number of plants collecting and processing waste chemical energy sources. The report includes information on the quantity and weight of batteries placed on the market as well as collected and treated battery waste, distinguishing between three categories: portable, automotive and industrial batteries and accumulators. In the register of chemical energy sources introduced to the Polish market, data on the mass and quantity of batteries and accumulators of the following types are collected separately: nickel-cadmium (Ni-Cd), lead-acid (Pb-Acid), button cells with mercury, button cells without mercury, zinc-carbon (Zn-C), zinc-manganese (Zn-Mn) and zinc-air (Zn-O_2). Batteries and accumulators that do not belong to any of these groups together form the last, sixth group—other batteries and accumulators. The register of collected waste batteries contains separate data on the quantity and weight of nickel-cadmium and lead-acid batteries and accumulators; all other types of batteries are counted together. In Poland, data on the number of lithium-ion batteries introduced to the market and collected are not gathered. In the annual reports of the Chief Inspector of Environmental Protection, the stream of the accumulators discussed forms part of the "other batteries and accumulators" group, which also includes silver, lithium and nickel-hydrogen accumulators.

According to the latest available report of the Chief Inspectorate of Environmental Protection of 2018, at the end of 2017 there were 3648 registered entrepreneurs placing batteries or accumulators on the market (1035 entrepreneurs placing batteries and accumulators on the market and 2613 entrepreneurs placing batteries or accumulators together with electrical and electronic equipment on the market) and 25 entrepreneurs operating in the field of waste batteries and accumulators processing. In total, 134,951.6 MT of batteries and accumulators were placed on the market in 2017, including 13,269.9 MT of portable batteries and accumulators, 30,306.5 MT of industrial batteries and accumulators, and 91,375.3 MT of automotive batteries and accumulators. Detailed data on the types of chemical energy sources placed on the market are broken down into three categories: portable, automotive and industrial. Both automotive and industrial batteries are mostly lead-acid batteries. The data collected in the reports of the Chief Inspectorate of Environmental Protection [20–27] show that they account for over 99% of automotive batteries and between 90% and 98% of industrial batteries. The largest number of lithium-ion batteries is used in mobile phones and personal computers and belongs to the category of portable batteries and accumulators [28].

Table 1 shows the number of portable batteries and accumulators introduced to the Polish market according to the data contained in the reports on the functioning of battery and accumulator management for the years 2010–2017 [20–27].

Table 1. Mass of batteries and accumulators introduced to Polish market in 2010–2017, in MT [20–27].

Year	2010	2011	2012	2013	2014	2015	2016	2017
Zn-Mn, Zn-C, Zn-O$_2$	5976.3	5451.3	5715.3	6515.3	7371.5	7343.4	7540.4	8259.9
Ni-Cd	662.9	660.1	489.4	449.4	606.2	375.2	390.5	212.6
Pb	348.9	369.5	296.3	413.6	409.7	416.5	281.7	413.4
Button-type batteries and accumulators without Hg	179.3	153.3	113.9	143.6	141.8	186.9	232.0	279.6
Button-type batteries and accumulators from Hg	2.4	3.6	5.6	22.5	8.27	7.8	16.3	8.4
Other batteries and accumulators	2696.5	3133.4	3978.6	3719.4	3261.3	3875.5	4124.3	4096.0

Table 1 shows that in 2017 more than 13,000 tons of portable batteries and accumulators were introduced to the market; 62.2% of this amount were zinc-carbon, alkaline and zinc-air batteries. The other specified types, i.e., button cell batteries and nickel-cadmium and lead-acid batteries, accounted for a small share of 6.9%. Whereas 30.9%, in 2017, i.e., almost 4.1 thousand tonnes, are other cells, i.e., silver, nickel-hydride, lithium and lithium-ion cells. The use of silver and lithium cells is relatively small, while nickel and hydrogen batteries are used mainly in hybrid cars and are currently being replaced by lithium ion batteries.

3. Li-Ion Battery and Accumulator Stream in Poland

In order to determine the size of the waste stream of Li-ion batteries, the necessary information is the level of effectiveness of their collection. The Chief Inspectorate of Environmental Protection [20–27] reports contain data on the amount of portable batteries and accumulators collected. Collection rates for waste batteries are set by the Directive on batteries and accumulators [29] and are: 25% from 2012 to 45% from 2016. In Poland, the required collection rates are specified in the Ordinance of 3 December 2009 on annual collection rates of waste portable batteries and waste portable accumulators [30], which in subsequent years were as follows: 2010—18%, 2011—22%, 2012—25%, 2013—30%, 2014—35%, 2015—40%, 2016 and 2017—45%. The required levels of collection of used energy sources were achieved by 2013 (2010—18.00%, 2011—22.72%, 2012—29.10%). In 2014 and 2015 to achieve an appropriate level of collection of this type of waste, approximately 2.0% (2014—33.06%, 2015—38.35%) were missing. The last three years have seen a significant increase in the collection of waste cells and so in 2015 54.92%, in 2016 78.14% and 2017 65.74% of waste portable batteries and accumulators were collected from the market. The lack of detailed information on the collection efficiency of waste Li-ion batteries makes it clear that, in order to estimate the amount of the waste stream, the collection rate of Li-ion batteries is equal to the collection rate of all portable batteries and accumulators. Assuming also that lithium-ion batteries constitute 80% of batteries and accumulators other than zinc-carbon, alkaline, zinc-air, nickel-cadmium, lead-acid and button batteries and accumulators, it is possible to calculate the stream of used Li-ion cells in the years 2010–2017 (Figure 1). Calculations show that from 2010 to 2017 the mass of Li-ion batteries introduced to the Polish market increased by 34.17% from 2157.24 MT to 3276.79 MT. However, the total stream of collected waste lithium cells in the last seven years could amount to as much as 10 076.15 MT and it should be noted that Poland is a country where there is no single technology for processing waste batteries and accumulators. Processing methods of portable batteries concern only used cells of the first type of Zn-C, Zn-Mn and Zn-O$_2$ cells and end with mechanical processing and separation of three ferromagnetic, diamagnetic and paramagnetic material fractions for management [31]. Based on the value of masses of Li-ion batteries and accumulators introduced into and collected from the Polish market, it can be concluded that in 2020 these streams may amount to 4361.40 MT and 1962.63 MT respectively. In 2030, the volume of these streams may increase 2.6 times to 11,312.36 MT and 5090.56 MT in relation to 2020. These forecasts are consistent with the data presented in the works of Rogulski and Czerwiński [32] and Rogulski and Dłubak [33], in which the authors presented a detailed analysis of the portable batteries market in Europe in terms of the management of electrochemical energy sources.

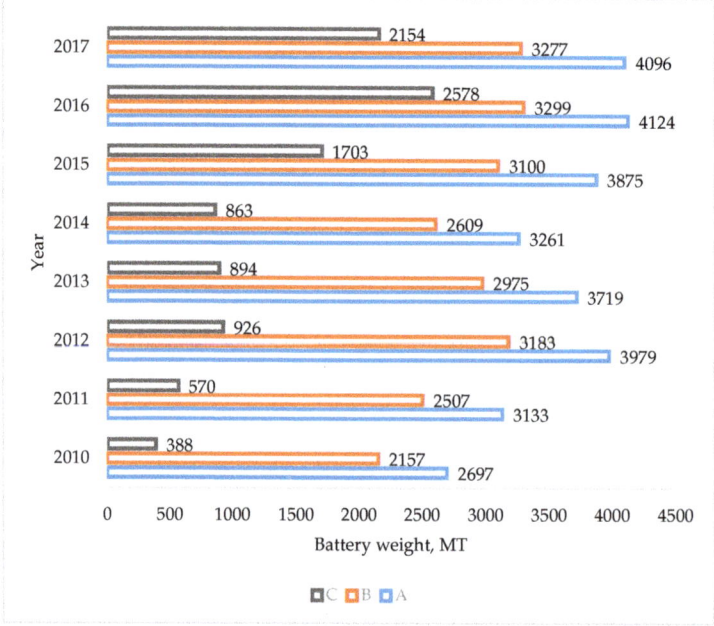

Figure 1. Estimated volume of Li-ion battery and accumulator streams in Poland in 2010–2017: A—weight of other batteries and accumulators placed on the market; B—weight of Li-ion batteries and accumulators placed on the market; C—mass of accumulated used Li-ion batteries and accumulators, in MT.

4. Recycling of Used Li-Ion Batteries and Accumulators in Poland

The reference system for the recycling of used batteries and accumulators should be based on three complementary and successive unit processes. The first of these is a mechanical treatment most often used for large cells (industrial type) and as a preliminary operation in most processing technologies. Separation processes involve the mechanical loosening of the structure (body) of the battery and separation of components with characteristic physical properties (density, size, magnetic properties). These activities are usually simple and cheaper than other processes, and for that reason they should be used to prepare the material stream for further processing [31–34]. The second main processes are pyrometallurgical and/or hydrometallurgical processes. Pyrometallurgical methods rely on the recovery of materials (in particular metals) by carrying them out at sufficiently high temperatures to specific condensed phases (including a metallic alloy) or to the gas phase with subsequent condensation. In general, these methods are more appropriate for phases rich in recoverable components, possibly concentrating at elevated temperatures in the gas phase (this applies to e.g., mercury removal, cadmium or zinc extraction). However, it should be remembered that this division is arbitrary and is not suitable for strict chemical or technological considerations. With regard to batteries, these processes can be carried out both in a traditional way, i.e., using the oxidation-reduction equilibria of the HCO system (hydrogen, carbon, oxygen), and in an extended manner, which is characteristic of advanced chemical metallurgy, where for example chlorination processes are utilized. The advantage of pyrometallurgical methods is the possibility of recycling various types of cells, including those containing various organic materials [35]. In contrast, hydrometallurgical methods usually rely on acid or alkaline leaching of properly prepared battery waste (after machining processes). They are followed by a series of physicochemical operations that lead to the separation and concentration of valuable or burdensome components between the respective phases, up to commercial products and semi-finished products for separate technological processes (pyrometallurgical or hydrometallurgical) or waste. It is believed that hydrometallurgical processes are less energy-consuming than pyrometallurgical ones, but the waste from them is more burdensome. The advantage of hydrometallurgical processes is also that they allow in most cases the processing of a mixture of different types of batteries simultaneously [36–38]. The third stage of recycling used cells should be the processes of managing all post-process waste in such a way that they are not harmful to the environment.

In 2017, there were 25 registered companies operating in the field of processing waste batteries and accumulators in Poland [27]. However, in reality only a few companies conduct raw material recovery from waste cells—Table 2 [39–47]. Processing of used zinc–carbon, zinc–manganese and zinc–air batteries is mainly mechanical, which results in creation of ferromagnetic fractions, diamagnetic fraction and paramagnetic fraction. The ferromagnetic fraction consisting of metals such as iron, chromium and nickel is a secondary raw material for steel works. The diamagnetic fraction in which plastics and paper accumulate, when mixed with sawdust and used cleaning cloth, is used as a substrate for the production of alternative fuel, the recipients of which are cement plants or combined heat and power plants. The paramagnetic fraction is mainly dominated by graphite and non-ferrous metals with the remainder of the other two fractions [48]. However, zinc and manganese contained in the last fraction that could be obtained by pyrometallurgical [49] or hydrometallurgical [50] methods are not recovered. The lack of a complete domestic technology for recycling used cells of the first type means that secondary raw materials containing valuable metals are resold to foreign companies based in Finland, Germany or Slovenia [27]. Only lead-acid cells are recycled in high-efficiency pyrometallurgical installations. Baterpol S.A. and Orzeł Biały S.A. Process Pb-acid cells by thermal methods to obtain lead alloys and polypropylene, which are again used in the production of car batteries [45,46].

Lithium-ion cells are not processed in any domestic installation, they are probably sorted out of the entire stream of used batteries and accumulators, collected and then resold at a price of 300 to 1000$/MT [51] to foreign companies having installations for their recycling. Therefore, assuming the average price of 600$/MT of used lithium-ion batteries collected in Poland, it can be calculated that the market value of the discussed waste in 2017 was approximately 1.30 million USD and in 2030 reached the value of 3.05 million USD. It seems that the estimated amounts, apart from the costs of designing and constructing the installation and its current maintenance, could give the potential investor real profits in the next 10 years. Of course, press reports show that several companies alone or in cooperation with research institutions are planning or are implementing research projects related to the creation of recycling technologies for used Li-ion batteries and accumulators [52–54]. However, it is difficult to deduce from this information in what time perspective a complete recycling installation could be built in Poland.

Table 2. Technologies for processing used batteries and accumulators in Poland [39–47].

Name of Technology	Recycling Method	Type of Remanufactured Battery	Recovered Metals and Compounds
BatEko	Mechanical processing	Zn–C Zn–Mn Zn–Air	— ferromagnetic fraction — diamagnetic fraction — paramagnetic fraction
Grupa Eneris	Mechanical processing	Zn–C Zn–Mn Zn–Air	— metallic Zn — brass — steel — RDF (foil and paper)

Table 2. Cont.

Name of Technology	Recycling Method	Type of Remanufactured Battery	Recovered Metals and Compounds
MB Recycling Sp. z o. o.	Mechanical processing	Zn–C Zn–Mn Zn–Air	— ferromagnetic fraction — diamagnetic fraction — paramagnetic fraction
Biosystem S. A.	Mechanical processing	Zn–C Zn–Mn Zn–Air Ni–Cd	— ferromagnetic fraction — diamagnetic fraction — paramagnetic fraction
MarCo Ltd. Sp. z o. o.	Mechanical processing	Ni–Cd	— iron-nickel electrodes — iron-cadmium electrodes — steel scrap — potassium lye — plastic waste
Eco Harpoon Recycling Sp. z o.o.	Mechanical processing	Zn–C Zn–Mn Zn–Air	— ferromagnetic fraction — diamagnetic fraction — paramagnetic fraction
		Ni–Cd Batteries containing mercury	— mercury — cadmium — ferrous and non-ferrous metals
Baterpol S.A.	Pyrometallurgical	Pb–acid	— alloys Pb — solid Na_2SO_4 — polypropylene
Orzeł Biały S.A.	Pyrometallurgical	Pb–acid	— alloys Pb — polypropylene pellets
ZM Silesia S.A.	Pyrometallurgical	Ni–Cd	— metallic Cd — ferro-alloys containing Ni

5. Conclusions

The above quantitative and qualitative analysis of the Polish market for batteries and accumulators as well as waste batteries and accumulators shows that in the next 10 years in the country the stream of discussed waste will change dynamically. In addition to the used cells of the first type, the number of which on the market is still high, one should expect an additional significant growth in number of cells of the secondary type, especially the lithium-ion type. That is why it is so important that the current system of managing chemical energy sources undergo versification and modification taking into account the changes associated with the technology of producing new cells, which in the near future will become a valuable source of secondary raw materials. Meanwhile despite an increase in the level of metal recovery from used chemical energy sources, the amount of batteries placed on the market is growing so rapidly that the problem of recycling of secondary type cells is constantly worsening. This is influenced in particular by the development of personal device technology, which mostly uses lithium-ion cells. Despite a number of publications and reports raising the high level of risk related to pollution of the environment by poorly managed cells containing highly harmful substances, the first overriding problem limiting the effectiveness of the recovery of raw materials from lithium-ion cells is low social awareness resulting in a small stream of selectively collected waste batteries. As a result, only every tenth cell is recycled, which has a direct impact on the high technological costs associated with the recovery of metals from smaller waste material stream. The task of states and properly appointed institutions should be to take care not only of the interests of the present but also future generations, who may have major problems with maintaining stable development and rational raw material management. Greater efforts should be made to improve public education and raise consumer awareness of the need to return raw materials for re-use, not only through regulations imposing minimum recycling rates, which usually involve higher costs, but also through improvements in the raw material situation. Also important is improvement of social awareness, especially in developed countries, where the number of devices containing cells introduced to the market is the first barrier to achieving rational levels of recovery of raw materials from electronic devices. Technologies for recovering metals from battery cells are slowly surpassing the technologies used to produce them. This is a signal that they should have a real impact on battery manufacturers to look for solutions that allow easier recovery of components and materials used in their production. There is now a clear tendency to optimize the production of cells, not only lithium-ion cells, in terms of their cost of production in relation to their efficiency without considering recovery and reuse. As a result, the resulting batteries are not prepared for proper treatment after use, and the costs associated with the recovery of materials are increasing.

Author Contributions: Conceptualization, A.S.-T. and W.U.; methodology, A.S.-T.; formal analysis, A.S.-T.; investigation, A.S.-T.; resources, A.S.-T.; data curation, A.S.-T. and W.U.; writing—original draft preparation, A.S.-T.; writing—review and editing, A.S.-T. and W.U.; visualization, A.S.-T.; supervision, A.S.-T.; project administration, A.S.-T.; funding acquisition, A.S.-T.

Funding: This study was funded by the National Science Centre of Poland (grant number 2017/01/X/ST10/00267).

Conflicts of Interest: The authors declare no conflict of interest.

References

1. Liu, C.; Lin, J.; Cao, H.; Zhang, Y.; Sun, Z. Recycling of spent lithium-ion batteries in view of lithium recovery: A critical review. *J. Clean. Prod.* **2019**, *228*, 801–813. [CrossRef]
2. Zubi, G.; Dufo-López, R.; Carvalho, M.; Pasaoglu, G. The lithium-ion battery: State of the art and future perspectives. *Renew. Sustain. Energy Rev.* **2018**, *89*, 292–308. [CrossRef]
3. Huang, B.; Pan, Z.; Su, X.; An, L. Recycling of lithium-ion batteries: Recent advances and perspectives. *J. Power Sources* **2018**, *399*, 274–286. [CrossRef]
4. He, L.-P.; Sun, S.-Y.; Song, X.-F.; Yu, J.-G. Leaching process for recovering valuable metals from the LiNi1/3Co1/3Mn1/3O2 cathode of lithium-ion batteries. *Waste Manag.* **2017**, *64*, 171–181. [CrossRef] [PubMed]
5. Huang, Y.; Han, G.; Liu, J.; Chai, W.; Wang, W.; Yang, S.; Su, S. A stepwise recovery of metals from hybrid cathodes of spent Li-ion batteries with leaching-flotation-precipitation process. *J. Power Sources* **2016**, *325*, 555–564. [CrossRef]
6. Peng, C.; Hamuyuni, J.; Wilson, B.P.; Lundström, M. reductive leaching of cobalt and lithium from industrially crushed waste Li-ion batteries in sulfuric acid system. *Waste Manag.* **2018**, *76*, 582–590. [CrossRef]
7. Metalary. Available online: https://www.metalary.com/ (accessed on 25 October 2019).
8. Lv, W.; Wang, Z.; Cao, H.; Sun, Y.; Zhang, Y.; Sun, Z. A critical review and analysis on the recycling of spent lithium-ion batteries. *ACS Sustain. Chem. Eng.* **2018**, *6*, 1504–1521. [CrossRef]
9. Swain, B. Recovery and recycling of lithium: A review. *Sep. Purif. Technol.* **2017**, *172*, 388–403. [CrossRef]
10. Langkau, S.; Espinoza, L.A.T. Technological change and metal demand over time: What can we learn from the past? *Sustain. Mater. Technol.* **2018**, *16*, 54–59. [CrossRef]
11. Li, L.; Biana, Y.; Zhanga, X.; Xuea, Q.; Fana, E.; Wua, F.; Chen, R. Economical recycling process for spent lithium-ion batteries and macro- and micro-scale mechanistic study. *J. Power Sources* **2018**, *377*, 70–79. [CrossRef]
12. Porvali, A.; Aaltonen, M.; Ojanen, S.; Velazquez-Martinez, O.; Eronen, E.; Liu, F.; Wilson, B.P.; Serna-Guerrero, R.; Lundström, M. Mechanical and hydrometallurgical processes in HCl media for the recycling of valuable metals from Li-ion battery waste. *Resour. Conserv. Recycl.* **2019**, *142*, 257–266. [CrossRef]

13. Dhiman, S.; Gupta, B. Partition studies on cobalt and recycling of valuable metals from waste Li-ion batteries via solvent extraction and chemical precipitation. *J. Clean. Prod.* **2019**, *225*, 820–832. [CrossRef]
14. Ordoñez, J.; Gago, E.J.; Girard, A. Processes and technologies for the recycling and recovery of spent lithium-ion batteries. *Renew. Sustain. Energy Rev.* **2016**, *60*, 195–205. [CrossRef]
15. Eurostat. Available online: http://appsso.eurostat.ec.europa.eu/nui/submitViewTableAction.do/ (accessed on 4 November 2018).
16. Accurec Recycling GmbH. Available online: https://accurec.de/battery-market (accessed on 16 November 2018).
17. Greenfish. Available online: https://www.greenfish.eu/lithium-ion-batteries-delivering-transition-solutions-to-different-sectors/ (accessed on 4 November 2018).
18. Waste Management World. Available online: https://waste-management-world.com/a/in-depth-recycling-to-supply-9-of-global-lithium-demand-by/ (accessed on 4 November 2018).
19. Journal of Laws of the Republic of Poland: Dziennik Ustaw Rzeczypospolitej Polskiej: Ustawa z dnia 24 kwietnia 2009 r. o bateriach i akumulatorach, Dz.U. 2009 nr 89 poz. 666 z późn. zm. Available online: http://prawo.sejm.gov.pl/isap.nsf/download.xsp/WDU20090790666/U/D20090666Lj.pdf (accessed on 11 December 2019). (In Polish)
20. Chief Inspectorate of Environmental Protection (CIPE). *Raport O Funkcjonowaniu Gospodarki Bateriami I Akumulatorami I Zużytymi Akumulatorami Za Rok 2010*. Available online: http://www.gios.gov.pl/images/dokumenty/raporty/raport_luty2012.pdf (accessed on 11 December 2019). (In Polish)
21. Chief Inspectorate of Environmental Protection (CIPE). *Raport O Funkcjonowaniu Gospodarki Bateriami I Akumulatorami I Zużytymi Akumulatorami Za Rok 2011*. Available online: http://www.gios.gov.pl/images/dokumenty/raporty/raport_baterie_2011.pdf (accessed on 11 December 2019). (In Polish)
22. Chief Inspectorate of Environmental Protection (CIPE). *Raport O Funkcjonowaniu Gospodarki Bateriami I Akumulatorami I Zużytymi Akumulatorami Za Rok 2012*. Available online: http://www.gios.gov.pl/images/dokumenty/raporty/Raport_2012.pdf (accessed on 11 December 2019). (In Polish)
23. Chief Inspectorate of Environmental Protection (CIPE). *Raport O Funkcjonowaniu Gospodarki Bateriami I Akumulatorami I Zużytymi Akumulatorami Za Rok 2013*. Available online: http://www.gios.gov.pl/images/dokumenty/raporty/Raport_2013_20140708.pdf (accessed on 11 December 2019). (In Polish)
24. Chief Inspectorate of Environmental Protection (CIPE). *Raport O Funkcjonowaniu Gospodarki Bateriami I Akumulatorami I Zużytymi Akumulatorami Za Rok 2014*. Available online: http://www.gios.gov.pl/images/dokumenty/raporty/raport2014.pdf (accessed on 11 December 2019). (In Polish)
25. Chief Inspectorate of Environmental Protection (CIPE). *Raport O Funkcjonowaniu Gospodarki Bateriami I Akumulatorami I Zużytymi Akumulatorami Za Rok 2015*. Available online: http://www.gios.gov.pl/images/dokumenty/raporty/raport_baterie_2015.pdf (accessed on 11 December 2019). (In Polish)
26. Chief Inspectorate of Environmental Protection (CIPE). *Raport O Funkcjonowaniu Gospodarki Bateriami I Akumulatorami I Zużytymi Akumulatorami Za Rok 2016*. Available online: http://www.gios.gov.pl/images/dokumenty/raporty/Raport_bateryjny_2016.pdf (accessed on 11 December 2019). (In Polish)
27. Chief Inspectorate of Environmental Protection (CIPE). *Raport O Funkcjonowaniu Gospodarki Bateriami I Akumulatorami I Zużytymi Akumulatorami Za Rok 2017*. Available online: http://www.gios.gov.pl/images/dokumenty/raporty/Raport_bateryjny_2017.pdf (accessed on 11 December 2019). (In Polish)
28. European Council. European Directive on Batteries and Accumulators 2006/66/EC. Available online: https://eur-lex.europa.eu/legal-content/EN/ALL/?uri=CELEX%3A32006L0066 (accessed on 11 December 2019).
29. Statista. Available online: https://www.statista.com/statistics/272595/global-shipments-forecast-for-tablets-laptops-and-desktop-pcs/ (accessed on 31 October 2018).
30. Journal of Laws of the Republic of Poland: Rozporządzenie Ministra Środowiska z dnia 3 grudnia 2009 r. w sprawie rocznych poziomów zbierania zużytych baterii przenośnych i zużytych akumulatorów przenośnych, Dz.U. 2009 nr 215 poz. 1671. Available online: http://prawo.sejm.gov.pl/isap.nsf/download.xsp/WDU20092151671/O/D20091671.pdf (accessed on 11 December 2019). (In Polish)
31. Sobianowska-Turek, A. Hydrometallurgical recovery of metals: Ce, La, Co, Fe, Mn, Ni and Zn from the stream of used Ni-MH cells. *Waste Manag.* **2018**, *77*, 213–219. [CrossRef]
32. Rogulski, Z.; Czerwiński, A. Market of portable batteries and accumulators. *Przem. Chem.* **2014**, *93*, 709–712. (In Polish) [CrossRef]

33. Rogulski, Z.; Dłubak, J. Batteries and accumulators in Europe. *Przem. Chem.* **2014**, *93*, 704–708. (In Polish) [CrossRef]
34. Velázquez-Martínez, O.; Valio, J.; Santasalo-Aarnio, A.; Reuter, M.; Serna-Guerrero, R. A critical review of lithium-ion battery recycling processes from a circular economy perspective. *Batteries* **2019**, *5*, 68. [CrossRef]
35. Petranikova, M.; Ebin, B.; Mikhailova, S.; Steenari, B.-M.; Ekberg, C. Investigation of the effects of thermal treatment on the leachability of Zn and Mn from discarded alkaline and ZnC batteries. *J. Clean. Prod.* **2018**, *170*, 1195–1205. [CrossRef]
36. Winiarska, K.; Klimkiewicz, R.; Tylus, W.; Sobianowska-Turek, A.; Winiarski, J.; Szczygieł, B.; Szczygieł, I. Study of the catalytic activity and surface properties of manganese-zinc ferrite prepared from used batteries. *J. Chem.* **2019**. [CrossRef]
37. Szczygieł, I.; Winiarska, K.; Sobianowska-Turek, A. The study of thermal, microstructural and magnetic properties of manganese-zinc ferrite prepared by co-precipitation method using different precipitants. *J. Anal. Calorim.* **2018**, *134*, 51–57. [CrossRef]
38. Meshrama, P.; Mishra, A.; Sahub, R. Environmental impact of spent lithium ion batteries and green recycling perspectives by organic acids—A review. *Chemosphere* **2020**, *242*, 125291. [CrossRef]
39. BatEko. Available online: http://www.bateko.com.pl/ (accessed on 3 December 2019).
40. Grupa Eneris. Available online: http://www.eneris.pl/ (accessed on 3 December 2019).
41. MB Recycling. Available online: https://mbrecycling.pl/ (accessed on 3 December 2019).
42. Biosystem S.A. Available online: http://www.biosystem.pl/ (accessed on 3 December 2019).
43. MarCo Ltd Sp. z o.o. Available online: http://marcoltd.pl/ (accessed on 3 December 2019).
44. EKO HARPOONRecykling Sp. z o.o. Available online: http://www.ekoharpoon-recykling.pl/ (accessed on 3 December 2019).
45. Baterpol S.A. Available online: http://www.baterpol.pl/ (accessed on 3 December 2019).
46. Orzeł Biały S.A. Available online: http://www.orzel-bialy.com.pl/pl/ (accessed on 3 December 2019).
47. ZM Silesia S.A. Available online: http://hutaolawa.pl/ (accessed on 3 December 2019).
48. Sobianowska-Turek, A.; Szczepaniak, W.; Maciejewski, P.; Gawlik-Kobylińska, M. Recovery of zinc and manganese, and other metals (Fe, Cu, Ni, Co, Cd, Cr, Na, K) from Zn-MnO$_2$ and Zn-C waste batteries: Hydroxyl and carbonate co-precipitation from solution after reducing acidic leaching with use of oxalic acid. *J. Power Sources* **2016**, *325*, 220–228. [CrossRef]
49. Institute of Non-Ferrous Metals (IMN). Method of Recycling Used Batteries, Particularly Carbon-Zinc Batteries. PL Patent 198,317 B1, 23 April 2002.
50. KGHM Metraco S.A. Method for Disposal of Waste Zn-C and Zn-Mn Batteries Remaining after Separation of Ferromagnetic Fraction. PL Patent 220,853 B1, 25 August 2011.
51. Metalary. Available online: http://www.metalary.com/scrap-metal-prices/ (accessed on 5 December 2019).
52. University of Warsaw. Available online: http://www.uott.uw.edu.pl/rdls-pge-s-a-dotacja-opracowanie-technologii-recyklingu-baterii-litowo-jonowych/ (accessed on 5 December 2019).
53. Business Insider Polska. Available online: https://businessinsider.com.pl/technologie/nowe-technologie/recykling-starych-baterii-szansa-na-nowy-przemysl-dla-polski/244r9tq (accessed on 5 December 2019).
54. Stena Recycling. Available online: https://www.stenarecycling.pl/zrownowazony-recyklingu/innowacje-w-recyklingu/badania-nad-akumulatorami-litowo-jonowymi/ (accessed on 5 December 2019).

© 2019 by the authors. Licensee MDPI, Basel, Switzerland. This article is an open access article distributed under the terms and conditions of the Creative Commons Attribution (CC BY) license (http://creativecommons.org/licenses/by/4.0/).

MDPI
St. Alban-Anlage 66
4052 Basel
Switzerland
Tel. +41 61 683 77 34
Fax +41 61 302 89 18
www.mdpi.com

Batteries Editorial Office
E-mail: batteries@mdpi.com
www.mdpi.com/journal/batteries

www.ingramcontent.com/pod-product-compliance
Lightning Source LLC
LaVergne TN
LVHW070415100526
838202LV00014B/1464